The Statistics and Machine Learning with R Workshop

Unlock the power of efficient data science modeling with this hands-on guide

Liu Peng

BIRMINGHAM—MUMBAI

The Statistics and Machine Learning with R Workshop

Group Product Manager: Ali Abidi
Book Project Manager: Farheen Fathima
Senior Editor: Nazia Shaikh
Technical Editor: Devanshi Ayare
Copy Editor: Safis Editing
Proofreader: Safis Editing
Indexer: Tejal Daruwale Soni
Production Designer: Joshua Misquitta
DevRel Marketing Coordinator: Vinshika Kalra

First published: September 2023
Production reference: 1290923

Published by Packt Publishing Ltd
Grosvenor House
11 St Paul's Square
Birmingham
B3 1RB

ISBN 978-1-80324-030-5
www.packtpub.com

This book is dedicated to my family, particularly my wife, Zheng, and my children, Jiaran, Jiaxin, and Jiayu. Jiaran comes first this time, as her older sister (Jiaxin) already declared victory in my other book.

Contributors

About the author

Liu Peng is an assistant professor of quantitative finance (practice) at Singapore Management University and an adjunct researcher at the National University of Singapore. He holds a Ph.D. in statistics from the National University of Singapore and has 10 years of working experience as a data scientist across the banking, technology, and hospitality industries.

This volume encapsulates a decade-long odyssey through the multifaceted landscape of data science, a journey that began as a spark of personal curiosity and evolved into an integrated blend of theoretical and practical insights. I owe a debt of gratitude to my esteemed mentors—Teo Chung Piaw, Chen Ying, and Ian Wilson—who have been instrumental in shaping my academic and professional trajectory, providing unwavering support every step of the way.

About the reviewer

Usha Rengaraju currently heads the data science research at Exa Protocol, and she is the first female triple Kaggle Grandmaster worldwide. She specializes in deep learning and probabilistic graphical models and was also one of the judges of the TigerGraph Graph for All Million Dollar Challenge. She was ranked as one of the top 10 data scientists in India by Analytics India Magazine and also ranked as one of the top 150 AI leaders and influencers by 3AI magazine. She is one of the winners of the ML in Action competition organized by the ML developer programs team at Google, and her team won first place at the WiDS Datathon 2022 organized by Stanford University. She was also the winner of the Kaggle ML Research Spotlight for 2022 and the winner of the TensorFlow Community Spotlight 2023.

Vybhavreddy KC is a dedicated data science practitioner by profession. He has fortified his passion for data with a Bachelor's degree in Computer Science and a Master's degree in Analytics. Vybhav's expertise includes leading the development of innovative ML/AI driven solutions for Compliance and Regulatory product suite. When he's not immersed in the realm of numbers and algorithms, Vybhav cherishes his free time, and loves playing with his children.

I would like to thank my wife Srilakshmi, my lovely kids Varshil and Reyansh for their unwavering support in achieving my academic and professional goals.

Table of Contents

3

Intermediate Data Processing   69

4

Data Visualization with ggplot2   97

Part 2: Fundamentals of Linear Algebra and Calculus in R

7

Part 3: Fundamentals of Mathematical Statistics in R

10

Probability Basics   301

11

Statistical Estimation   345

Preface

This workshop book is a comprehensive resource designed to offer a deep dive into the essential aspects of both statistics and machine learning. It provides explanations of key concepts with practical examples and hands-on exercises to enable a holistic learning experience. Starting with the fundamentals, the book walks you through the complete model development process, covering everything from data preprocessing to model development.

In addition to its focus on machine learning, the book also delves into R's statistical capabilities. You'll learn how to manipulate various data types and tackle complex mathematical challenges, from algebra and calculus to probability and Bayesian statistics. The text even guides you through linear regression techniques and more advanced statistical methodologies.

By the conclusion of this workshop book, you'll not only have a robust foundational understanding of statistics and machine learning, but you will also be proficient in using R's extensive libraries for tasks such as data processing and model training. Through this integrated approach, you'll emerge well equipped to leverage the full power of R in your future projects.

Who this book is for

Beginner- to intermediate-level data scientists will get a lot out of this book. So will undergraduate- to masters-level students, and early- to mid-senior data scientists or people in analytics-related roles.

Basic knowledge of linear algebra and modeling will be helpful in understanding the concepts covered in this book.

What this book covers

Chapter 1, *Getting Started with R*, introduces the basics of R programming, including basic data structures such as vectors, matrices, factors, DataFrames, and lists and control logic such as loops, function writing, and so on.

Chapter 2, *Data Processing with dplyr*, introduces common data manipulation and processing techniques using the dplyr library, covering data transformation, aggregation, selection, and merging.

Chapter 3, *Intermediate Data Processing*, introduces common data processing challenges such as converting data types, filling in missing values, and matching for strings. This chapter also covers advanced techniques to ensure data quality, including categorical and textual data.

Chapter 4, *Data Visualization with ggplot2*, introduces common plotting techniques using ggplot2, covering the beginning-level aesthetics, geometrics, and themes of the library, as well as intermediate techniques such as overlaying the graphics with statistical models, coordinate systems, and facets.

Chapter 5, *Exploratory Data Analysis*, introduces different ways to work with and explore different types of data, including categorical data and numerical data, and different ways to summarize the data. This chapter also covers a case study that starts from data cleaning all the way to different visualization and analysis.

Chapter 6, *Effective Reporting with R Markdown*, introduces dynamic documents using R Markdown. Different from static contents, the outputs built using the R Markdown ecosystem offer interactivity covering graphs and tables. This chapter covers the fundamentals of R Markdown reports, including how to add, fine-tune, and customize figures and tables to make interactive and effective reports.

Chapter 7, *Linear Algebra in R*, covers beginner-level linear algebra with illustrated examples in R, including linear equations, vector spaces, and matrix basics such as common matrix operations, such as multiplication, inversion, and transposition.

Chapter 8, *Intermediate Linear Algebra in R*, introduces intermediate topics in linear algebra and implementations in R, including determinants of a matrix – the norm, rank, and trace of a matrix, and the eigenvalues and eigenvectors values.

Chapter 9, *Calculus in R*, introduces the basics of calculus and implementations in R, including fitting a function to data and plotting, derivatives and numerical differentiation, and integrals and integration.

Chapter 10, *Probability Basics*, introduces the basic concepts of probability and implementation in R, including common discrete probability distributions such as geometric distribution, binomial distribution, and Poisson distribution, as well as common continuous distributions such as normal distribution and exponential distribution.

Chapter 11, *Statistics Estimation*, introduces common statistical estimation and inference procedures for both numerical and categorical data. Key concepts such as hypothesis testing and confidence intervals will also be covered.

Chapter 12, *Linear Regression in R*, introduces simple and multiple linear regression models, covering topics such as model estimation, closed-form solutions, evaluation, and linear regression assumptions.

Chapter 13, *Logistic Regression in R*, introduces logistic regression and its connection to linear regression and the loss function and its application to modeling imbalanced datasets.

Chapter 14, *Bayesian Statistics*, introduces the Bayesian inference framework, covering topics such as posterior updates and uncertainty quantification.

To get the most out of this book

To get the most out of this book, it's advisable to have a basic understanding of programming, ideally in R, although a strong foundation in any programming language should suffice. Familiarity with elementary statistics and mathematical concepts will also be beneficial, as the book delves into statistical methods and mathematical models. While the book is structured to guide you from foundational to advanced topics, prior exposure to data analysis techniques will enhance your learning experience. If you're new to R, you may want to spend some extra time on the initial chapters to become comfortable with the programming environment and syntax.

Software covered in the book	Operating system requirements
R	Windows, macOS, or Linux

If you are using the digital version of this book, we advise you to type the code yourself or access the code from the book's GitHub repository (a link is available in the next section). Doing so will help you avoid any potential errors related to the copying and pasting of code.

Download the example code files

You can download the example code files for this book from GitHub at `https://github.com/PacktPublishing/The-Statistics-and-Machine-Learning-with-R-Workshop`. If there's an update to the code, it will be updated in the GitHub repository.

We also have other code bundles from our rich catalog of books and videos available at `https://github.com/PacktPublishing/`. Check them out!

Conventions used

There are a number of text conventions used throughout this book.

`Code in text`: Indicates code words in text, database table names, folder names, filenames, file extensions, pathnames, dummy URLs, user input, and Twitter handles. Here is an example: "Finally, we plot the function to show the characteristic S-shaped curve of the sigmoid function and add the gridlines using the `grid()` function."

A block of code is set as follows:

```
lm_model = lm(Class_num ~ Duration, data=GermanCredit)
coefs = coefficients(lm_model)
intercept = coefs[1]
slope = coefs[2]
```

Bold: Indicates a new term, an important word, or words that you see onscreen. For instance, words in menus or dialog boxes appear in **bold**. Here is an example: "To create an R Markdown file, we can select **File | New File | R Markdown** in RStudio."

> Tips or important notes
> Appear like this.

Get in touch

Feedback from our readers is always welcome.

General feedback: If you have questions about any aspect of this book, email us at customercare@packtpub.com and mention the book title in the subject of your message.

Errata: Although we have taken every care to ensure the accuracy of our content, mistakes do happen. If you have found a mistake in this book, we would be grateful if you would report this to us. Please visit www.packtpub.com/support/errata and fill in the form.

Piracy: If you come across any illegal copies of our works in any form on the internet, we would be grateful if you would provide us with the location address or website name. Please contact us at copyright@packt.com with a link to the material.

If you are interested in becoming an author: If there is a topic that you have expertise in and you are interested in either writing or contributing to a book, please visit authors.packtpub.com.

Share your thoughts

Once you've read *The Statistics and Machine Learning with R Workshop*, we'd love to hear your thoughts! Scan the QR code below to go straight to the Amazon review page for this book and share your feedback.

https://packt.link/r/1-803-24030-X

Your review is important to us and the tech community and will help us make sure we're delivering excellent quality content.

Download a free PDF copy of this book

Thanks for purchasing this book!

Do you like to read on the go but are unable to carry your print books everywhere?

Is your eBook purchase not compatible with the device of your choice?

Don't worry, now with every Packt book you get a DRM-free PDF version of that book at no cost.

Read anywhere, any place, on any device. Search, copy, and paste code from your favorite technical books directly into your application.

The perks don't stop there, you can get exclusive access to discounts, newsletters, and great free content in your inbox daily

Follow these simple steps to get the benefits:

1. Scan the QR code or visit the link below

https://packt.link/free-ebook/9781803240305

2. Submit your proof of purchase

3. That's it! We'll send your free PDF and other benefits to your email directly

Part 1:
Statistics Essentials

This part is designed to equip you with knowledge of statistical and programming fundamentals, focusing particularly on the versatile R language, which will serve as the cornerstone for more advanced topics in subsequent parts.

By the end of this part, you'll have a strong grasp of the core statistical and programming concepts essential for any data science practitioner to understand. With these foundational skills in hand, you'll be well prepared to delve into the more specialized topics that await you in subsequent parts of this book.

This part has the following chapters:

- *Chapter 1, Getting Started with R*
- *Chapter 2, Data Processing with dplyr*
- *Chapter 3, Intermediate Data Processing*
- *Chapter 4, Data Visualization with ggplot2*
- *Chapter 5, Exploratory Data Analysis*
- *Chapter 6, Effective Reporting with R Markdown*

1

Getting Started with R

In this chapter, we will cover the basics of R, the most widely used open source language for statistical analysis and modeling. We will start with an introduction to RStudio, how to perform simple calculations, the common data structures and control logic, and how to write functions in R.

By the end of the chapter, you will be able to do basic computations in R using common data structures such as vectors, lists and data frames in the RStudio **integrated development environment** (IDE). You will also be able to wrap these calculations in functions using different methods.

In this chapter, we will cover the following:

- Introducing R
- Covering the R and RStudio basics
- Common data structures in R
- Control logic in R
- Exploring functions in R

Technical requirements

To complete the exercises in this chapter, you will need to have the following:

- The latest version of R, which is 4.1.2 at the time of writing
- The latest version of RStudio Desktop, which is 2021.09.2+382

All the code for this chapter is available at `https://github.com/PacktPublishing/The-Statistics-and-Machine-Learning-with-R-Workshop/blob/main/Chapter_1/Chapter_1.R`.

Introducing R

R is a popular open source language that supports statistical analysis and modeling, and it is most widely used by statisticians developing statistical models and performing data analysis. One question commonly asked by learners is how to choose between Python and R. For those new to both and needing a simple model for a not-so-big dataset, R would be a better choice. It has rich resources to support modeling and plotting tasks that were developed by statisticians long before Python was born. Besides its many off-the-shelf graphing and statistical modeling offerings, the R community is also catching up in advanced machine learning such as deep learning, which the Python community currently dominates.

There are many differences between the two languages, and recent years have witnessed increasing convergence in many aspects. This book aims to equip you with the essential knowledge to understand and use statistics and calculus via R. We hope that at some point, you will be able to extract from the inner workings of the language itself and think at the methodological level when performing some analysis. After cultivating the essential skills from the fundamentals, it will just be a matter of personal preference regarding the specific language in use. To this end, R provides dedicated utility functions to automatically "convert" Python code to be used within the R context, which gives us another reason not to worry about choosing a specific language.

Covering the R and RStudio basics

It is easy to confuse R with **RStudio** if you are a first-time user. In a nutshell, R is the engine that supports all sorts of backend computations, and RStudio is a convenient tool for navigating and managing related coding and reference resources. Specifically, RStudio is an **IDE** where the user writes R code, performs analysis, and develops models without worrying much about the backend logistics required by the R engine. The interface provided by RStudio makes the development work much more convenient and user-friendly than the vanilla R interface.

First, we need to install R on our computer, as the RStudio will ship with the computation horsepower upon installation. We can choose the corresponding version of R at `https://cloud.r-project.org/`, depending on the specific type of operating system we use. RStudio can then be downloaded at `https://www.rstudio.com/products/rstudio/download/` and installed accordingly. When launching the RStudio application after installing both software, the R engine will be automatically detected and used. Let's go through an exercise to get familiar with the interface.

Exercise 1.01 – exploring RStudio

RStudio provides a comprehensive environment for working with R scripts and exploring the data simultaneously. In this exercise, we will look at a basic example of how to write a simple script to store a string and perform a simple calculation using RStudio.

Perform the following steps to complete this exercise:

1. Launch the RStudio application and observe the three panes:

   - The **Console** pane is used to execute R commands and display the immediate result.

   - The **Environment** pane stores all the global variables in the current **session**.

   - The **Files** pane lists all the files within the current working directory along with other tabs, as shown in *Figure 1.1*.

     Note that the R version is printed as a message in the console (highlighted in the dashed box):

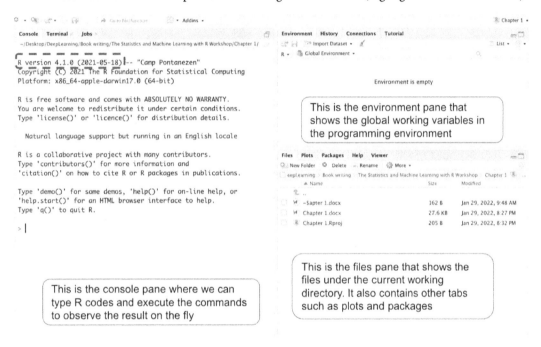

Figure 1.1 – A screenshot of the RStudio upon the first launch

We can also type `R.version` in the console to retrieve more detailed information on the version of the R engine in use, as shown in *Figure 1.2*. It is essential to check the R version, as different versions *may* produce different results when running the same code.

```
> R.version
                _
platform        x86_64-apple-darwin17.0
arch            x86_64
os              darwin17.0
system          x86_64, darwin17.0
status
major           4
minor           1.0
year            2021
month           05
day             18
svn rev         80317
language        R
version.string  R version 4.1.0 (2021-05-18)
nickname        Camp Pontanezen
```

Figure 1.2 – Typing a command in the console to check the R version

2. Build a new **R script** by clicking on the plus sign in the upper-left corner or via **File | New File | R Script**. An R script allows us to write longer R code that involves functions and chunks of code executed in sequence. We will build an R script and name it test.R upon saving the file. See the following figure for an illustration:

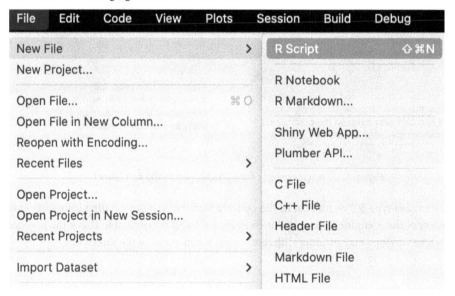

Figure 1.3 – Creating a new R script

3. Running the script can be achieved by placing the cursor at the current line and pressing *Cmd + Enter* for macOS or *Ctrl + Enter* for Windows; alternatively, click on the **Run** button at the top of the R script pane, as shown in the following figure:

Figure 1.4 – Executing the script by clicking on the Run button

4. Type the following commands in the script editing pane and observe the output in the console as well as the changes in the other panes. First, we create a **variable** named test by assigning "I am a string". A variable can be used to store an object, which could take the form of a string, number, data frame, or even function (more on this later). Strings consist of characters, a common data type in R. The test variable created in the script is also reflected in the **Environment** pane, which is a convenient check as we can also observe the content in the variable. See *Figure 1.5* for an illustration:

```
# String assignment
test = "I am a string"
print(test)
```

Figure 1.5 – Creating a string-type variable

We also assign a simple addition operation to test2 and print it out in the console. These commands are also annotated via the # sign, where the contents after the sign are not executed and are only used to provide an explanation of the following code. See *Figure 1.6* for an illustration:

```
# Simple calculation
test2 = 1 + 2
print(test2)
```

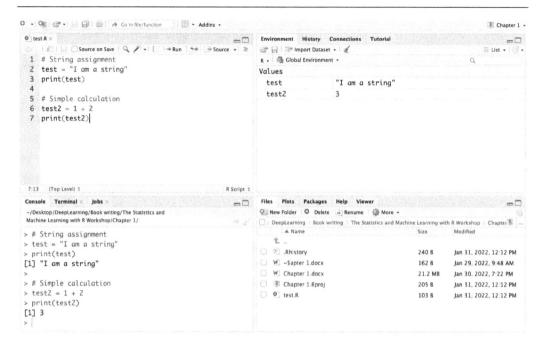

Figure 1.6 – Assigning a string and performing basic computation

5.   We can also check the contents of the environment workspace via the `ls()` function:

```
>>> ls()
"test"   "test2"
```

In addition, note that the newly created R script is also reflected in the **Files** pane. RStudio is an excellent one-stop IDE for working with R and will be the programming interface for this book. We will introduce more features of RStudio in a more specific context along the way.

> **Note**
>
> The canonical way of assigning some value to a variable is via the `<-` operator instead of the `=` sign as in the example. However, the author chose to use the `=` sign as it is faster to type on the screen and has an equivalent effect as the `<-` sign in the majority of cases.
>
> In addition, note that the output message in the **Console** pane has a preceding `[1]` sign, which indicates that the result is a one-dimensional output. We will ignore this sign in the output message unless otherwise specified.

The exercise in the previous section provides an additional example, which is an essential operation in R. As with other modern programming languages, R also ships with many standard arithmetic operators, including subtraction (`-`), multiplication (`*`), division (`/`), exponentiation (`^`), and modulo (`%%`) operators. The modulo operator returns the remainder of the numerator in the division operation.

Let's look at an exercise to go through some common arithmetic operations.

Exercise 1.02 – common arithmetic operations in R

This exercise will perform different arithmetic operations (addition, subtraction, multiplication, division, exponentiation, and modulo) between two numbers: 5 and 2.

Type the commands under the **EXERCISE 1.02** comment section in the **R Script** pane and observe the output message in the console shown in *Figure 1.7*. Note that we removed the print () function, as directly executing the command will also print out the result as highlighted in the console:

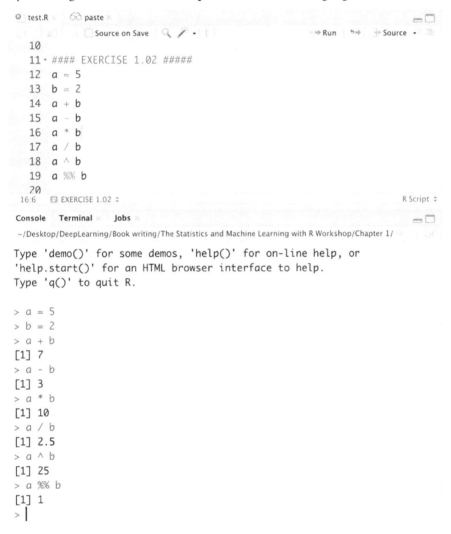

Figure 1.7 – Performing common arithmetic operations in R

Note that these elementary arithmetic operations can jointly form complex operations. When evaluating a complex operation that consists of multiple operators, the general rule of thumb is to use parentheses to enforce the execution of a specific component according to the desired sequence. This follows in most numeric analyses using any programming language.

But, what forms can we expect the data to take in R?

Common data types in R

There are five most basic data types in R: **numeric**, **integer**, **character**, **logical**, and **factor**. Any complex R object can be decomposed into individual elements that fall into one of these five data types and, therefore, contain one or more data types. The definition of these five data types is as follows:

- **Numeric** is the default data type in R and represents a decimal value, such as 1.23. A variable is treated as a numeric even if we assign an integer value to it in the first place.

- **Integer** is a whole number and so a subset of the numeric data type.

- **Character** is the data type used to store a sequence of characters (including letters, symbols, or even numbers) to form a string or a piece of text, surrounded by double or single quotes.

- **Logical** is a Boolean data type that only takes one of two values: TRUE or FALSE. It is often used in a conditional statement to determine whether specific codes after the condition should be executed.

- **Factor** is a special data type used to store categorical variables that contain a limited number of categories (or **levels**), ordered or unordered. For example, a list of student heights classified as low, medium, and high can be represented as a factor type to encode the inherent ordering, which would not be available when represented as a character type. On the other hand, unordered lists such as male and female can also be represented as factor types.

Let's go through an example to understand these different data types.

Exercise 1.03 – understanding data types in R

R has strict rules on the data types when performing arithmetic operations. In general, the data types of all variables should be the same when evaluating a particular **statement** (a piece of code). Performing an arithmetic operation on different data types may give an error. In this exercise, we will look at how to check the data type to ensure the type consistency and different ways to convert the data type from one into another:

1. We start by creating five variables, each belonging to a different data type. Check the data type using the class() function. Note that we can use the semicolon to separate different actions:

```
>>> a = 1.0; b = 1; c = "test"; d = TRUE; e = factor("test")
>>> class(a); class(b); class(c); class(d); class(e)
```

```
"numeric"
"numeric"
"character"
"logical"
"factor"
```

As expected, the data type of the b variable is converted into numeric even when it is assigned an integer in the first place.

2. Perform addition on the variables. Let's start with the a and b variables:

```
>>> a + b
2
>>> class(a + b)
"numeric"
```

Note that the decimal point is ignored when displaying the result of the addition, which is still numeric as verified via the class() function.

Now, let's look at the addition between a and c:

```
>>> a + c
Error in a + c : non-numeric argument to binary operator
```

This time, we received an *error* message due to a mismatch in data types when evaluating an addition operation. This is because the + addition operator in R is a binary operator designed to take in two values (operands) and produce another, all of which need to be numeric (including integer, of course). The error pops up when any of the two input *arguments* are non-numeric.

3. Let's trying adding a and d:

```
>>> a + d
2
>>> class(a + d)
"numeric"
```

Surprisingly, the result is the same as a + b, suggesting that the Boolean b variable taking a TRUE value is converted into a value of one under the hood. Correspondingly, a Boolean value of FALSE, obtained by adding an exclamation mark before the variable, would be treated as zero when performing an arithmetic operation with a numeric:

```
>>> a + !d
1
```

Note that the implicit Boolean conversion occurs in settings when such conversion is necessary to proceed in a specific statement. For example, d is converted into a numeric value of one when evaluating whether a equals d:

```
>>> a == d
TRUE
```

4. Convert the data types using the as.(datatype) family of functions in R.

    For example, the as.numeric() function converts the input parameter into a numeric, as.integer() returns the integer part of the input decimal, as.character() converts all inputs (including numeric and Boolean) into strings, and as.logical() converts any non-zero numeric into TRUE and zero into FALSE. Let's look at a few examples:

    ```
    >>> class(as.numeric(b))
    "numeric"
    ```

This suggests that the b variable is successfully converted into numeric. Note that type conversion is a standard data processing operation in R, and type incompatibility is a popular source of error that may be difficult to trace:

```
>>> as.integer(1.8)
1
>>> round(1.8)
2
```

Since as.integer() only returns the integer part of the input, the result is always "floored" to the lower bound integer. We could use the round() function to round it up or down, depending on the value of the first digit after the decimal point:

```
>>> as.character(a)
"1"
>>> as.character(d)
"TRUE"
```

The as.character() function converts all input parameters into strings as represented by the double quotes, including numeric and Boolean. The converted value no longer maintains the original arithmetic property. For example, a numeric converted into a character would not go through the addition operation. Also, a Boolean converted into a character would no longer be evaluated via a logical statement and treated as a character:

```
>>> as.factor(a)
1
Levels: 1
>>> as.factor(c)
test
Levels: test
```

Since there is only one element in the input parameter, the resulting number of levels is only 1, meaning the original input itself.

> **Note**
>
> A categorical variable is called a **nominal** variable when there is no natural ordering among the categories, and an **ordinal** variable if there is natural ordering. For example, the temperature variable valued as either `high`, `medium`, or `low` has an inherent ordering in nature, while a gender variable valued as either `male` or `female` has no order.

Common data structures in R

Data structures provide an organized way to store various data points that follow either the same or different types. This section will look at the typical data structures used in R, including the vector, matrix, data frame, and list.

Vector

A **vector** is a one-dimensional array that can hold a series of elements of any consistent data type, including numeric, integer, character, logical, or factor. We can create a vector by filling in comma-separated elements in the input argument of the combine function, `c()`. The arithmetic operations between two vectors are similar to the single-element example earlier, provided that their lengths are equal. There needs to be a one-to-one correspondence between the elements of the two vectors; if not, the calculation *may* give an error. Let's look at an exercise.

Exercise 1.04 – working with vectors

We will create two vectors of the same length in this exercise and add them up. As an extension, we will also attempt the same addition using a vector of a different length. We will also perform a pairwise comparison between the two vectors:

1. Create two vectors named `vec_a` and `vec_b` and extract simple summary statistics such as mean and sum:

    ```
    >>> vec_a = c(1,2,3)
    >>> vec_b = c(1,1,1)
    >>> sum(vec_a)
    6
    >>> mean(vec_a)
    2
    ```

    The sum and mean of a vector can be generated using the `sum()` and `mean()` function, respectively. We will cover more ways to summarize a vector later.

2. Add up vec_a and vec_b:

    ```
    >>> vec_a + vec_b
    2 3 4
    ```

The addition between two vectors is performed element-wise. The result can also be saved into another variable for further processing. How about adding a single element to a vector?

3. Add vec_a and 1:

    ```
    >>> vec_a + 1
    2 3 4
    ```

Under the hood, element one is broadcasted into vector c(1,1,1), whose length is decided by vec_a. **Broadcasting** is a unique mechanism that replicates the elements of the short vector into the required length, as long as the length of the longer vector is a multiple of the short vector's length. The same trick may not apply when it is not a multiple.

4. Add vec_a and c(1,1):

    ```
    >>> vec_a + c(1,1)
    2 3 4
    Warning message:
    In vec_a + c(1, 1) :
    longer object length is not a multiple of shorter object length
    ```

We still get the same result, except for a warning message saying that the longer vector's length of three is not a multiple of the shorter vector length of two. Pay attention to this warning message. It is not recommended to follow such practice as the warning may become an explicit error or become the implicit cause of an underlying bug in an extensive program.

5. Next, we will perform a pairwise comparison between the two vectors:

    ```
    vec_a > vec_b
    FALSE   TRUE    TRUE
    vec_a == vec_b
    TRUE FALSE FALSE
    ```

Here, we have used evaluation operators such as > (greater than) and == (equal to), returning logical results (TRUE or FALSE) for each pair.

Note, there are multiple logical comparison operators in R. The common ones include the following:

- < for less than
- <= for less than or equal to

- • > for greater than
- • >= for greater than or equal to
- • == for equal to
- • != for not equal to

Besides the common arithmetic operations, we may also be interested in selected partial components of a vector. We can use square brackets to select specific elements of a vector, which is the same way to select elements in other data structures such as in a matrix or a data frame. In between the square brackets are indices indicating what elements to select. For example, we can use vec_a[1] to select the first element of vec_a. Let's go through an exercise to look at different ways to subset a vector.

Exercise 1.05 – subsetting a vector

We can pass in the select index (starting from 1) to select the corresponding element in the vector. We can wrap the indices via the c() combine function and pass in the square brackets to select multiple elements. Selecting multiple sequential indices can also be achieved via a shorthand notation by writing the first and last index with a colon in between. Let's run through different ways of subsetting a vector:

1. Select the first element in vec_a:

    ```
    >>> vec_a[1]
    1
    ```

2. Select the first and third elements in vec_a:

    ```
    >>> vec_a[c(1,3)]
    1 3
    ```

3. Select all three elements in vec_a:

    ```
    >>> vec_a[c(1,2,3)]
    1 2 3
    ```

    Selecting multiple elements in this way is not very convenient since we need to type every index. When the indices are sequential, a nice shorthand trick is to use the starting and end index separated by a colon. For example, 1:3 would be the same as c(1,2,3):

    ```
    >>> vec_a[1:3]
    1 2 3
    ```

We can also perform more complex subsetting by adding a conditional statement within the square brackets as the selection criteria. For example, the logical evaluation introduced earlier returns either True or False. An element whose index is marked as true in the square bracket would be selected. Let's see an example.

4. Select elements in vec_a that are bigger than the corresponding elements in vec_b:

```
>>> vec_a[vec_a > vec_b]
2 3
```

The result contains the last two elements since only the second and third indices are set as true.

Matrix

Like a vector, a **matrix** is a two-dimensional array consisting of a collection of elements of the same data type arranged in a fixed number of rows and columns. It is often faster to work with a data structure exclusively containing the same data type since the program does not need to differentiate between different types of data. This makes the matrix a popular data structure in scientific computing, especially in an optimization procedure that involves intensive computation. Let's get familiar with the matrix, including different ways to create, index, subset, and enlarge a matrix.

Exercise 1.06 – creating a matrix

The standard way to create a matrix in R is to call the matrix() function, where we need to supply three input arguments:

- The elements to be filled in the matrix
- The number of rows in the matrix
- The filling direction (either by row or by column)

We will also rename the rows and columns of the matrix:

1. Use vec_a and vec_b to create a matrix called mtx_a:

```
>>> mtx_a = matrix(c(vec_a,vec_b), nrow=2, byrow=TRUE)
>>> mtx_a
     [,1] [,2] [,3]
[1,]    1    2    3
[2,]    1    1    1
```

First, the input vectors, vec_a and vec_b, are combined via the c() function to form a long vector, which then gets sequentially arranged into two rows (nrow=2) row-wise (byrow=TRUE). Feel free to try out different dimension configurations, such as setting three rows and two columns when creating the matrix.

Pay attention to the row and column names in the output. The rows are indexed by the first index in the square bracket, while the second indexes the columns. We can also rename the matrix as follows.

2. Rename the matrix mtx_a via the rownames() and colnames() functions:

```
>>> rownames(mtx_a) = c("r1", "r2")
>>> colnames(mtx_a) = c("c1", "c2", "c3")
>>> mtx_a
   c1 c2 c3
r1  1  2  3
r2  1  1  1
```

Let's look at how to select elements from the matrix.

Exercise 1.07 – subsetting a matrix

We can still use the square brackets to select one or more matrix elements. The colon shorthand trick also applies to matrix subsetting:

1. Select the element at the first row and second column of the mtx_a matrix:

```
>>> mtx_a[1,2]
2
```

2. Select all elements of the last two columns across all rows in the mtx_a matrix:

```
>>> mtx_a[1:2,c(2,3)]
   c2 c3
r1  2  3
r2  1  1
```

3. Select all elements of the second row of the mtx_a matrix:

```
>>> mtx_a[2,]
c1 c2 c3
 1  1  1
```

In this example, we have used the fact that the second (column-level) index indicates that all columns are selected when left blank. The same applies to the first (row-level) index as well.

We can also select the second row using the row name:

```
>>> mtx_a[rownames(mtx_a)=="r2",]
c1 c2 c3
 1  1
```

Selecting elements by matching the row name using a conditional evaluation statement offers a more precise way of subsetting the matrix, especially when counting the exact index becomes troublesome. Name-based indexing also applies to columns.

4. Select the third row of the mtx_a matrix:

```
>>> mtx_a[,3]
r1 r2
 3  1
>>> mtx_a[,colnames(mtx_a)=="c3"]
r1 r2
 3  1
```

Therefore, we have multiple ways to select the specific elements of interest from a matrix.

Working with a matrix requires similar arithmetic operations compared to a vector. In the next exercise, we will look at summarizing a matrix both row-wise and column-wise and performing basic operations such as addition and multiplication.

Exercise 1.08 – arithmetic operations with a matrix

Let's start by making a new matrix:

1. Create another matrix named mtx_b whose elements are double those in mtx_a:

```
>>> mtx_b = mtx_a * 2
>>> mtx_b
   c1 c2 c3
r1  2  4  6
r2  2  2  2
```

Besides multiplication, all standard arithmetic operators (such as +, -, and /) apply in a similar element-wise fashion to a matrix, backed by the same broadcasting mechanism. Operations between two matrices of the same size are also performed element-wise.

2. Divide mtx_a by mtx_b:

```
>>> mtx_a / mtx_b
    c1  c2   c3
r1 0.5 0.5  0.5
r2 0.5 0.5  0.5
```

3. Calculate the row-wise and column-wise sum and mean of mtx_a using rowSums(), colSums(), rowMeans(), and colMeans() respectively:

```
>>> rowSums(mtx_a)
r1 r2
 6  3
>>> colSums(mtx_a)
c1 c2 c3
 2  3  4
```

```
>>> rowMeans(mtx_a)
r1 r2
 2  1
>>> colMeans(mtx_a)
c1  c2  c3
1.0 1.5 2.0
```

When running an optimizing procedure, we often need to save some intermediate metrics, such as model loss and accuracy, for diagnosis. These metrics can be saved in a matrix form by gradually appending new data to the current matrix. Let's look at how to expand a matrix both row-wise and column-wise.

Exercise 1.09 – expanding a matrix

Adding a column or multiple columns to a matrix can be achieved via the cbind() function, which merges a new matrix or vector column-wise. Similarly, an additional matrix or vector can be concatenated row-wise via the rbind() function:

1. Append mtx_b to mtx_a column-wise:

    ```
    >>> cbind(mtx_a, mtx_b)
       c1 c2 c3 c1 c2 c3
    r1  1  2  3  2  4  6
    r2  1  1  1  2  2  2
    ```

    We may need to rename the columns since some of them overlap. This also applies to the row-wise concatenation as follows.

2. Append mtx_b to mtx_a row-wise:

    ```
    >>> rbind(mtx_a, mtx_b)
       c1 c2 c3
    r1  1  2  3
    r2  1  1  1
    r1  2  4  6
    r2  2  2  2
    ```

So, we've seen the matrix in operation. How about data frames next?

Data frame

A **data frame** is a standard data structure where variables are stored as columns and observations as rows in an **object**. It is an advanced version of a matrix in that the elements for each column can be of different data types.

The R engine comes with several default datasets stored as data frames. In the next exercise, we will look at different ways to examine and understand the structure of a data frame.

Exercise 1.10 – understanding data frames

The data frame is a famous data structure representing rectangular-shaped data similar to Excel. Let's examine a default dataset in R as an example:

1. Load the `iris` dataset:

    ```
    >>> data("iris")
    >>> dim(iris)
    150   5
    ```

    Checking the dimension using the `dim()` function suggests that the `iris` dataset contains 150 rows and five columns. We can initially understand its contents by looking at the first and last few observations (rows) in the dataset.

2. Examine the first and last five rows using `head()` and `tail()`:

    ```
    >>> head(iris)
      Sepal.Length Sepal.Width Petal.Length Petal.Width Species
    1          5.1         3.5          1.4         0.2  setosa
    2          4.9         3.0          1.4         0.2  setosa
    3          4.7         3.2          1.3         0.2  setosa
    4          4.6         3.1          1.5         0.2  setosa
    5          5.0         3.6          1.4         0.2  setosa
    6          5.4         3.9          1.7         0.4  setosa
    >>> tail(iris)
        Sepal.Length Sepal.Width Petal.Length Petal.Width   Species
    145          6.7         3.3          5.7         2.5 virginica
    146          6.7         3.0          5.2         2.3 virginica
    147          6.3         2.5          5.0         1.9 virginica
    148          6.5         3.0          5.2         2.0 virginica
    149          6.2         3.4          5.4         2.3 virginica
    150          5.9         3.0          5.1         1.8 virginica
    ```

    Note that the row names are sequentially indexed by integers starting from one by default. The first four columns are numeric, and the last is a character (or factor). We can look at the structure of the data frame more systematically.

3. Examine the structure of the `iris` dataset using `str()`:

    ```
    >>> str(iris)
    'data.frame':    150 obs. of  5 variables:
     $ Sepal.Length: num  5.1 4.9 4.7 4.6 5 5.4 4.6 5 4.4 4.9 ...
     $ Sepal.Width : num  3.5 3 3.2 3.1 3.6 3.9 3.4 3.4 2.9 3.1 ...
     $ Petal.Length: num  1.4 1.4 1.3 1.5 1.4 1.7 1.4 1.5 1.4 1.5
     ...
    ```

```
$ Petal.Width : num   0.2 0.2 0.2 0.2 0.2 0.4 0.3 0.2 0.2 0.1
...
$ Species      : Factor w/ 3 levels "setosa","versicolor",..: 1
1 1 1 1 1 1 1 1 1 ...
```

The `str()` function summarizes the data frame structure, including the total number of observations and variables, the complete list of variable names, data type, and the first few observations. The number of categories (levels) is also shown if the column is a factor.

We can also create a data frame by passing in vectors as columns of the same length to the `data.frame()` function.

4. Create a data frame called `df_a` with two columns that correspond to `vec_a` and `vec_b` respectively:

    ```
    >>> df_a = data.frame("a"=vec_a, "b"=vec_b)
    >>> df_a
      a b
    1 1 1
    2 2 1
    3 3 1
    ```

Selecting the elements of a data frame can be done in a similar fashion to matrix selection. Other functions such as `subset()` make the selection more flexible. Let's go through an example.

Exercise 1.11 – selecting elements in a data frame

In this exercise, we will first look at different ways to select a particular set of elements and then introduce the `subset()` function to perform customized conditional selection:

1. Select the second column of the `df_a` data frame:

    ```
    >>> df_a[,2]
    1 1 1
    ```

    The row-level indexing is left blank to indicate that all rows will be selected. We can also make it explicit by referencing all row-level indices:

    ```
    >>> df_a[1:3,2]
    1 1 1
    ```

    We can also select by using the name of the second column as follows:

    ```
    >>> df_a[,"b"]
    1 1 1
    ```

    Alternatively, we can use the shortcut $ sign to reference the column name directly:

    ```
    >>> df_a$b
    1 1 1
    ```

The `subset ()` function provides an easy and structured way to perform row-level filtering and column-level selection. Let's see how it works in practice.

2. Select the rows of `df_a` where column a is greater than two:

```
>>> subset(df_a, a>2)
  a b
3 3 1
```

Note that row index three is also shown as part of the output.

We can directly use column a within the context of the `subset ()` function, saving us from using the $ sign instead. We can also select the column by passing the column name to the `select` argument.

3. Select column b where column a is greater than two in `df_a`:

```
>>> subset(df_a, a>2, select="b")
  b
3 1
```

Another typical operation in data analysis is sorting one or more variables of a data frame. Let's see how it works in R.

Exercise 1.12 – sorting vectors and data frames

The `order ()` function can be used to return the ranked position of the elements in the input vector, which can then be used to sort the elements via updated indexing:

1. Create the `c (5, 1, 10)` vector in `vec_c` and sort it in ascending order:

```
>>> vec_c = c(5,1,10)
>>> order(vec_c)
2 1 3
>>> vec_c[order(vec_c)]
1  5 10
```

Since the smallest element in `vec_c` is 1, the corresponding ranked position is 1. Similarly, 5 is set as the second rank and 10 as the third and highest rank. The ranked positions are then used to reshuffle and sort the original vector, the same as how we would select its elements via **positional indexing**.

The `order ()` function ranks the elements in ascending order by default. What if we want to sort by descending order? We could simply add a minus sign to the input vector.

2. Sort the `df_a` data frame by column a in descending order:

```
>>> df_a[order(-df_a$a),]
  a b
```

```
3  3  1
2  2  1
1  1  1
```

Data frames will be the primary structures we will work with in this book. Let's look at the last and most complex data structure: list.

List

A **list** is a flexible data structure that can hold different data types (numeric, integer, character, logical, factor, or even list itself), each possibly having a different length. It is the most complex structure we have introduced so far, gathering various objects in a structured way. To recap, let's compare the four data structures in terms of the contents, data type, and length in *Figure 1.8*. In general, all four structures can store elements of any data type. Vectors (one-dimensional array) and matrices (two-dimensional array) require the contents to be homogeneous data types. A data frame contains one or more vectors whose data types could differ, and a list could contain entries of different data types. Matrices and data frames follows a rectangular shape and so require the same length for each column. However, the entries in a list could be of arbitrary lengths (subject to memory constraint) different from each other.

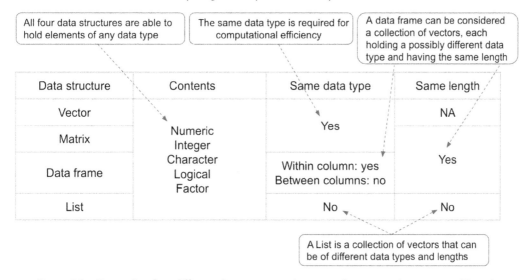

Figure 1.8 – Comparing four different data structures in terms of contents, data type, and length

Let's look at how to create a list.

Exercise 1.13 – creating a list

In this exercise, we will go through different ways to manipulate a list, including creating and renaming a list, and accessing, adding, and removing elements in a list:

1. Create a list using the previous a, vec_a, and df_a variables:

```
>>> ls_a = list(a, vec_a, df_a)
>>> ls_a
[[1]]
[1] 1

[[2]]
[1] 1 2 3

[[3]]
  a b
1 1 1
2 2 1
3 3 1
```

The output shows that the list elements are indexed by double square brackets, which can be used to access the entries in the list.

2. Access the second entry in the list, ls_a:

```
>>> ls_a[[2]]
1 2 3
```

The default indices can also be renamed to enable entry selection by name.

3. Rename the list based on the original names and access the vec_a variable:

```
>>> names(ls_a) <- c("a", "vec_a", "df_a")
ls_a
$a
[1] 1

$vec_a
[1] 1 2 3

$df_a
  a b
1 1 1
2 2 1
3 3 1
>>> ls_a[['vec_a']]
```

```
1 2 3
>>> ls_a$vec_a
1 2 3
```

We can access a specific entry in the list by using the name either in square brackets or via the $ sign.

4. Add a new entry named new_entry with the content "test" in the ls_a list:

```
>>> ls_a[['new_entry']] = "test"
>>> ls_a
$a
[1] 1
$vec_a
[1] 1 2 3
$df_a
  a b
1 1 1
2 2 1
3 3 1
$new_entry
[1] "test"
```

The result shows that "test" is now added to the last entry of ls_a. We can also remove a specific entry by assigning NULL to it.

5. Remove the entry named df_a in ls_a:

```
>>> ls_a[['df_a']] = NULL
>>> ls_a
$a
[1] 1
$vec_a
[1] 1 2 3
$new_entry
[1] "test"
```

The entry named df_a is now successfully removed from the list. We can also update an existing entry in the list.

6. Update the entry named vec_a to be c(1,2):

```
>>> ls_a[['vec_a']] = c(1,2)
>>> ls_a
$a
[1] 1
$vec_a
```

```
[1] 1 2
$new_entry
[1] "test"
```

The entry named `vec_a` is now successfully updated.

The flexibility and scalability of the list structure make it a popular choice for storing heterogeneous data elements, similar to the dictionary in Python. In the next section, we will extend our knowledge base by going over the control logic in R, which gives us more flexibility and precision when writing long programs.

Control logic in R

Relational and logical operators help compare statements as we add logic to the program. We can also add to the complexity by evaluating multiple conditional statements via loops that repeatedly iterate over a sequence of actions. This section will cover the essential relational and logical operators that form the building blocks of conditional statements.

Relational operators

We briefly covered a few relational operators such as `>=` and `==` earlier. This section will provide a detailed walkthrough on the use of standard relational operators. Let's look at a few examples.

Exercise 1.14 – practicing with standard relational operators

Relational operators allow us to compare two quantities and obtain the single result of the comparison. We will go over the following steps to learn how to express and use standard relational operators in R:

1. Execute the following evaluations using the equality operator (`==`) and observe the output:

    ```
    >>> 1 == 2
    FALSE
    >>> "statistics" == "calculus"
    FALSE
    >>> TRUE == TRUE
    TRUE
    >>> TRUE == FALSE
    FALSE
    ```

    The equality operator performs by strictly evaluating the two input arguments on both sides (including logical data) and only returns `TRUE` if they are equal.

2. Execute the same evaluations using the inequality operator (`!=`) and observe the output:

    ```
    >>> 1 != 2
    TRUE
    ```

```
>>> "statistics" != "calculus"
TRUE
>>> TRUE != TRUE
FALSE
>>> TRUE != FALSE
TRUE
```

The inequality operator is the exact opposite of the equality operator.

3. Execute the following evaluations using the greater than and less than operators (> and <) and observe the output:

```
>>> 1 < 2
TRUE
>>> "statistics" > "calculus"
TRUE
>>> TRUE > FALSE
TRUE
```

In the second evaluation, the comparison between character data follows the pairwise alphabetical order of both strings starting from the leftmost character. In this case, the letter s comes after c and is encoded as a higher-valued numeric. In the third example, TRUE is converted into one and FALSE into zero, so returning a logical value of TRUE.

4. Execute the following evaluations using the greater-than-or-equal-to operator (>=) and less-than-or-equal-to operator (<=) and observe the output:

```
>>> 1 >= 2
FALSE
>>> 2 <= 2
TRUE
```

Note that these operators consist of two conditional evaluations connected via an OR operator (|). We can, therefore, break it down into two evaluations in brackets, resulting in the same output as before:

```
>>> (1 > 2) | (1 == 2)
FALSE
>>> (2 < 2) | (2 == 2)
TRUE
```

The relational operators also apply to vectors, which we encountered earlier, such as row-level filtering to subset a data frame.

5. Compare vec_a with 1 using the greater-than operator:

```
>>> vec_a > 1
FALSE   TRUE   TRUE
```

We would get the same result by separately comparing each element and combining the resulting using c ( ).

Logical operators

A **logical operator** is used to combine the results of multiple relational operators. There are three basic logical operators in R, including AND (&), OR (|), and NOT (!). The AND operator returns TRUE only if both operands are TRUE, and the OR operator returns TRUE if at least one operand is TRUE. On the other hand, the NOT operator flips the evaluation result to the opposite.

Let's go through an exercise on the use of these logical operators.

Exercise 1.15 – practicing using standard logical operators

We will start with the AND operator, the most widely used control logic to ensure a specific action only happens if multiple conditions are satisfied at the same time:

1. Execute the following evaluations using the AND operator and observe the output:

```
>>> TRUE & FALSE
FALSE
>>> TRUE & TRUE
TRUE
>>> FALSE & FALSE
FALSE
>>> 1 > 0 & 1 < 2
TRUE
```

The result shows that both conditions need to be satisfied to obtain a TRUE output.

2. Execute the following evaluations using the OR operator and observe the output:

```
>>> TRUE | FALSE
TRUE
>>> TRUE | TRUE
TRUE
>>> FALSE | FALSE
FALSE
>>> 1 < 0 | 1 < 2
TRUE
```

The result shows that the output is TRUE if at least one condition is evaluated as TRUE.

3. Execute the following evaluations using the NOT operator and observe the output:

```
>>> !TRUE
FALSE
```

```
>>> !FALSE
TRUE
>>> !(1<0)
TRUE
```

In the third example, the evaluation is the same as 1 >= 0, which returns TRUE. The NOT operator, therefore, reverses the evaluation result after the exclamation sign.

These operators can also be used to perform pairwise logical evaluations in vectors.

4.   Execute the following evaluations and observe the output:

```
>>> c(TRUE, FALSE) & c(TRUE, TRUE)
TRUE FALSE
>>> c(TRUE, FALSE) | c(TRUE, TRUE)
TRUE TRUE
>>> !c(TRUE, FALSE)
FALSE   TRUE
```

There is also a long-form for the AND (&&) and the OR (||) logical operators. Different from the element-wise comparison in the previous short-form, the long-form is used to evaluate only the first element of each input vector, and such evaluation continues only until the result is determined. In other words, the long-form only returns a single result when evaluating two vectors of multiple elements. It is most widely used in modern R programming control flow, especially in the conditional if statement.

Let's look at the following example:

```
>>> c(TRUE, FALSE) && c(FALSE, TRUE)
FALSE
>>> c(TRUE, FALSE) || c(FALSE, TRUE)
TRUE
```

Both evaluations are based on the first element of each vector. That is, the second element of each vector is ignored in both evaluations. This offers computational benefit, especially when the vectors are large. Since there is no point in continuing the evaluation if the final result can be obtained by evaluating the first element, we can safely discard the rest.

In the first evaluation using &&, comparing the first element of the two vectors (TRUE and FALSE) returns FALSE, while continuing the comparison of the second element will also return FALSE, so the second comparison is unnecessary. In the second evaluation using ||, comparing the first element (TRUE | FALSE) gives TRUE, saving the need to make the second comparison, as the result will always be evaluated as TRUE.

Conditional statements

A **conditional statement**, or more specifically, the `if-else` statement, is used to combine the result of multiple logical operators and decide the flow of follow-up actions. It is commonly used to increase the complexity of large R programs. The `if-else` statement follows a general structure as follows, where the evaluation condition is first validated. If the validation returns TRUE, the expression within the curve braces of the `if` clause would be executed and the rest of the code is ignored. Otherwise, the expression within the `else` clause would be executed:

```
if(evaluation condition){
some expression
} else {
other expression
}
```

Let's go through an exercise to see how to use the `if-else` control statement.

Exercise 1.16 – practicing using the conditional statement

Time for another exercise! Let's practice using the conditional statement:

1. Initialize an x variable with a value of 1 and write an `if-else` condition to determine the output message. Print out `"positive"` if x is greater than zero, and `"not positive"` otherwise:

    ```
    >>> x = 1
    >>> if(x > 0){
    >>>    print("positive")
    >>> } else {
    >>>    print("not positive")
    >>> }
    "positive"
    ```

    The condition within the `if` clause evaluates to be TRUE, and the code inside is executed, printing out `"positive"` in the console. Note that the `else` branch is optional and can be removed if we only intend to place one check to the input. Additional `if-else` control can also be embedded within a branch.

    We can also add additional branches using the `if-else` conditional control statement, where the middle part can be repeated multiple times.

2. Initialize an x variable with 0 and write a control flow to determine and print out its sign:

    ```
    >>> x = 0
    >>> if(x > 0){
    >>>    print("positive")
    >>> } else if(x == 0){
    ```

```
>>>    print("zero")
>>> } else {
>>>    print("negative")
>>> }
"zero"
```

As the conditions are sequentially evaluated, the second statement returns TRUE and so prints out "zero".

Loops

A **loop** is similar to the if statement; the codes will only be executed if the condition evaluates to be TRUE. The only difference is that a loop will continue to iteratively execute the code as long as the condition is TRUE. There are two types of loops: the while loop and the for loop. The while loop is used when the number of iterations is unknown, and the termination relies on either the evaluation condition or a separated condition within the running expression using the break control statement. The for loop is used when the number of iterations is known.

The while loop follows a general structure as follows, where condition 1 first gets evaluated to determine the expression within the outer curly braces that should be executed. There is an (optional) if statement to decide whether the while loop needs to be terminated based on condition 2. These two conditions control the termination of the while loop, which exits the execution as long as any one condition evaluates as TRUE. Inside the if clause, condition 2 can be placed anywhere within the while block:

```
while(condition 1){
some expression
if(condition 2){
        break
}
}
```

Note that condition 1 within the while statement needs to be FALSE at some point; otherwise, the loop will continue indefinitely, which may cause a session expiry error within RStudio.

Let's go through an exercise to look at how to use the while loop.

Exercise 1.17 – practicing the while loop

Let's try out the while loop:

1. Initialize an x variable with a value of 2 and write a while loop. If x is less than 10, square it and print out its value:

    ```
    >>> x = 2
    >>> while(x < 10){
    ```

```
>>>    x = x^2
>>>    print(x)
>>> }
4
16
```

The `while` loop is executed twice, bringing the value of x from 2 to 16. During the third evaluation, x is above 10 and the conditional statement evaluates to be `FALSE`, thus exiting the loop. We can also print out x to double-check its value:

```
>>> x
16
```

2. Add a condition after the squaring to exit the loop if x is greater than 10:

```
>>> x = 2
>>> while(x < 10){
>>>    x = x^2
>>>    if(x > 10){
>>>       break
>>>    }
>>>    print(x)
>>> }
4
```

Only one number is printed out this time. The reason is that when x is changed to `16`, the `if` condition evaluates to be `TRUE`, thus triggering the `break` statement to exit the `while` loop and ignore the `print()` statement. Let's verify the value of x:

```
>>> x
16
```

Let's look at the `for` loop, which assumes the following general structure. Here, `var` is a placement to sequentially reference the contents in `sequence`, which can be a vector, a list, or another data structure:

```
for(var in sequence){
some expression
}
```

The same expression will be evaluated for each unique variable in `sequence`, unless an explicit `if` condition is triggered to either exit the loop using `break`, or skip the rest of the code and immediately jump to the next iteration using `next`. Let's go through an exercise to put these in perspective.

Exercise 1.18 – practicing using the for loop

Next, let's try the `for` loop:

1. Create a vector to store three strings (`statistics, and,` and `calculus`) and print out each element:

    ```
    >>> string_a = c("statistics","and","calculus")
    >>> for(i in string_a){
    >>>    print(i)
    >>> }
    "statistics"
    "and"
    "calculus"
    ```

    Here, the `for` loop iterates through each element in the `string_a` vector by sequentially assigning the element value to the `i` variable at each iteration. We can also choose to iterate using the vector index, as follows:

    ```
    >>> for(i in 1:length(string_a)){
    >>>    print(string_a[i])
    >>> }
    "statistics"
    "and"
    "calculus"
    ```

    Here, we created a series of integer indexes from 1 up to the length of the vector and assigned them to the `i` variable in each iteration, which is then used to reference the element in the `string_a` vector. This is a more flexible and versatile way of referencing elements in a vector since we can also use the same index to reference other vectors. Directly referencing the element as in the previous approach is more concise and readable. However, it lacks the level of control and flexibility without the looping index.

2. Add a condition to break the loop if the current element is `"and"`:

    ```
    >>> for(i in string_a){
    >>>    if(i == "and"){
    >>>      break
    >>>    }
    >>>    print(i)
    >>> }
    "statistics"
    ```

    The loop is exited upon satisfying the `if` condition when the current value in `i` is `"and"`.

3.   Add a condition to jump to the next iteration if the current element is `"and"`:

```
>>> for(i in string_a){
>>>     if(i == "and"){
>>>         next
>>>     }
>>>     print(i)
>>> }
"statistics"
"calculus"
```

When the next statement is evaluated, the following `print()` function is ignored, and the program jumps to the next iteration, printing only `"statistics"` and `"calculus"` with the `"and"` element.

So far, we have covered some of the most fundamental building blocks in R. We are now ready to come to the last and most widely used building block: functions.

Exploring functions in R

A **function** is a collection of statements in the form of an object that receives an (optional) input, completes a specific task, and (optionally) generates an output. We may or may not be interested in how a function achieves the task and produces the output. When we only care about utilizing an existing function, which could be built-in and provisioned by R itself or pre-written by someone else, we can treat it as a black box and pass the required input to obtain the output we want. Examples include the `sum()` and `mean()` functions we used in the previous exercise. We can also define our own function to operate as an interface that processes a given input signal and produces an output. See *Figure 1.9* for an illustration:

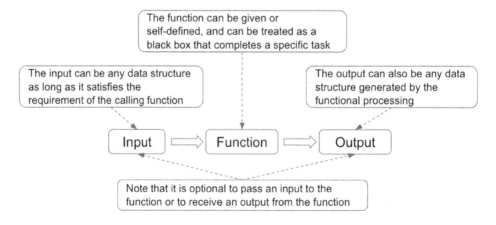

Figure 1.9 – Illustration of a function's workflow

A function can be created using the `function` keyword with the following format:

```
function_name = function(argument_1, argument_2, …){
  some statements
}
```

A function can be decomposed into the following parts:

- **Function name**: The name of the functional object registered and stored in the R environment. We use this name followed by a pair of parentheses and (optionally) input arguments within the parentheses to call the function.

- **Input argument**: A placeholder used to receive input value when calling the function. An argument can be optional (with a default value assigned) or compulsory (with no default value assigned). Setting all arguments as optional is the same as requiring no compulsory input arguments for the function. However, we will need to pass a specific value to a compulsory argument in order to call the function. In addition, the optional argument can also appear after the compulsory argument, if any.

- **Function body**: This is the area where the main statement is executed to complete a specific action and fulfill the purpose of the function.

- **Return value**: The last statement to be evaluated within the function body, usually explicitly wrapped within the `return()` function.

Let's go through an exercise on creating a user-defined function.

Exercise 1.19 – creating a user-defined function

Now, let's try it out:

1. Create a function named `test_func` to receive an input and print out `"(input) is fun"`. Allow the option to print the message in uppercase:

    ```
    test_func = function(x, cap=FALSE){
      msg = paste(x,"is fun!")
      if(cap){
        msg = toupper(msg)
      }
      return(msg)
    }
    ```

    Note that we used the = sign instead of < - to assign the functional object to the `test_func` variable. However, the latter is more commonly observed when creating functions in R. In the input, we created two arguments: the compulsory argument, x, to receive the message to be printed, and the optional argument, cap, to determine whether the message needs to be converted into uppercase. The optional argument means that the user can either go with the default setting (that is, a lowercase message) by not supplying anything to this argument or overwrite the default behavior by explicitly passing in a value.

In the function body, we first create a `msg` variable and assign the message content by calling the `paste()` function, a built-in function to concatenate the two input arguments. If the `cap` argument is `FALSE`, the `if` statement will evaluate to `FALSE` and `msg` will be directly returned as the function's output. Otherwise, the statement within the `if` clause will be triggered to convert the `msg` variable into uppercase using the `toupper()` function, another built-in function in R.

2. Let's see what happens after calling the function in different ways:

```
>>> test_func("r")
"r is fun!"
>>> test_func("r",cap=TRUE)
"R IS FUN!"
>>> test_func()
Error in paste(x, "is fun!") : argument "x" is missing, with no
default
```

The first two cases work as expected. In the third case, we did not supply any value to the `x` argument, defined as a compulsory argument. This leads to an error and fails to call the function.

Summary

In this chapter, we covered the essential building blocks in R, including how to leverage and navigate the RStudio IDE, basic arithmetic operations (addition, subtraction, multiplication, division, exponentiation, and modulo), common data structures (vectors, matrices, data frames, and lists), control logic, including relational operators (>, ==, <, >=, <=, and !=) and logical operators (&, |, !, &&, and ||), conditional statements using `ifelse`, the `for` and `while` loops, and finally, functions in R. Understanding these fundamental aspects will greatly benefit our learning in later chapters as we gradually introduce more challenging topics.

In the next chapter, we will cover `dplyr`, one of the most widely used libraries for data processing and manipulation. Tapping into the various utility functions provided by `dplyr` will make it much easier to handle most data processing tasks.

2

Data Processing with dplyr

In the previous chapter, we covered the basics of the R language itself. Grasping these fundamentals will help us better tackle the challenges in the most common task in data science projects: **data processing**. Data processing refers to a series of data wrangling and massaging steps that transform the data into its intended format for downstream analysis and modeling. We can consider it as a function that accepts the raw data and outputs the desired data. However, we need to explicitly specify how the function executes the cooking recipe and processes the data.

By the end of this chapter, you will be able to perform common data wrangling steps such as filtering, selection, grouping, and aggregation using `dplyr`, one of the most widely used data processing libraries in R.

In this chapter, we will cover the following topics:

- Introducing `tidyverse` and `dplyr`
- Data transformation with `dplyr`
- Data aggregation with `dplyr`
- Data merging with `dplyr`
- Case study – working with the Stack Overflow dataset

Technical requirements

To complete the exercises in this chapter, you will need the following:

- The latest version of the `tidyverse` package, which is 1.3.1 at the time of writing

All the code and data for this chapter is available at `https://github.com/PacktPublishing/The-Statistics-and-Machine-Learning-with-R-Workshop/tree/main/Chapter_2`.

Introducing tidyverse and dplyr

One of the most widely used R libraries that contains a set of individual packages is `tidyverse`; it includes `dplyr` and `ggplot2` (to be covered in *Chapter 4*). It can support most data processing and visualization needs and comes with an easy and fast implementation compared to base R commands. Therefore, it is recommended to outsource a specific data processing or visualization task to `tidyverse` instead of implementing it ourselves.

Before we dive into the world of data processing, there is one more data structure that's used in the ecosystem of `tidyverse`: `tibble`. A `tibble` is an advanced version of a DataFrame and offers much better format control, leading to clean expressions in code. It is the central data structure in `tidyverse`. A DataFrame can be converted into a `tibble` object and vice versa. Let's go through an exercise on this.

Exercise 2.01 – converting between tibble and a DataFrame

First, we will explore the `tidyverse` ecosystem by installing this package and then converting the `iris` DataFrame into `tibble` format:

1. Install the `tidyverse` package and load the `dplyr` package:

    ```
    install.packages("tidyverse")
    library(dplyr)
    ```

    Installing the `tidyverse` package will automatically install `dplyr`, which can be loaded into our working environment via the `library()` function.

2. Load the `iris` dataset and check its data structure:

    ```
    >>> data("iris")
    >>> class(iris)
    "data.frame"
    ```

    The `data()` function loads the `iris` dataset, a default dataset provided by base R, and a DataFrame that's checked using the `class()` function.

3. Convert the dataset into `tibble` format and verify its data structure:

    ```
    >>> iris_tbl = as_tibble(iris)
    >>> class(iris_tbl)
    "tbl_df"      "tbl"           "data.frame"
    ```

    There are three class attributes in `iris_tbl`, which means that the object can be used as both a `tibble` and a DataFrame. Having multiple class attributes in one object supports better compatibility since we can treat it differently.

A `tibble` object also supports smart printing by listing the top few rows, the shape of the dataset (150 rows and 5 columns), and the data type of each column. On the other hand, a DataFrame would just display all its contents to the console when printed:

```
>>> iris_tbl
# A tibble: 150 x 5
   Sepal.Length Sepal.Width Petal.Length Petal.Width Species
          <dbl>       <dbl>        <dbl>       <dbl> <fct>
 1          5.1         3.5          1.4         0.2 setosa
 2          4.9         3            1.4         0.2 setosa
 3          4.7         3.2          1.3         0.2 setosa
 4          4.6         3.1          1.5         0.2 setosa
 5          5           3.6          1.4         0.2 setosa
 6          5.4         3.9          1.7         0.4 setosa
 7          4.6         3.4          1.4         0.3 setosa
 8          5           3.4          1.5         0.2 setosa
 9          4.4         2.9          1.4         0.2 setosa
10          4.9         3.1          1.5         0.1 setosa
# ... with 140 more rows
```

Multiple utility functions for data transformation are provided by `tidyverse` and `dplyr`. Let's look at a few commonly used functions, such as `filter()` and `arrange()`.

Data transformation with dplyr

Data transformation refers to a collection of techniques for performing row-level treatment on the raw data using `dplyr` functions. In this section, we will cover five fundamental functions for data transformation: `filter()`, `arrange()`, `mutate()`, `select()`, and `top_n()`.

Slicing the dataset using the filter() function

One of the biggest highlights of the `tidyverse` ecosystem is the **pipe operator**, `%>%`, which provides the statement before it as the contextual input for the statement after it. Using the pipe operator gives us better clarity in terms of code structuring, besides saving the need to type multiple repeated contextual statements. Let's go through an exercise on how to use the pipe operator to slice the `iris` dataset using the `filter()` function.

Exercise 2.02 – filtering using the pipe operator

For this exercise, we have been asked to keep only the setosa species in the iris dataset using the pipe operator and the filter() function:

1. Print all unique species in the iris dataset:

    ```
    >>> unique(iris_tbl$Species)
    setosa      versicolor virginica
    Levels: setosa versicolor virginica
    ```

    The result shows that the Species column is a factor with three levels.

2. Keep only the "setosa" species in iris_tbl using filter() and save the result in iris_tbl_subset:

    ```
    iris_tbl_subset = iris_tbl %>%
      filter(Species == "setosa")
    >>> iris_tbl_subset
    # A tibble: 50 x 5
       Sepal.Length Sepal.Width Petal.Length Petal.Width Species
          <dbl>        <dbl>        <dbl>        <dbl> <fct>
    1      5.1          3.5          1.4          0.2 setosa
    2      4.9          3            1.4          0.2 setosa
    3      4.7          3.2          1.3          0.2 setosa
    4      4.6          3.1          1.5          0.2 setosa
    5      5            3.6          1.4          0.2 setosa
    6      5.4          3.9          1.7          0.4 setosa
    7      4.6          3.4          1.4          0.3 setosa
    8      5            3.4          1.5          0.2 setosa
    9      4.4          2.9          1.4          0.2 setosa
    10     4.9          3.1          1.5          0.1 setosa
    # ... with 40 more rows
    ```

    The pipe operator indicates that the following filtering operation is applied to the iris_tbl object. Given this context, we could directly reference the Species column (instead of using iris_tbl$Species) and use the == logical operator to set the equality condition to be evaluated row-wise. The result shows a total of 50 rows stored in iris_tbl_subset.

3. To double-check the filtering result, we could print out the unique species in iris_tbl_subset:

    ```
    >>> unique(iris_tbl_subset$Species)
    setosa
    Levels: setosa versicolor virginica
    ```

4. Now, the dataset only contains the "setosa" species. However, the Species column still encodes the previous information as a factor that has three levels. This is a unique feature

for the factor data type, where information about the total levels is encoded in all individual elements of a factor-typed column. We can remove such information by converting it into a character, as follows:

```
>>> unique(as.character(iris_tbl_subset$Species))
"setosa"
```

Note that we are chaining together two functions that are evaluated from the innermost `as.character()` to the outermost `unique()`.

The `filter()` function makes it easy to add multiple filtering conditions by separating them using a comma. For example, we can add another condition to set the maximum value of `Sepal.Length` as 5, as follows:

```
iris_tbl_subset = iris_tbl %>%
  filter(Species == "setosa",
        Sepal.Length <= 5)
>>> max(iris_tbl_subset$Sepal.Length)
5
>>> dim(iris_tbl_subset)
28   5
```

The result shows that the maximum `Sepal.Length` is now 5 and there are 28 rows left out of the original 150.

Next, we will look at how to sort a `tibble` object (or a DataFrame) based on a specific column(s).

Sorting the dataset using the arrange() function

Another common data transformation operation is sorting, which leads to a dataset with one or multiple columns arranged in increasing or decreasing order. This can be achieved via the `arrange()` function, which is provided by `dplyr`. Let's go through an exercise to look at different ways of sorting a dataset.

Exercise 2.03 – sorting using the arrange() function

In this exercise, we will look at how to sort columns of a dataset in either ascending or descending order, as well as combine the sorting operation with filtering via the pipe operator:

1. Sort the `Sepal.Length` column of the `iris` dataset in ascending order using `arrange()`:

```
iris_tbl_sorted = iris_tbl %>%
  arrange(Sepal.Length)
>>> iris_tbl_sorted
# A tibble: 150 x 5
   Sepal.Length Sepal.Width Petal.Length Petal.Width Species
```

	<dbl>	<dbl>	<dbl>	<dbl>	<fct>
1	4.3	3	1.1	0.1	setosa
2	4.4	2.9	1.4	0.2	setosa
3	4.4	3	1.3	0.2	setosa
4	4.4	3.2	1.3	0.2	setosa
5	4.5	2.3	1.3	0.3	setosa
6	4.6	3.1	1.5	0.2	setosa
7	4.6	3.4	1.4	0.3	setosa
8	4.6	3.6	1	0.2	setosa
9	4.6	3.2	1.4	0.2	setosa
10	4.7	3.2	1.3	0.2	setosa

```
# … with 140 more rows
```

The result shows that the `arrange()` function sorts the specific column in ascending order by default. Now, let's look at how to sort in descending order.

2. Sort the same column in descending order:

```
iris_tbl_sorted = iris_tbl %>%
  arrange(desc(Sepal.Length))
>>> iris_tbl_sorted
# A tibble: 150 x 5
   Sepal.Length Sepal.Width Petal.Length Petal.Width Species
```

	<dbl>	<dbl>	<dbl>	<dbl>	<fct>
1	7.9	3.8	6.4	2	virginica
2	7.7	3.8	6.7	2.2	virginica
3	7.7	2.6	6.9	2.3	virginica
4	7.7	2.8	6.7	2	virginica
5	7.7	3	6.1	2.3	virginica
6	7.6	3	6.6	2.1	virginica
7	7.4	2.8	6.1	1.9	virginica
8	7.3	2.9	6.3	1.8	virginica
9	7.2	3.6	6.1	2.5	virginica
10	7.2	3.2	6	1.8	virginica

```
# … with 140 more rows
```

Adding the `desc()` function to the column before passing in `arrange()` can flip the ordering and achieve sorting in descending order. We can also pass in multiple columns to sort them sequentially.

Besides this, the `arrange()` function can also be used together with other data processing steps, such as filtering.

3. Sort both `Sepal.Length` and `Sepal.Width` in descending order after keeping `Species` set to `"setosa"` only and `Sepal.Length` up to a maximum value of 5:

```
iris_tbl_subset_sorted = iris_tbl %>%
  filter(Species == "setosa",
```

```
             Sepal.Length <= 5) %>%
       arrange(desc(Sepal.Length),desc(Sepal.Width))
>>> iris_tbl_subset_sorted
# A tibble: 28 x 5
   Sepal.Length Sepal.Width Petal.Length Petal.Width Species
          <dbl>       <dbl>        <dbl>       <dbl> <fct>
 1            5         3.6          1.4         0.2 setosa
 2            5         3.5          1.3         0.3 setosa
 3            5         3.5          1.6         0.6 setosa
 4            5         3.4          1.5         0.2 setosa
 5            5         3.4          1.6         0.4 setosa
 6            5         3.3          1.4         0.2 setosa
 7            5         3.2          1.2         0.2 setosa
 8            5         3            1.6         0.2 setosa
 9          4.9         3.6          1.4         0.1 setosa
10          4.9         3.1          1.5         0.1 setosa
# ... with 18 more rows
```

The result shows a two-layer sorting, where for the same value of Sepal.Length, Sepal. Width is further sorted in descending order. These two sorting criteria are separated by a comma, just like separating multiple conditions in filter().

In addition, the pipe operator connects and evaluates multiple functions sequentially. In this case, we start by setting the context for working with iris_tbl, followed by filtering and then sorting, both of which are connected via the pipe operator.

Adding or changing a column using the mutate() function

A tibble object or DataFrame essentially consists of multiple columns stored together as a list of lists. We may want to edit an existing column by changing its contents, type, or format; such editing could also make us end up with a new column appended to the original dataset. Column-level editing can be achieved via the mutate() function. Let's go through an example of how to use this function in conjunction with other functions.

Exercise 2.04 – changing and adding columns using the mutate() function

In this exercise, we will look at how to change the type of an existing column and add a new column to support the filtering operation:

1. Change the Species column to the character type:

    ```
    >>> paste("Before:", class(iris_tbl$Species))
    iris_tbl = iris_tbl %>%
      mutate(Species = as.character(Species))
    >>> paste("After:", class(iris_tbl$Species))
    ```

```
"Before: factor"
"After: character"
```

Here, we used the `mutate()` function to change the type of `Species`, which is referenced directly within the context of the `iris_tbl` object via the pipe operator.

2. Create a column called `ind` to indicate whether `Sepal.Width` is bigger than `Petal.Length`:

```
iris_tbl = iris_tbl %>%
  mutate(ind = Sepal.Width > Petal.Length)
>>> iris_tbl
# A tibble: 150 x 6
   Sepal.Length Sepal.Width Petal.Length Petal.Width Species ind
          <dbl>       <dbl>        <dbl>       <dbl>
   <chr>  <lgl>
1           5.1         3.5          1.4         0.2
   setosa TRUE
2           4.9         3            1.4         0.2
   setosa TRUE
3           4.7         3.2          1.3         0.2
   setosa TRUE
4           4.6         3.1          1.5         0.2
   setosa TRUE
5           5           3.6          1.4         0.2
   setosa TRUE
6           5.4         3.9          1.7         0.4
   setosa TRUE
7           4.6         3.4          1.4         0.3
   setosa TRUE
8           5           3.4          1.5         0.2
   setosa TRUE
9           4.4         2.9          1.4         0.2
   setosa TRUE
10          4.9         3.1          1.5         0.1
   setosa TRUE
# … with 140 more rows
```

The result shows that we have added an indicator column with logical values. We can get the counts of TRUE and FALSE values via the `table()` function:

```
>>> table(iris_tbl$ind)
FALSE  TRUE
100    50
```

3. Keep only rows whose `Sepal.Width` is bigger than `Petal.Length`:

```
iris_tbl_subset = iris_tbl %>%
  filter(ind==TRUE)
>>> table(iris_tbl_subset$ind)
```

```
TRUE
50
```

Since we are essentially performing a filtering operation, this two-step process of first creating an indicator column and then filtering could be combined into a single step by directly setting the filtering condition within the `filter()` function:

```
iris_tbl_subset2 = iris_tbl %>%
  filter(Sepal.Width > Petal.Length)
>>> nrow(iris_tbl_subset2)
50
```

The result is the same as the two-step approach.

Now, let's cover the last commonly used utility function – `select()`.

Selecting columns using the select() function

The `select()` function works by selecting the columns specified by the input argument, a vector of strings representing one or multiple columns. When using `select()` in the context of the pipe operator, it means that all the following statements are evaluated based on the selected columns. When the `select` statement comes last, it returns the selected columns as the output `tibble` object.

Let's go through an exercise on different ways of selecting columns from a dataset.

Exercise 2.05 – selecting columns using select()

In this exercise, we will look at different ways of selecting columns from a `tibble` dataset:

1. Select the first three columns from the `iris` dataset:

```
rst = iris_tbl %>%
  select(Sepal.Length, Sepal.Width, Petal.Length)
>>> rst
# A tibble: 150 x 3
   Sepal.Length Sepal.Width Petal.Length
          <dbl>       <dbl>        <dbl>
 1          5.1         3.5          1.4
 2          4.9         3            1.4
 3          4.7         3.2          1.3
 4          4.6         3.1          1.5
 5          5           3.6          1.4
 6          5.4         3.9          1.7
 7          4.6         3.4          1.4
 8          5           3.4          1.5
 9          4.4         2.9          1.4
10          4.9         3.1          1.5
# … with 140 more rows
```

When you need to increase the number of columns to be selected, typing them one by one would become tedious. Another way to do this is to pass the first and last columns separated by a colon (:), as follows:

```
rst = iris_tbl %>%
  select(Sepal.Length:Petal.Length)
>>> rst
# A tibble: 150 x 3
   Sepal.Length Sepal.Width Petal.Length
          <dbl>       <dbl>        <dbl>
1           5.1         3.5          1.4
2           4.9         3            1.4
3           4.7         3.2          1.3
4           4.6         3.1          1.5
5           5           3.6          1.4
6           5.4         3.9          1.7
7           4.6         3.4          1.4
8           5           3.4          1.5
9           4.4         2.9          1.4
10          4.9         3.1          1.5
# … with 140 more rows
```

This approach selects all columns contained between Sepal.Length and Petal.Length. Using a colon helps us select multiple consecutive columns in one shot. Besides, we can also combine it with other individual columns via the c() function.

2. Select columns that contain "length":

```
rst = iris_tbl %>%
  select(contains("length"))
>>> rst
# A tibble: 150 x 2
   Sepal.Length Petal.Length
          <dbl>        <dbl>
1           5.1          1.4
2           4.9          1.4
3           4.7          1.3
4           4.6          1.5
5           5            1.4
6           5.4          1.7
7           4.6          1.4
8           5            1.5
9           4.4          1.4
10          4.9          1.5
# … with 140 more rows
```

Here, we used the contains () function to perform a case-insensitive string match. Other utility functions that support string matching include starts_with () and ends_with (). Let's look at an example.

3. Select columns that start with "petal":

```
rst = iris_tbl %>%
   select(starts_with("petal"))
>>> rst
# A tibble: 150 x 2
   Petal.Length Petal.Width
          <dbl>       <dbl>
1           1.4         0.2
2           1.4         0.2
3           1.3         0.2
4           1.5         0.2
5           1.4         0.2
6           1.7         0.4
7           1.4         0.3
8           1.5         0.2
9           1.4         0.2
10          1.5         0.1
# … with 140 more rows
```

Next, we will look at selecting the top rows using the top_n () function, which comes in handy when we want to examine a few rows after sorting the DataFrame based on a specific column.

Selecting the top rows using the top_n() function

The top_n () function can be helpful when we are interested in the top few observations for a particular column. It expects two input arguments: the number of top observations (implicitly sorted in descending order) returned and the specific column sorted. The mechanism would be the same if we were to sort a column in descending order using arrange () and return the top few rows using head (). Let's try it out.

Exercise 2.06 – selecting the top rows using top_n()

In this exercise, we will demonstrate how to use top_n () in combination with other verbs:

1. Return the observation with the biggest Sepal.Length:

```
rst = iris_tbl %>%
   top_n(1, Sepal.Length)
>>> rst
# A tibble: 1 x 6
```

```
      Sepal.Length Sepal.Width Petal.Length Petal.Width
  Species    ind
              <dbl>        <dbl>        <dbl>        <dbl>
  <chr>      <lgl>
  1            7.9          3.8          6.4            2 virginica
  FALSE
```

We can see that the result is a full row whose `Sepal.Length` is the biggest. This can also be achieved by explicitly sorting the dataset using this column and returning the first row, illustrated as follows:

```
rst = iris_tbl %>%
  arrange(desc(Sepal.Length)) %>%
  head(1)
>>> rst
# A tibble: 1 x 6
      Sepal.Length Sepal.Width Petal.Length Petal.Width
  Species    ind
              <dbl>        <dbl>        <dbl>        <dbl>
  <chr>      <lgl>
  1            7.9          3.8          6.4            2 virginica
  FALSE
```

We can also apply `top_n()` in a `group_by()` context, which aggregates the data into different groups. We will cover more details on data aggregation in the next section.

2. Return the biggest `Sepal.Length` for each category of `Species`:

```
rst = iris_tbl %>%
  group_by(Species) %>%
  top_n(1, Sepal.Length) %>%
  select(Species, Sepal.Length)
>>> rst
# A tibble: 3 x 2
# Groups:    Species [3]
  Species      Sepal.Length
  <chr>               <dbl>
1 setosa                5.8
2 versicolor            7
3 virginica             7.9
```

We can also use the `max()` function to achieve the same purpose:

```
rst = iris_tbl %>%
  group_by(Species) %>%
  summarize(max_sepal_length = max(Sepal.Length))
>>> rst
# A tibble: 3 x 2
  Species      max_sepal_length
  <chr>                   <dbl>
```

```
1 setosa                      5.8
2 versicolor                  7
3 virginica                   7.9
```

The `summarize()` function compresses the dataset into one row (with the maximum `Sepal.Length`) for each group of `Species`. More on this later.

3.   Return the biggest `Sepal.Length` and its category:

```
rst = iris_tbl %>%
  group_by(Species) %>%
  summarize(max_sepal_length = max(Sepal.Length)) %>%
  top_n(1, max_sepal_length)
>>> rst
# A tibble: 1 x 2
  Species     max_sepal_length
  <chr>                  <dbl>
1 virginica                7.9
```

This example shows that we can use `top_n()` in multiple contexts together with other verbs.

Now, let's combine the five verbs we've covered here.

Combining the five verbs

The five utility functions we have covered so far can be combined, thus offering a flexible and concise way of processing data. Let's go through an exercise that involves all five functions.

Exercise 2.07 – combining the five utility functions

The example we'll cover in this exercise is a somewhat contrived one so that all five verb functions can be used. In this exercise, we have been asked to find the average absolute difference between `Sepal.Length` and `Petal.Length` for the top 100 rows with the highest `Sepal.Length` and whose `Sepal.Width` is bigger than `Petal.Length`.

When performing a complex query like this one, it is helpful to work through the requirements backward, starting from the conditions for subsetting the dataset and then working on the metric. In this case, we will start by sorting `Sepal.Length` in a descending order using the `arrange()` function and keep the top 100 rows using the `head()` function. Another filtering condition that uses the `filter()` function then comes in to retain the rows whose `Sepal.Width` is bigger than `Petal.Length`. Next, we must create a new column using the `mutate()` function to represent the absolute difference between `Sepal.Length` and `Petal.Length`. Finally, we must apply the `select()` function to focus on the new column and calculate its average. See the following code block for a detailed implementation:

```
rst = iris_tbl %>%
  top(80, Sepal.Length) %>%
  filter(Sepal.Width > Petal.Length) %>%
```

```
    mutate(Diff = abs(Sepal.Length - Petal.Length)) %>%
    select(Diff) %>%
    colMeans()
>>> rst
Diff
 4.266667
```

Next, we will look at additional two verbs: rename() and transmute().

Introducing other verbs

Two other verbs are also commonly used: rename() and transmute(). The rename() function changes the name of a specific column. For example, when using the count() function, a column named n is automatically created. We can use rename(Count = n) within the pipe context to change its default name from n to Count.

There is another way to change the column's name. When selecting a column of a dataset, we can pass the same statement that we did to rename() to the select() function. For example, the following code shows selecting the Sepal.Length and Sepal.Width columns from the iris dataset while renaming the second column Sepal_Width:

```
rst = iris_tbl %>%
    select(Sepal.Length, Sepal_Width=Sepal.Width)
>>> rst
# A tibble: 150 x 2
   Sepal.Length Sepal_Width
          <dbl>       <dbl>
 1          5.1         3.5
 2          4.9         3
 3          4.7         3.2
 4          4.6         3.1
 5          5           3.6
 6          5.4         3.9
 7          4.6         3.4
 8          5           3.4
 9          4.4         2.9
10          4.9         3.1
# ... with 140 more rows
```

On the other hand, the transmute() function is a combination of select() and mutate(). It will return a subset of columns where some could be transformed. For example, suppose we would like to calculate the absolute between Sepal.Length and Petal.Length and return the result together with Species. We can achieve both tasks using transmute(), as follows:

```
rst = iris_tbl %>%
    transmute(Species, Diff = abs(Sepal.Length - Petal.Length))
```

```
>>> rst
# A tibble: 150 x 2
   Species  Diff
   <chr>    <dbl>
 1 setosa     3.7
 2 setosa     3.5
 3 setosa     3.4
 4 setosa     3.1
 5 setosa     3.6
 6 setosa     3.7
 7 setosa     3.2
 8 setosa     3.5
 9 setosa     3
10 setosa     3.4
# … with 140 more rows
```

Although these verbs could be used interchangeably, there are a few technical differences. As shown in *Figure 2.1*, the select() function returns the specified columns without changing the value of these columns, which can be achieved via either mutate() or transmutate(). Both mutate() and rename() keep the original columns in the returned result when creating additional new columns, while select() and transmute() only return specified columns in the result:

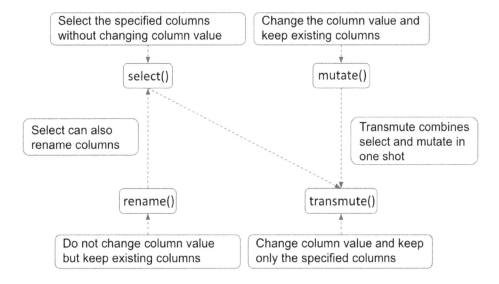

Figure 2.1 – Summarizing the four verbs in terms of their purposes and connections

Now that we know how to transform the data, we can go a step further to make it more interpretable and presentable via aggregation and summarization. We will cover different ways to aggregate the data in the next section.

Data aggregation with dplyr

Data aggregation refers to a set of techniques that summarizes the dataset at an aggregate level and characterizes the original dataset at a higher level. Compared to data transformation, it operates at the row level for the input and the output.

We have already encountered a few aggregation functions, such as calculating the mean of a column. This section will cover some of the most widely used aggregation functions provided by dplyr. We will start with the count () function, which returns the number of observations/rows for each category of the specified input column.

Counting observations using the count() function

The count () function automatically groups the dataset into different categories according to the input argument and returns the number of observations for each category. The input argument could include one or more columns of the dataset. Let's go through an exercise and apply it to the iris dataset.

Exercise 2.08 – counting observations by species

This exercise will use the count () function to get the number of observations for each unique species, followed by adding filtering conditions using the filter () function:

1. Count the number of observations for each unique type of species in the iris dataset:

```
rst = iris_tbl %>%
  count(Species)
>>> rst
# A tibble: 3 x 2
  Species       n
  <chr>     <int>
1 setosa       50
2 versicolor   50
3 virginica    50
```

The output is a tibble dataset with two columns, where the first column contains the unique category in Species and the second column (named n by default) is the corresponding count of rows.

Let's look at how to perform filtering before the counting operation.

2. Perform the exact counting for those observations whose absolute difference between `Sepal.Length` and `Sepal.Width` is greater than that of `Petal.Length` and `Petal.Width`. Return the result in descending order:

```
rst = iris_tbl %>%
  filter(abs(Sepal.Length-Sepal.Width) > abs(Petal.Length-Petal.
Width)) %>%
  count(Species, sort=TRUE)
>>> rst
# A tibble: 3 x 2
  Species        n
  <chr>      <int>
1 setosa        45
2 versicolor    33
3 virginica     28
```

Here, we added a filtering condition to keep rows that meet the specified criterion before counting. We enabled the `sort` argument to arrange the results in descending order.

The `count()` function essentially combines two steps: grouping by each category of a specified column and then counting the number of observations. It turns out that we can achieve the same task using the `group_by()` and `summarize()` functions, as introduced in the next section.

Aggregating data via group_by() and summarize()

`count()` is a helpful way to aggregate data. However, it is a particular case of two more general aggregation functions, `group_by()` and `summarize()`, which are often used together. The `group_by()` function splits the original dataset into different groups according to one or more columns in the input argument, while the `summarize()` function summarizes and collapses all observations within a specific category into one metric, which could be the count of rows in the case of `count()`.

Multiple summarization functions can be used in the `summarize()` function. Typical ones include the following:

- `sum()`: Sums all observations of a particular group

- `mean()`: Calculates the average of all observations

- `median()`: Calculates the median of all observations

- `max()`: Calculates the maximum of all observations

- `min()`: Calculates the minimum of all observations

Let's go through an exercise on calculating different summary statistics using `group_by()` and `summarize()`.

Exercise 2.09 – summarizing a dataset using group_by() and summarize()

This exercise covers using the `group_by()` and `summarize()` functions to extract the count and mean statistics, combined with some of the verbs introduced earlier, including `filter()`, `mutate()`, and `arrange()`:

1. Get the count of observations for each unique type of `Species`:

```
rst = iris_tbl %>%
  group_by(Species) %>%
  summarise(n=n())
>>> rst
# A tibble: 3 x 2
  Species        n
  <chr>      <int>
1 setosa        50
2 versicolor    50
3 virginica     50
```

In the preceding code, we used the `n()` function to get the number of observations and assigned the result to a column named n. The counting comes after grouping the observations based on the unique type of `Species`.

2. Add the same filter as in the previous exercise and sort the result in descending order:

```
rst = iris_tbl %>%
  filter(abs(Sepal.Length-Sepal.Width) > abs(Petal.Length-Petal.
Width)) %>%
  group_by(Species) %>%
  summarise(n=n()) %>%
  arrange(desc(n))
>>> rst
# A tibble: 3 x 2
  Species        n
  <chr>      <int>
1 setosa        45
2 versicolor    33
3 virginica     28
```

In this code block, the filtering condition is applied first to limit the grouping operations to a subset of observations. In the `arrange()` function, we directly used the n column to sort in descending order.

3. Create a logical column based on the same filter and perform a two-level grouping using `Species`. Then, create a logical column to calculate the average `Sepal.Length`:

```
rst = iris_tbl %>%
  mutate(ind = abs(Sepal.Length-Sepal.Width) > abs(Petal.Length-
Petal.Width)) %>%
  group_by(Species, ind) %>%
  summarise(mean_sepal_length=mean(Sepal.Length))
>>> rst
# A tibble: 6 x 3
# Groups:   Species [3]
  Species     ind    mean_sepal_length
  <chr>       <lgl>            <dbl>
1 setosa      FALSE          5
2 setosa      TRUE           5.01
3 versicolor  FALSE          5.78
4 versicolor  TRUE           6.02
5 virginica   FALSE          6.39
6 virginica   TRUE           6.74
```

We can put multiple categorical columns in the `group_by()` function to perform a multi-level grouping. Note that the result contains a `Groups` attribute based on `Species`, suggesting that the `tibble` object has a group structure. Let's learn how to remove the structure.

4. Remove the group structure in the returned `tibble` object using `ungroup()`:

```
rst = iris_tbl %>%
  mutate(ind = abs(Sepal.Length-Sepal.Width) > abs(Petal.Length-
Petal.Width)) %>%
  group_by(Species, ind) %>%
  summarise(mean_sepal_length=mean(Sepal.Length)) %>%
  ungroup()
>>> rst
# A tibble: 6 x 3
  Species     ind    mean_sepal_length
  <chr>       <lgl>            <dbl>
1 setosa      FALSE          5
2 setosa      TRUE           5.01
3 versicolor  FALSE          5.78
4 versicolor  TRUE           6.02
5 virginica   FALSE          6.39
6 virginica   TRUE           6.74
```

Now, the result contains a normal `tibble` object with the average sepal length for each unique combination of `Species` and `ind`.

Now that we know how to transform and aggregate one dataset, we will cover how to work with multiple datasets via merging and joining.

Data merging with dplyr

In practical data analysis, the information we need is not necessarily confined to one table but is spread across multiple tables. Storing data in separate tables is memory-efficient but not analysis-friendly. **Data merging** is the process of merging different datasets into one table to facilitate data analysis. When joining two tables, there need to be one or more columns, or **keys**, that exist in both tables and serve as the common ground for joining.

This section will cover different ways to join tables and analyze them in combination, including inner join, left join, right join, and full join. The following list shows the verbs and their definitions for these four types of joining:

- `inner_join()`: Returns common observations in both tables according to the matching key.
- `left_join()`: Returns all observations from the left table and matched observations from the right table. Note that in the case of a duplicate key value in the right table, an additional row will be automatically created and added to the left table. Empty cells are filled with NA. More on this in the exercise.
- `right_join()`: Returns all observations from the right table and matched observations from the left table. Empty cells are filled with NA.
- `full_join()`: Returns all observations from both tables. Empty cells are filled with NA.

Figure 2.2 illustrates the four joins using Venn diagrams:

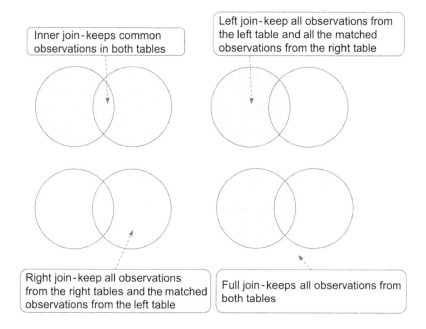

Figure 2.2 – Four different joins commonly used in practice

Let's go through an exercise on these four joins.

Exercise 2.10 – joining datasets

This exercise will create two dummy `tibble` datasets and apply different joining verbs to merge them:

1. Create two dummy datasets by following the steps in this code block:

```
a = 1:3
tbl_A = tibble(key_A=a, col_A=2*a)
tbl_B = tibble(key_B=a+1, col_B=3*a)
>>> tbl_A
# A tibble: 3 x 2
  key_A col_A
  <int> <dbl>
1     1     2
2     2     4
3     3     6
>>> tbl_B
# A tibble: 3 x 2
  key_B col_B
```

```
     <dbl> <dbl>
1      2     3
2      3     6
3      4     9
```

Both dummy datasets have three rows and two columns, with the first column being the key column to be used for joining.

2. Perform an inner join of the two datasets:

```
rst = tbl_A %>%
  inner_join(tbl_B, by=c("key_A"="key_B"))
>>> rst
# A tibble: 2 x 3
  key_A col_A col_B
  <dbl> <dbl> <dbl>
1    2     4     3
2    3     6     6
```

The preceding code shows that the matching keys are specified in the by argument via the c() function. Since "key_A" and "key_B" have only two values in common, the resulting table after the inner join operation is a 2x3 tibble that keeps only "key_A" as the key column and all other non-key columns from both tables. It only keeps observations that have an exact match and works the same way with either table in either direction.

We can also pass in additional match keys (on which the tables will be joined) in the by argument while following the same format to perform a multi-level merging.

Let's look at how to perform a left join.

3. Perform a left join of the two datasets:

```
rst = tbl_A %>%
  left_join(tbl_B, by=c("key_A"="key_B"))
>>> rst
# A tibble: 3 x 3
  key_A col_A col_B
  <dbl> <dbl> <dbl>
1    1     2    NA
2    2     4     3
3    3     6     6
```

Note that the resulting tables contain the whole tbl_A and an additional column, col_B, that's referenced from tbl_B. Since col_B does not have 1, the corresponding cell in col_B shows NA. In general, any cell that cannot be matched will assume a value of NA in the resulting table.

Note that when there are multiple rows with duplicate values in col_B, the resulting table after a left join will also have a duplicate row automatically created since it is now a one-to-two mapping from left to right. Let's see an example.

4. Create another table with a duplicate key value and perform a left join with tbl_A:

```
tbl_C = tbl_B %>%
  bind_rows(tbl_B[1,])
tbl_C[nrow(tbl_C),"col_B"] = 10
>>> tbl_C
# A tibble: 4 x 2
  key_B col_B
  <dbl> <dbl>
1   2     3
2   3     6
3   4     9
4   2    10
```

Here, we used the bind_rows() function to append a new row with the same value in "key_B" as the first row and a different value for col_B. Let's see what happens when we join it to tbl_A:

```
rst = tbl_A %>%
  left_join(tbl_C, by=c("key_A"="key_B"))
>>> rst
# A tibble: 4 x 3
  key_A col_A col_B
  <dbl> <dbl> <dbl>
1   1     2    NA
2   2     4     3
3   2     4    10
4   3     6     6
```

Having a duplicate value in the key column of the right table is a common source of bugs that could be difficult to trace. Seasoned data scientists should pay particular attention to this potential unintended result by checking the dimension of the dataset *before* and *after* the left join. Now, let's look at how to perform a right join.

5. Perform a right join of tbl_A to tbl_B:

```
rst = tbl_A %>%
right_join(tbl_B, by=c("key_A"="key_B"))
>>> rst
# A tibble: 3 x 3
  key_A col_A col_B
  <dbl> <dbl> <dbl>
```

```
1      2      4      3
2      3      6      6
3      4      NA     9
```

Similarly, all observations in tbl_B are kept and the missing value in col_A is filled with NA. In addition, the key column is named "key_A" instead of "key_B".

6.   Perform a full join of tbl_A and tbl_B:

```
rst = tbl_A %>%
  full_join(tbl_B, by=c("key_A"="key_B"))
>>> rst
# A tibble: 4 x 3
  key_A col_A col_B
  <dbl> <dbl> <dbl>
1    1     2    NA
2    2     4     3
3    3     6     6
4    4    NA     9
```

Using a full join, all matched results from both tables are maintained, with missing values filled with NA. We can use this when we do not want to leave out any observations from the source tables.

These four joining statements can be repeatedly used to join multiple tables and combined with any data transformation verbs covered earlier. For example, we can remove the rows with NA values and keep only the complete rows, which should give us the same result in the inner join. This can be achieved using drop_na(), a utility function provided by the tidyr package that's designed specifically for data cleaning within the tidyverse ecosystem:

```
library(tidyr)
rst = tbl_A %>%
  full_join(tbl_B, by=c("key_A"="key_B")) %>%
  drop_na()
>>> rst
# A tibble: 2 x 3
  key_A col_A col_B
  <dbl> <dbl> <dbl>
1    2     4     3
2    3     6     6
```

We may also want to replace NA values with 0, which can be achieved via the replace_na() function provided by tidyr. In the following code, we are specifying the replacement values for each column of interest and wrapping them in a list to be passed into replace_na():

```
rst = tbl_A %>%
  full_join(tbl_B, by=c("key_A"="key_B")) %>%
```

```
    replace_na(list(col_A=0, col_B=0))
>>> rst
# A tibble: 4 x 3
  key_A col_A col_B
  <dbl> <dbl> <dbl>
1     1     2     0
2     2     4     3
3     3     6     6
4     4     0     9
```

Note that there are other merging options, such as the semi-join and the anti-join, which correspond to the semi_join() and anti_join() functions, respectively. Semi-join returns only the rows from the table in the first argument where there is a match in the second table. Although similar to a full join operation, a semi-join only keeps *the columns in the first table*. The anti-join operation, on the other hand, is the opposite of a semi-join, returning only *the unmatched rows in the first table*. Since many merging operations, including these two, can be derived using the four basic operations we introduced in this section, we will not cover these slightly more complicated joining operations in detail. Instead, we encourage you to explore using these four fundamental joining functions to achieve complicated operations instead of relying on other shortcut joining functions.

Next, we will go through a case study and observe how to transform, merge, and aggregate a dataset using the functions we've covered in this chapter.

Case study – working with the Stack Overflow dataset

This section will cover an exercise to help you practice different data transformation, aggregation, and merging techniques based on the public Stack Overflow dataset, which contains a set of tables related to technical questions and answers posted on the Stack Overflow platform. The supporting raw data has been uploaded to the accompanying Github repository of this book. We will directly download it from the source GitHub link using the readr package, another tidyverse offering that provides an easy, fast, and friendly way to read a wide range of data sources, including those from the web.

Exercise 2.11 – working with the Stack Overflow dataset

Let's begin this exercise:

1. Download three data sources on questions, tags, and their mapping table from GitHub:

    ```
    library(readr)
    df_questions = read_csv("https://raw.githubusercontent.com/
    PacktPublishing/The-Statistics-and-Machine-Learning-with-R-
    Workshop/main/Chapter_2/data/questions.csv")
    >>> df_questions
    # A tibble: 294,735 x 3
            id creation_date score
    ```

```
       <dbl> <date>       <dbl>
 1 22557677 2014-03-21        1
 2 22557707 2014-03-21        2
 3 22558084 2014-03-21        2
 4 22558395 2014-03-21        2
 5 22558613 2014-03-21        0
 6 22558677 2014-03-21        2
 7 22558887 2014-03-21        8
 8 22559180 2014-03-21        1
 9 22559312 2014-03-21        0
10 22559322 2014-03-21        2
# … with 294,725 more rows
```

The questions dataset contains the question ID, date of creation, and score, which indicates the number of (positive) upvotes and (negative) downvotes. We can examine the range of scores using the summary() function:

```
>>> summary(df_questions$score)
   Min.  1st Qu.   Median     Mean  3rd Qu.      Max.
-21.000    0.000    1.000    1.904    2.000  2474.000
```

The range of scores is quite broad. Indeed, many answers wait for excellent and proper questions to be asked. Do remember to ask questions when in doubt and search for similar questions other people have asked.

Let's import the other two data sources:

```
df_question_tags = read_csv("https://raw.githubusercontent.
com/PacktPublishing/The-Statistics-and-Machine-Learning-with-R-
Workshop/main/Chapter_2/data/question_tags.csv")
>>> df_question_tags
# A tibble: 497,153 x 2
   question_id tag_id
         <dbl>  <dbl>
 1    22557677     18
 2    22557677    139
 3    22557677  16088
 4    22557677   1672
 5    22558084   6419
 6    22558084  92764
 7    22558395   5569
 8    22558395    134
 9    22558395   9412
10    22558395  18621
# … with 497,143 more rows
```

This dataset contains the mapping table between `question_id` and `tag_id`:

```
df_tags = read_csv("https://raw.githubusercontent.com/
PacktPublishing/The-Statistics-and-Machine-Learning-with-R-
Workshop/main/Chapter_2/data/tags.csv")
>>> df_tags
# A tibble: 48,299 x 2
        id tag_name
     <dbl> <chr>
 1 124399 laravel-dusk
 2 124402 spring-cloud-vault-config
 3 124404 spring-vault
 4 124405 apache-bahir
 5 124407 astc
 6 124408 simulacrum
 7 124410 angulartics2
 8 124411 django-rest-viewsets
 9 124414 react-native-lightbox
10 124417 java-module
# … with 48,289 more rows
```

This dataset contains the ID and the content of each tag. To analyze the tags, we need to merge the three datasets into one using the relevant mapping keys.

2. Reference the tag ID from `df_question_tags` into `df_questions` via a left join:

```
df_all = df_questions %>%
  left_join(df_question_tags, by=c("id"="question_id"))
>>> df_all
# A tibble: 545,694 x 4
          id creation_date score tag_id
       <dbl> <date>        <dbl>  <dbl>
 1 22557677 2014-03-21         1     18
 2 22557677 2014-03-21         1    139
 3 22557677 2014-03-21         1  16088
 4 22557677 2014-03-21         1   1672
 5 22557707 2014-03-21         2     NA
 6 22558084 2014-03-21         2   6419
 7 22558084 2014-03-21         2  92764
 8 22558395 2014-03-21         2   5569
 9 22558395 2014-03-21         2    134
10 22558395 2014-03-21         2   9412
# … with 545,684 more rows
```

Note that the total number of rows almost doubled when comparing `df_questions` to `df_all`. You may have noticed that this is due to the one-to-many relationship: a question often has multiple tags, so each tag gets appended as a separate row to the left table during the left join operation.

3. Let's continue to reference the tags from `df_tags`:

```
df_all = df_all %>%
  left_join(df_tags, by=c("tag_id"="id"))
>>> df_all
# A tibble: 545,694 x 5
          id creation_date score tag_id tag_name
       <dbl> <date>        <dbl>  <dbl> <chr>
 1 22557677 2014-03-21         1     18 regex
 2 22557677 2014-03-21         1    139 string
 3 22557677 2014-03-21         1  16088 time-complexity
 4 22557677 2014-03-21         1   1672 backreference
 5 22557707 2014-03-21         2     NA NA
 6 22558084 2014-03-21         2   6419 time-series
 7 22558084 2014-03-21         2  92764 panel-data
 8 22558395 2014-03-21         2   5569 function
 9 22558395 2014-03-21         2    134 sorting
10 22558395 2014-03-21         2   9412 vectorization
# … with 545,684 more rows
```

Next, we will do some analysis of the tags, starting with counting their frequency.

4. Count the occurrence of each non-`NA` tag in descending order:

```
df_all = df_all %>%
  filter(!is.na(tag_name))
rst = df_all %>%
  count(tag_name, sort = TRUE)
>>> rst
# A tibble: 7,840 x 2
   tag_name       n
   <chr>      <int>
 1 ggplot2    28228
 2 dataframe  18874
 3 shiny      14219
 4 dplyr      14039
 5 plot       11315
 6 data.table  8809
 7 matrix      6205
 8 loops       5149
```

```
 9 regex       4912
10 function    4892
# … with 7,830 more rows
```

Here, we first used `filter()` to remove rows if `tag_name` was NA, then calculated the counts using the `count()` function. The result shows that `dplyr` is among one of the most popular R-related tags on Stack Overflow, which is a good sign as it shows we are learning about something useful and trendy.

5. Count the number of tags in each year:

```
library(lubridate)
rst = df_all %>%
  mutate(year = year(creation_date)) %>%
  count(year)
>>> rst
# A tibble: 12 x 2
    year      n
   <dbl> <int>
 1  2008     18
 2  2009    874
 3  2010   3504
 4  2011   8787
 5  2012  18251
 6  2013  34998
 7  2014  50749
 8  2015  66652
 9  2016  76056
10  2017  90462
11  2018  96819
12  2019  49983
```

The result shows an increasing number of tags per year, with 2019 being a special case as the data ends in mid-2019 (verified here). Note that we used the `year()` function from the `lubricate` package from `tidyverse` to convert a date-formatted column into the corresponding year:

```
>>> max(df_all$creation_date)
"2019-07-01"
```

6. Calculate the average occurrence of tags per month.

We need to derive the monthly occurrence of tags to calculate their average. First, we must create two columns to indicate the month and year-month for each tag:

```
df_all = df_all %>%
  mutate(month = month(creation_date),
```

```
            year_month = format(creation_date, "%Y%m"))
>>> df_all
# A tibble: 497,153 x 7
         id creation_date score tag_id tag_name                month
year_month
      <dbl> <date>        <dbl>  <dbl> <chr>                   <dbl>
<chr>
 1 22557677 2014-03-21        1     18 regex                       3
201403
 2 22557677 2014-03-21        1    139 string                      3
201403
 3 22557677 2014-03-21        1  16088 time-complexity             3
201403
 4 22557677 2014-03-21        1   1672 backreference               3
201403
 5 22558084 2014-03-21        2   6419 time-series                 3
201403
 6 22558084 2014-03-21        2  92764 panel-data                  3
201403
 7 22558395 2014-03-21        2   5569 function                    3
201403
 8 22558395 2014-03-21        2    134 sorting                     3
201403
 9 22558395 2014-03-21        2   9412 vectorization               3
201403
10 22558395 2014-03-21        2  18621 operator-precedence         3
201403
# … with 497,143 more rows
```

Then, we must count the occurrence of tags per month for each year-month:

```
rst1 = df_all %>%
  count(year_month, month)
>>> rst1
# A tibble: 130 x 3
   year_month month     n
   <chr>      <dbl> <int>
 1 200809         9    13
 2 200811        11     4
 3 200812        12     1
 4 200901         1     8
 5 200902         2    10
 6 200903         3     7
 7 200904         4    24
 8 200905         5     3
 9 200906         6    12
```

```
10 200907            7    100
# … with 120 more rows
```

Finally, we must average over all years for each month:

```
rst2 = rst1 %>%
  group_by(month) %>%
  summarise(avg_num_tag = mean(n))
>>> rst2
# A tibble: 12 x 2
   month avg_num_tag
   <dbl>       <dbl>
 1     1       3606.
 2     2       3860.
 3     3       4389.
 4     4       4286.
 5     5       4178.
 6     6       4133.
 7     7       3630.
 8     8       3835.
 9     9       3249.
10    10       3988.
11    11       3628.
12    12       3125.
```

The result shows that March has the highest occurrence of tags on average. Maybe school just got started and people are actively learning and asking questions more in March.

7. Calculate the count, minimum, average score, and maximum score for each tag and sort the result by count in descending order:

```
rst = df_all %>%
  group_by(tag_name) %>%
  summarise(count = n(),
            min_score = min(score),
            mean_score = mean(score),
            max_score = max(score)) %>%
  arrange(desc(count))
>>> rst
# A tibble: 7,840 x 5
   tag_name    count min_score mean_score max_score
   <chr>       <int>     <dbl>      <dbl>     <dbl>
 1 ggplot2     28228        -9       2.61       666
 2 dataframe   18874       -11       2.31      1241
 3 shiny       14219        -7       1.45        79
```

```
 4 dplyr        14039      -9      1.95       685
 5 plot         11315     -10      2.24       515
 6 data.table    8809      -8      2.97       685
 7 matrix        6205     -10      1.66       149
 8 loops         5149      -8      0.743      180
 9 regex         4912      -9      2          242
10 function      4892     -14      1.39       485
# … with 7,830 more rows
```

Here, we used multiple summary functions in the group_by() and summarize() context to calculate the metrics.

Summary

In this chapter, we covered essential functions and techniques for data transformation, aggregation, and merging. For data transformation at the row level, we learned about common utility functions such as filter(), mutate(), select(), arrange(), top_n(), and transmute(). For data aggregation, which summarizes the raw dataset into a smaller and more concise summary view, we introduced functions such as count(), group_by(), and summarize(). For data merging, which combines multiple datasets into one, we learned about different joining methods, including inner_join(), left_join(), right_join(), and full_join(). Although there are other more advanced joining functions, the essential tools we covered in our toolkit are enough for us to achieve the same task. Finally, we went through a case study based on the Stack Overflow dataset. The skills we learned in this chapter will come in very handy in many data analysis tasks.

In the next chapter, we will cover a more advanced topic on natural language processing, taking us one step further to work with textual data using tidyverse.

3

Intermediate Data Processing

The previous chapter covered a suite of commonly used functions offered by `dplyr` for data processing. For example, when characterizing and extracting the statistics of a dataset, we can follow the split-apply-combine procedure using `group_by()` and `summarize()`. This chapter continues from the previous one and focuses on intermediate data processing techniques, including transforming categorical and numeric variables and reshaping DataFrames. Besides that, we will also introduce string manipulation techniques for working with textual data, whose format is fundamentally different from the neatly shaped tables we have been working with so far.

By the end of this chapter, you will be able to perform more advanced data manipulation and extend your data massaging skills to string-based texts, which are fundamental to the field of natural language processing.

In this chapter, we will cover the following topics:

- Transforming categorical and numeric variables
- Reshaping the DataFrame
- Manipulating string data
- Working with `stringr`
- Introducing regular expressions
- Working with tidy text mining

Technical requirements

To complete the exercises in this chapter, you will need to have the following:

- The latest version of the `rebus` package, which is 0.1-3 at the time of writing
- The latest version of the `tidytext` package, which is 0.3.2 at the time of writing
- The latest version of the `tm` package, which is 0.7-8 at the time of writing

All the code and data for this chapter is available at `https://github.com/PacktPublishing/The-Statistics-and-Machine-Learning-with-R-Workshop/tree/main/Chapter_3`.

Transforming categorical and numeric variables

As covered in the previous chapter, we can use the `mutate()` function from `dplyr` to transform existing variables and create new ones. The specific transformation depends on the type of the variable and the resulting shape we would like it to be. For example, we may want to change the value of a categorical variable according to a mapping dictionary, create a new variable based on a combination of filtering conditions of existing variables, or group a numeric variable into different ranges in a new variable. Let us look at these scenarios in turn.

Recoding categorical variables

There are many cases when you would want to recode the values of a variable, such as mapping countries' short names to the corresponding full names. Let's create a dummy `tibble` dataset to illustrate this.

In the following code, we have created a `students` variable that stores information on age, country, gender, and height. This is a small dummy dataset but it's good enough for demonstration purposes:

```
students = tibble(age = c(26, 30, 28, 31, 25, 29, 30, 29),
                  country = c('SG', 'CN', 'US', 'UK','CN', 'SG', 'IN',
'SG'),
                  gender = c('F', 'F', 'M', 'M', 'M', 'F', 'F', 'M'),
height = c(168, 169, 175, 178, 170, 170, 172, 180))
```

Now, let's go through an example of converting the values of the `country` variable into their full names.

Exercise 3.1 – converting the country variable values into their full names

This exercise will use the `recode()` function from the `dplyr` package to map the existing short country names to the corresponding full names:

1. Add a new column that converts the short country names into the corresponding full names by providing a mapping table using `recode()`:

    ```
    students_new = students %>%
      mutate(country_fullname = recode(country,
                                       "SG"="Singapore",
                                       "CN"="China",
                                       "UK"="United Kingdom",
                                       "IN"="India"))
    ```

```
>>> students_new
# A tibble: 8 x 5
    age country gender height country_fullname
  <dbl> <chr>   <chr>   <dbl> <chr>
1    26 SG      F         168 Singapore
2    30 CN      F         169 China
3    28 UK      M         175 United Kingdom
4    31 UK      M         178 United Kingdom
5    25 CN      M         170 China
6    29 SG      F         170 Singapore
7    30 IN      F         172 India
8    29 SG      M         180 Singapore
```

Here, we provided a mapping dictionary as an argument in the recode() function, which searches for the keys in the left column and assigns the corresponding values in the right column to country_fullname. Note that the newly created column assumes a character type.

2. Perform the same conversion and store the result as a factor type:

```
students_new = students_new %>%
  mutate(country_fullname2 = recode_factor(country,
                             "SG"="Singapore",
                             "CN"="China",
                             "UK"="United Kingdom",
                             "IN"="India"))
>>> students_new
# A tibble: 8 x 6
    age country gender height country_fullname country_fullname2
  <dbl> <chr>   <chr>   <dbl> <chr>            <fct>
1    26 SG      F         168 Singapore        Singapore
2    30 CN      F         169 China            China
3    28 UK      M         175 United Kingdom   United Kingdom
4    31 UK      M         178 United Kingdom   United Kingdom
5    25 CN      M         170 China            China
6    29 SG      F         170 Singapore        Singapore
7    30 IN      F         172 India            India
8    29 SG      M         180 Singapore        Singapore
```

We can see that the resulting variable, country_fullname2, is a factor by using recode_factor().

When the new column we want to create depends on a complex combination of existing ones, we can resort to the case_when() function, as introduced in the following section.

Creating variables using case_when()

The case_when() function provides a convenient way to set multiple if-else conditions when creating a new variable. It takes a sequence of two-sided formulas, where the left-hand side contains the filtering conditions and the right-hand side provides the replacement value that matches the preceding conditions. The syntax inside the function follows a logical condition(s) ~ replacement value pattern that gets evaluated sequentially, where multiple variables can be used inside the logical conditioning. At the end of the sequence is a TRUE ~ default value case that gets assigned to the variable if all preceding conditions evaluate to FALSE.

Let's go through an exercise to create a new variable based on multiple if-else conditions involving multiple columns.

Exercise 3.2 – creating a new variable using multiple conditions and columns

In this exercise, we will create a new variable that indicates the age and region of the students.

Create a new variable type to identify whether the students come from Asia and are in their 20s or 30s by assuming asia_20+ and asia_30+ as values, respectively. Set the value to others if there is no match:

```
students_new = students %>%
  mutate(type = case_when(age >= 30 & country %in% c("SG","IN","CN") ~
"asia_30+",
                          age < 30 & age >= 20 & country %in%
c("SG","IN","CN") ~ "asia_20+",
                          TRUE ~ "others"))
>>> students_new
# A tibble: 8 x 5
    age country gender height type
  <dbl> <chr>   <chr>   <dbl> <chr>
1    26 SG      F         168 asia_20+
2    30 CN      F         169 asia_30+
3    28 UK      M         175 others
4    31 UK      M         178 others
5    25 CN      M         170 asia_20+
6    29 SG      F         170 asia_20+
7    30 IN      F         172 asia_30+
8    29 SG      M         180 asia_20+
```

Here, we used the & sign to combine multiple AND conditions that evaluate age and country. When neither of the preceding conditions in the sequence evaluate to TRUE, the function will fall into the all-encompassing TRUE case and assign others as the default value.

Next, we will look at converting a numeric column into different bins/categories.

Binning numeric variables using cut()

A numeric column can be partitioned into different categories using the cut () function. For a numeric column, it assigns the value to the corresponding predefined intervals and codes the value based on the assigned interval. The resulting column of intervals assumes an ordered factor type.

The cut () function has three key arguments: x to accept a numeric vector to be binned, breaks to accept a numeric vector of cut points, which could include negative infinity, -Inf, and positive infinity, Inf, and labels to indicate the labels of the resulting intervals.

Let's go through an exercise using cut () to convert the age column into different age groups.

Exercise 3.3 – binning the age column into three groups

In this exercise, we will use the cut () function to assign the value of the age column to one of the following ranges: (-infinity, 25), [26, 30], or [31, infinity]:

1. Segment the age column into three intervals with breakpoints at 25 and 30 (inclusive on the right) and store them in a new column named age_group:

```
students_new = students %>%
  mutate(age_group = cut(x = age,
                     breaks = c(-Inf, 25, 30, Inf),
                     labels = c("<=25", "26-30", ">30")))
>>> students_new
# A tibble: 8 x 5
    age country gender height age_group
  <dbl> <chr>   <chr>   <dbl> <fct>
1    26 SG      F         168 26-30
2    30 CN      F         169 26-30
3    28 UK      M         175 26-30
4    31 UK      M         178 >30
5    25 CN      M         170 <=25
6    29 SG      F         170 26-30
7    30 IN      F         172 26-30
8    29 SG      M         180 26-30
```

Here, we can see that age_group is an ordered factor with three levels.

A few cutting functions perform automatic binning when we do not have a specific cutoff point in mind. For example, cut_interval () cuts the original vector into a specified number of groups with equal intervals, while cut_number () converts the input vector into a specific number of groups, where each group has approximately the same number of observations. The tidyverse package provides both functions. Let's try them out.

2. Group the `age` column into three bins of equal length using `cut_interval()`:

```
students_new = students %>%
   mutate(age_group = cut_interval(age, n=3))
>>> students_new
# A tibble: 8 x 5
     age country gender height age_group
   <dbl> <chr>   <chr>   <dbl> <fct>
1     26 SG      F         168 [25,27]
2     30 CN      F         169 (29,31]
3     28 UK      M         175 (27,29]
4     31 UK      M         178 (29,31]
5     25 CN      M         170 [25,27]
6     29 SG      F         170 (27,29]
7     30 IN      F         172 (29,31)
8     29 SG      M         180 (27,29)
```

The `age_group` column now consists of three levels that represent equal-length intervals. Let's check out the counts of each level using the `summary()` function:

```
>>> summary(students_new$age_group)
[25,27] (27,29) (29,31)
      2       3       3
```

3. Group the `age` column into three bins of an equal number of observations using `cut_interval()`:

```
students_new = students %>%
   mutate(age_group = cut_number(age, n=3))
>>> students_new
# A tibble: 8 x 5
     age country gender height age_group
   <dbl> <chr>   <chr>   <dbl> <fct>
1     26 SG      F         168 [25,28.3]
2     30 CN      F         169 (29.7,31]
3     28 UK      M         175 [25,28.3]
4     31 UK      M         178 (29.7,31]
5     25 CN      M         170 [25,28.3]
6     29 SG      F         170 (28.3,29.7]
7     30 IN      F         172 (29.7,31)
8     29 SG      M         180 (28.3,29.7)
```

The cutoff points are now assuming decimal points to make the resulting count of observations approximately equal, as verified in the following code:

```
>>> summary(students_new$age_group)
  [25,28.3]  (28.3,29.7)    (29.7,31)
    3              2             3
```

So far, we have looked at different ways to transform an existing categorical or numeric variable and create a new variable based on specific conditions. Next, we will look at how to transform and reshape the whole DataFrame to facilitate our analysis.

Reshaping the DataFrame

A DataFrame that consists of a combination of categorical and numeric columns can be expressed in both wide and long formats. For example, the `students` DataFrame is considered a long format since all countries are stored in the `country` column. Depending on the specific purpose of processing, we may want to create a separate column for each unique country in the dataset, which adds more columns to the DataFrame and converts it into a wide format.

Converting between wide and long formats can be achieved via the `spread()` and `gather()` functions, both of which are provided by the `tidyr` package from the `tidyverse` ecosystem. Let's see how it works in practice.

Converting from long format into wide format using spread()

There will be times when we'll want to turn a long-formatted DataFrame into a wide format. The `spread()` function can be used to convert a categorical column with multiple categories into multiple columns, as specified by the `key` argument, with each category added to the DataFrame as a separate column. The column names will be the unique values of the categorical column. The `value` argument specifies the contents to be spread and filled in these additional columns upon calling the `spread()` function. Let's go through an exercise on this.

Exercise 3.4 – converting from long format into wide format

In this exercise, we will convert the `students` DataFrame into a wide format using `spread()`:

1. Use `country` as the `key` argument and `height` as the `value` argument to convert students into a wide format using `spread()`. Store the resulting DataFrame in `students_wide`:

```
students_wide = students %>%
  spread(key = country, value = height)
>>> students_wide
# A tibble: 7 x 6
   age gender    CN    IN    SG    UK
  <dbl> <chr>  <dbl> <dbl> <dbl> <dbl>
```

1	25 M	170	NA	NA	NA
2	26 F	NA	NA	168	NA
3	28 M	NA	NA	NA	175
4	29 F	NA	NA	170	NA
5	29 M	NA	NA	180	NA
6	30 F	169	172	NA	NA
7	31 M	NA	NA	NA	178

We can see that the original `height` column disappears, and four additional columns are added. These four columns correspond to the unique countries, and the values of these columns are filled in by the heights. If the corresponding height for a particular country is not available, NA is used to fill in the missing combination.

If we want to specify a default value for these NA values, we can set the `fill` argument in `spread()`.

2. Use the rounded average height to fill the NA values in the resulting wide format. Store the resulting DataFrame in `students_wide2`:

```
avg_height = round(mean(students$height))
students_wide2 = students %>%
  spread(key = country, value = height, fill = avg_height)
>>> students_wide2
# A tibble: 7 x 6
    age gender    CN    IN    SG    UK
  <dbl> <chr>  <dbl> <dbl> <dbl> <dbl>
1    25 M       170   173   173   173
2    26 F       173   173   168   173
3    28 M       173   173   173   175
4    29 F       173   173   170   173
5    29 M       173   173   180   173
6    30 F       169   172   173   173
7    31 M       173   173   173   178
```

Converting from long into wide could be helpful from an analysis and presentational perspective since we can visually compare the heights across all countries for a specific combination of age and gender. However, this comes with additional storage costs, as shown by the multiple NA values earlier.

Now, let's learn how to convert a wide-formatted DataFrame into a long format.

Converting from wide format into long format using gather()

When we are in the opposite situation, where the given data is in a wide format, we can use the `gather()` function to convert it into a long format for more convenient follow-up processing. For example, by compressing the four country columns into the `key` variable and storing all heights under the `value` variable specified in `gather()`, we can continue with the usual split-apply-combine treatments we introduced earlier based on just two columns instead of four or more.

The `gather()` function also uses the key and `value` arguments to specify the name of the resulting key and `value` columns in the long-formatted table. Besides, we need to specify the columns whose names are used to fill in the key column and the values in the `value` column. When there are many adjacent columns to specify, we can use the `:` operator by passing the starting and ending column names to select all columns in between. Let's go through a practice exercise.

Exercise 3.5 – converting from wide format into long format

This exercise will convert the wide-formatted `students_wide` DataFrame back into its original long format:

1. Convert `students_wide` into a long format by specifying the key column as `country` and the `value` column as `height`, and using the values of the CN, IN, SG, and UK columns to fill in the key and `value` columns, respectively:

```
students_long = students_wide %>%
  gather(key = "country", value = "height", CN:UK)
>>> students_long
# A tibble: 28 x 4
      age gender country height
   <dbl> <chr>  <chr>    <dbl>
 1    25 M      CN         170
 2    26 F      CN          NA
 3    28 M      CN          NA
 4    29 F      CN          NA
 5    29 M      CN          NA
 6    30 F      CN         169
 7    31 M      CN          NA
 8    25 M      IN          NA
 9    26 F      IN          NA
10    28 M      IN          NA
# … with 18 more rows
```

We can see that several rows with missing values in the `height` column have been added to `students_long`. This is because of their original presence in `students_wide`. Let's remove them using the `drop_na()` function from `dplyr`.

2. Remove the rows with NA values in the `height` column:

```
students_long = students_long %>%
  drop_na(height)
>>> students_long
# A tibble: 8 x 4
```

```
        age  gender  country  height
      <dbl>  <chr>   <chr>    <dbl>
  1    25    M       CN        170
  2    30    F       CN        169
  3    30    F       IN        172
  4    26    F       SG        168
  5    29    F       SG        170
  6    29    M       SG        180
  7    28    M       UK        175
  8    31    M       UK        178
```

With that, we have obtained the long-formatted DataFrame. Now, let's verify whether it is the same as the original students DataFrame.

3. Verify whether students_long is the same as students using all_equal():

    ```
    >>> all_equal(students, students_long, ignore_row_order = T,
    ignore_col_order = T)
    TRUE
    ```

The all_equal() function from dplyr compares two datasets and checks whether they are identical. It provides a flexible way to carry out an equality comparison and supports ignoring the ordering of rows and/or columns. The result shows that we have successfully converted back into the original dataset.

With that, we have looked at different ways to reshape the DataFrame. Next, we will cover how to deal with string data.

Manipulating string data

Character-typed strings are standard in real-life data, such as name and address. Analyzing string data requires properly cleaning the raw characters and converting the information embedded in a blob of textual data into a quantifiable numeric summary. For example, we may want to find the matching names of all students that follow a specific pattern.

This section will cover different ways to define patterns via regular expressions to detect, split, and extract string data. Let's start with the basics of strings.

Creating strings

A **string** is a character-typed variable that is represented by a sequence of characters (including punctuation) wrapped by a pair of double quotes (" "). Sometimes, a single quote (') is also used to denote a string, although it is generally recommended to use double quotes unless the characters themselves include double quotes.

There are multiple ways to create a string. The following exercise introduces a few different ways to initialize a character-typed string.

Exercise 3.6 – expressing strings in R

In this exercise, we will look at creating strings in R:

1. Try to type out the following strings in the R console:

    ```
    >>> "statistics workshop"
    "statistics workshop"
    ```

    The strings are printed without error. Let's see what happens if we wrap statistics with double quotes to highlight this word.

2. Add double quotes to statistics in the string:

    ```
    >>> ""statistics" workshop"
    Error: unexpected symbol in """statistics"
    ```

    This time, an error pops up because R takes the second double quote as the ending quote of the string. This error can be avoided by switching to using single quotes for the outside quotes when double quotes are used in a string.

3. Wrap the previous string with single quotes:

    ```
    >>> '"statistics" workshop'
    "\"statistics\" workshop"
    ```

    Now, R interprets the string correctly and considers all content within the pair of single quotes as a whole string. Note that the resulting string is still printed with double quotes in the console. The two double quotes within the string are also preceded by a backward slash (\). This is called an escape sequence and is used to indicate the literal interpretation of the double quotes as characters instead of the start of a string. The escape sequence is a useful way to include special characters in a string.

    We can also manually add the escape character inside the string to enforce the correct interpretation, which will print out the same result as before.

4. Add the escape sequence before the double quotes inside the string:

    ```
    >>> "\"statistics\" workshop"
    "\"statistics\" workshop"
    ```

    Printing out the string sequence with a backslash is not convenient for reading. To beautify the output, we can pass the exact string to the writeLines() function.

5. Print the same string using writeLines():

    ```
    >>> writeLines("\"statistics\" workshop")
    "statistics" workshop
    ```

Next, we will look at how to turn numbers into strings for better interpretation.

Converting numbers into strings

As we learned earlier, numbers can be converted into strings via the `as.character()` function. However, it would be inconvenient to directly read and report a big number such as `123000`. We would usually express it in a more readable manner such as 123,000 or a more concise way such as 1.23e+05, where the latter follows the scientific representation with e+05 equal to 10^5. Additionally, we may want to display a limited number of digits after the decimal point for a floating number.

All of these can be achieved via the `format()` function, which is useful when converting and printing numbers as strings while following different formats. Let's see how this is done in practice.

Exercise 3.7 – converting numbers into strings using format()

This exercise will use `format()` to convert numbers into pretty and easily readable strings:

1. Add a comma as a thousands separator to `123000` by specifying the `big.mark` argument in `format()`:

    ```
    >>> format(123000, big.mark = ",")
    "123,000"
    ```

    Notice that the result is now a character-type string with a comma added.

2. Convert `123000` into scientific format:

    ```
    >>> format(123000, scientific = TRUE)
    "1.23e+05"
    ```

    Using scientific format is a concise way to represent large numbers. We can also shorten a long floating number by specifying the number of digits to display.

3. Display only three digits of `1.256` by specifying the `digits` argument:

    ```
    >>> format(1.256, digits = 3)
    "1.26"
    ```

    The result is rounded and converted into a string, displaying only three digits as specified. We can also achieve the same rounding effect using `round()`.

4. Round `1.256` to two decimal points:

    ```
    >>> round(1.256, digits = 2)
    1.26
    ```

    This time, the result is still numeric since `round()` does not involve type conversion.

In the next section, we will look at connecting multiple strings.

Connecting strings

When there are multiple strings, we can use `paste()` to connect and join them into a single string. This becomes important if we want to print long and customized messages in our program instead of relying on manually typing them out.

The `paste()` function takes an arbitrary number of string inputs as arguments and combines them into one. Let's see how it works.

Exercise 3.8 – combining strings using paste()

In this exercise, we will look at different ways to combine multiple string inputs:

1. Connect the `statistics` and `workshop` strings to generate `statistics workshop`:

    ```
    >>> paste("statistics", "workshop")
    "statistics workshop"
    ```

    Here, we can see that a space is automatically added between the two strings. This is controlled by the `sep` argument, which specifies the filling content between strings and assumes a default value of a space. We can choose to override the default behavior by passing a separating character to this argument.

2. Remove the space in between and generate `statisticsworkshop`:

    ```
    >>> paste("statistics", "workshop", sep = "")
    "statisticsworkshop"
    ```

    We can also achieve the same result using the `paste0()` function, which directly connects two strings by default:

    ```
    >>> paste0("statistics", "workshop")
    "statisticsworkshop"
    ```

    Let's see what happens when we connect a single string to a vector of strings.

3. Connect a vector of `statistics` and `workshop` with `course`:

    ```
    >>> paste(c("statistics", "workshop"), "course")
    "statistics course" "workshop course"
    ```

The result shows that `course` is added to each element in the vector. This is completed via the recycling operation under the hood, where `course` is recycled so that it can be combined with each string in the vector. This is similar to the broadcasting mechanism in Python.

We can also remove the vector structure and combine all elements into a single string by specifying the `collapse` argument.

4. Compress the previous output into a single string separated by +:

    ```
    >>> paste(c("statistics", "workshop"), "course", collapse = " +
    ")
    "statistics course + workshop course"
    ```

    After plugging in all the components of the combined vector and separating them by the specified argument, the result is a single collapsed string.

So far, we have learned the basics when it comes to working with string data. The `stringr` package provided by the `tidyverse` ecosystem provides many handy functions if we want to have more flexible control of our strings. This will be covered in the following section.

Working with stringr

The `stringr` package provides a cohesive set of functions that all start with `str_` and are designed to make working with strings as easy as possible.

Let's start with the basic functions of `stringr` by replicating the same results from the previous exercise.

Basics of stringr

The `str_c()` function from the `stringr` package can concatenate multiple strings with similar functionalities as in `paste()`. Let's see its use in action.

Exercise 3.9 – combining strings using paste()

In this exercise, we will reproduce the same as what we did in *Exercise 3.8* using `str_c()`:

1. Concatenate `statistics` with `workshop` with a separating space in between:

    ```
    >>> str_c("statistics", "workshop", sep = " ")
    "statistics workshop"
    ```

    We can use the `sep` argument to specify the separator between strings.

2. Combine a vector of `statistics` and `workshop` with `course`:

    ```
    >>> str_c(c("statistics", "workshop"), "course", sep = " ")
    "statistics course" "workshop course"
    ```

    The same recycling behavior also appears here.

3. Compress the preceding output into a single string separated by +:

    ```
    >>> str_c(c("statistics", "workshop"), "course", sep = " ",
    collapse = " + ")
    "statistics course + workshop course"
    ```

There are two other common `stringr` functions: `str_length()`, which returns the length of the string, and `str_sub()`, which subtracts parts of the string:

- For example, we can get the length of each string in a vector, as shown in the following code snippet.

  ```
  >>> str_length(c("statistics", "workshop"))
  10   8
  ```

- Alternatively, we can use the `nchar()` function from base R to achieve the same result, as shown here:

  ```
  >>> nchar(c("statistics", "workshop"))
  10   8
  ```

- We can also use `str_sub()` to extract parts of the string by providing a starting and ending index:

  ```
  >>> str_sub(c("statistics", "workshop"), start = 1, end = 3)
  "sta" "wor"
  ```

Extracting parts of a string is one way to look for patterns in the string. In the next section, we will cover a more advanced approach for pattern matching beyond positional indexing.

Pattern matching in a string

Matching patterns in a string is a common way to extract intelligence from textual data. When a match is found, we could split or replace the string based on the match, add additional data such as the number of matches, or perform other text-based analyses. Let's go through a few exercises to get familiar with string matches.

Exercise 3.10 – locating matches in a string

In this exercise, we will cover three functions that are commonly used in locating a match in a string, including detecting a match using `str_detect()`, selecting the strings of a vector that have a match using `str_subset()`, and counting the number of matches in a string using `str_count()`:

1. Detect the occurrence of `stat` in a vector of strings containing `statistics` and `workshop`:

    ```
    >>> str_detect(c("statistics", "workshop"), "stat")
    TRUE FALSE
    ```

    The `str_detect()` function looks for a specified pattern in the input strings and returns a logical vector of the same length as the input vector, with `TRUE` indicating a match and `FALSE` otherwise.

2. Subset the string that contains `stat`:

    ```
    >>> str_subset(c("statistics", "workshop"), "stat")
    "statistics"
    ```

The `str_subset()` function completes detection and selection in one shot. It will return only the strings that match the specified pattern.

3. Count the occurrence of t in each string of the previous vector:

```
>>> str_count(c("statistics", "workshop"), "t")
 3  0
```

The `str_count()` function returns an integer vector of the same length as the input vector and shows each string's frequency of a particular match.

Next, we will look at how to split a string based on a particular match.

Splitting a string

Splitting a string based on a specific pattern can be achieved via the `str_split()` function, which assumes a similar naming and argument setting as in previous functions. The original string could then be decomposed into smaller pieces to support a more refined analysis. Let's see how it can be used.

Exercise 3.11 – splitting a string using str_split()

This exercise will use `str_split()` to decompose a string into smaller pieces based on a specific matching condition:

1. Separate the `statistics & machine learning workshop` string at the & sign:

```
>>> str_split(c("statistics & machine leaning workshop"), "&")
[[1]]
[1] "statistics "              " machine leaning workshop"
```

The result is a single-entry list that contains a vector of two elements in the first entry. Note that both the resulting substrings have a space inside, showing that the an exact pattern match is used to split the string. We could then include the space in the matching pattern to remove the spaces in the resulting substrings.

2. Include preceding and trailing spaces in the matching pattern:

```
>>> str_split(c("statistics & machine leaning workshop"), " & ")
[[1]]
[1] "statistics"              "machine leaning workshop"
```

With that, the spaces in the substrings have been removed. As shown in the following code snippet, since the result is wrapped in a list, we can follow the list indexing rule to access the corresponding element.

3. Access the second element from the previous result:

```
>>> str_split(c("statistics & machine leaning workshop"), " & ")
[[1]][2]
"machine leaning workshop"
```

The result shows that we have successfully accessed the second element of the first entry in the list.

As we learned in *Chapter 1*, *Getting Started with R*, a list is a flexible structure that can hold data of different types and lengths. This is particularly useful since we would not know how many substrings would be found in advance. See the following example:

```
>>> str_split(c("statistics & machine leaning workshop", "stats
& ml & workshop"), " & ")
[[1]]
[1] "statistics"              "machine leaning workshop"
[[2]]
[1] "stats"     "ml"          "workshop"
```

In this example, the first original string is split into two substrings while the second is split into three. Each original string corresponds to an entry in the list and can assume a different number of substrings.

We can convert the return into a more structured and rectangular format, such as a DataFrame, by specifying the simplify argument as TRUE:

```
>>> str_split(c("statistics & machine leaning workshop", "stats
& ml & workshop"), " & ", simplify = TRUE)
     [,1]          [,2]                         [,3]
[1,] "statistics" "machine leaning workshop" ""
[2,] "stats"      "ml"                         "workshop"
```

The resulting DataFrame will assume the same number of rows as the input vector and the same number of columns as the longest entry in the list.

Next, we will look at replacing a matched pattern in a string.

Replacing a string

The str_replace() and str_replace_all() functions replace the matches with new text specified by the replacement argument. The difference is that str_replace() only replaces the first match, while str_replace_all() replaces all matches as its name suggests.

Let's try replacing the & sign with and using both functions:

```
>>> str_replace(c("statistics & machine leaning workshop", "stats & ml
& workshop"), pattern = "&", replacement = "and")
"statistics and machine leaning workshop" "stats and ml & workshop"
>>> str_replace_all(c("statistics & machine leaning workshop", "stats
& ml & workshop"), pattern = "&", replacement = "and")
"statistics and machine leaning workshop" "stats and ml and workshop"
```

We can see that all & signs are replaced by and in the second string. Again, replacing a particular match involves a two-step process: locating the match, if any, and performing the replacement. The str_replace() and str_replace_all() functions complete both steps in one shot.

In the next section, we will go through a bit of a challenge that requires combining these `stringr` functions.

Putting it together

Often, a particular string processing task involves using more than one `stringr` function. Together, these functions could deliver useful transformations to the textual data. Let's go through an exercise that puts together what we have covered so far.

Exercise 3.12 – converting strings using multiple functions

In this exercise, we will use different string-based functions to convert `statistics and machine leaning workshop` into `stats & ml workshop`. First, we will replace and with the & sign, split the string, and work with the individual pieces. Let's see how this can be achieved:

1. Create a `title` variable to hold the string and replace and with &:

   ```
   >>> title = "statistics and machine leaning workshop"
   >>> title = str_replace(title, pattern = "and", replacement =
   "&")
   >>> title
   "statistics & machine leaning workshop"
   ```

   Here, we used `str_replace()` to replace and with &.

2. Split `title` into substrings using &:

   ```
   >>> a = str_split(title, " & ")
   >>> a
   [[1]]
   [1] "statistics"                   "machine leaning workshop"
   ```

   Here, we used `str_split()` to split the original string into smaller substrings. Note that additional spaces are added to the matching pattern as well. We will work with these individual pieces now.

3. Convert `statistics` into `stats`:

   ```
   >>> b = str_c(str_sub(a[[1]][1], 1, 4), str_sub(a[[1]][1], -1,
   -1))
   >>> b
   "stats"
   ```

   Here, we extracted the first four characters, namely `stat`, and the last character, `s`, using `str_sub()`, then concatenated them using `str_c()`. Note that -1 indicates the last positional indexing of a string.

Now, we can start to work on the second part.

4. Split the second element of the a variable with a space:

```
>>> c = unlist(str_split(a[[1]][2], " "))
>>> c
"machine"  "leaning"  "workshop"
```

Here, we used `str_split()` to split the `machine leaning workshop` string with a space and converted the result from a list into a vector using `unlist()`. We did this conversion to save some typing in the follow-up referencing since there is only one entry in the returned list.

Now, we can repeat a similar step by extracting the first characters of `machine` and `learning` and combining them to form `ml`.

5. Form `ml` based on the previous outputs:

```
>>> d = str_c(str_sub(c[1], 1, 1), str_sub(c[2], 1, 1))
>>> d
"ml"
```

Now, we can combine all the worked components into one string.

6. Use the previous outputs to form the final expected string:

```
>>> e = str_c(b, "&", d, c[3], sep = " ")
>>> e
"stats & ml workshop"
```

In the next section, we will learn about more advanced pattern matching techniques that use regular expressions.

Introducing regular expressions

A **regular expression** is a sequence of characters that bear a special meaning and are used for pattern matching in strings. Since the specific meaning of characters in a regular expression requires some memorization and can easily be forgotten if you do not use them often, we will avoid introducing its underlying syntax and focus on intuitive and more human-friendly programming using the `rebus` package. It is a good companion to `stringr` and provides utility functions that facilitate string manipulation and make building regular expressions much easier. Remember to install this package via `install.package("rebus")` when you use it for the first time.

The `rebus` package has a special operator called `%R%` that's used to concatenate matching conditions. For example, to detect whether a string starts with a particular character, such as `s`, we could specify the pattern as `START %R% "s"` and pass it to the pattern argument of the `str_detect()` function, where `START` is a special keyword that's used to indicate the start of a string. Similarly, the `END` keyword indicates the end of a string. Together, they are called anchors in the `rebus` library. Let's look at the following example:

```
>>> str_detect(c("statistics", "machine learning"), pattern = START
%R% "s")
TRUE FALSE
```

We can also type `START` in the console. The result, which is a carat sign, is exactly the character used in vanilla regression expressions to indicate the start of a string:

```
>>> START
<regex> ^
```

In addition, `str_view()` is another useful function that visualizes the matched parts of a string. Running the following command will bring up an HTML viewer panel with the matched parts highlighted:

```
>>> str_view(c("statistics", "machine learning"), pattern = START %R%
"s")
```

This is shown in *Figure 3.1*:

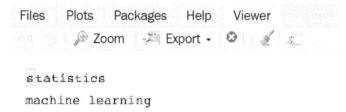

Figure 3.1 – Visualizing the matching result in the viewer pane using str_view()

Let's go through an exercise to learn more about the various pattern matching functions in rebus.

Exercise 3.13 – applying regular expressions using rebus

In this exercise, we will apply different regular expressions to match the expected pattern in the strings:

1. Run the following command to create a vector of strings. Note that the strings are designed to be simple but good enough to demonstrate the purpose of the matching functions we will introduce here:

    ```
    >>> texts = c("stats 101", "machine learning", "R 101 ABC
    workshop", "101 R workshop")
    ```

2. Search for a string from the vector that ends with `learning`:

    ```
    >>> str_subset(texts, pattern = "learning" %R% END)
    "machine learning"
    ```

    Here, we used the END keyword in the `pattern` argument to indicate that the string should end with `learning`.

3. Search for a string that contains any character followed by `101`:

    ```
    >>> str_subset(texts, pattern = ANY_CHAR %R% "101")
    "stats 101"          "R 101 ABC workshop"
    ```

    Note that ANY_CHAR is a special keyword, a wildcard that indicates any single character and corresponds to a dot in normal regression expressions, as shown in the following code:

    ```
    >>> ANY_CHAR
    <regex> .
    ```

    Since the pattern says any character followed by `101`, two strings have been selected due to the presence of `101`. `101 R workshop` has not been selected since there is no character before `101`.

4. Search for a string whose third character is `a`:

    ```
    >>> str_subset(texts, pattern = START %R% ANY_CHAR %R% ANY_CHAR
    %R% "a")
    "stats 101"
    ```

    Here, we specified the third character to be a by passing in two wildcard keywords at the beginning.

5. Search for a string that starts with `stats` or `R`:

    ```
    >>> str_subset(texts, pattern = START %R% or("stats", "R"))
    "stats 101"          "R 101 ABC workshop"
    ```

    The `or()` function is useful when specifying more than one matching condition.

6. Search for a string that contains one or more a or A characters:

```
>>> str_subset(texts, pattern = one_or_more(char_class("aA")))
"stats 101"          "machine learning"     "R 101 ABC workshop"
```

Two new functions were used here. The char_class() function enforces matching one and only one of the allowable characters specified in the input argument, while the one_or_more() function says that the pattern wrapped within the parentheses could be repeated more than once.

Next, we will cover the tidytext package, which allows us to conveniently work with unstructured textual data and the tidyverse ecosystem.

Working with tidy text mining

The tidytext package handles unstructured text by following the tidy data principle, which mandates that data is represented as a structured, rectangular-shaped, and tibble-like object. In the case of text mining, this requires converting a piece of text in a single cell into one token per row in the DataFrame.

Another commonly used representation for a collection of texts (called a **corpus**) is the **document-term matrix**, where each row represents one document (this could be a short sentence or a lengthy article) and each column represents one term (a unique word in the whole corpus, for example). Each cell in the matrix usually contains a representative statistic, such as frequency of occurrence, to indicate the number of times the term appears in the document.

We will dive into both representations and look at how to convert between a document-term matrix and a tidy data format for text mining in the following sections.

Converting text into tidy data using unnest_tokens()

Let's create a slightly different dummy dataset, as follows:

```
texts = c("stats 101", "Machine Learning", "R and ML workshop", "R
workshop & Statistics with R")
texts_df = tibble(id = 1:length(texts), text = texts)
>>> texts_df
# A tibble: 4 x 2
     id text
  <int> <chr>
1     1 stats 101
2     2 Machine Learning
3     3 R and ML workshop
4     4 R workshop & Statistics with R
```

The `texts` column in this dataset contains text of arbitrary length. Although it is stored as a tibble object, it is not quite suitable for tidy text analysis. For example, each row consists of multiple words in the `texts` column, making it challenging to derive statistical summaries such as the frequency of words. It would be much easier to get these statistics when each row corresponds to a single word for all the text.

Note that looking at word-level information is common in text mining, although we could extend to other variations such as pairs of words or even sentences. The unit of analysis used for text mining is called a **token**, and **tokenization** is the process of converting raw text into a series of tokens. Also, each row of text is often referred to as a document. In this case, we would like to have a data representation that has one token per row for each document, which can be achieved using the `unnest_tokens()` function from the `tidytext` package. Remember to install and load this package if you have not done so.

The `unnest_tokens()` function takes two inputs: the column used to host the resulting tokens from text, and the column whose text will be decomposed into tokens. There are also other aspects that the `unnest_tokens()` function takes care of when converting into a tidy text DataFrame. Let's go through an exercise to learn more about this.

Exercise 3.14 – building tidy text using unnest_tokens()

In this exercise, we will use `unnest_tokens()` to build tidy text and extract word frequency:

1. Convert `texts_df` into tidy text format using `unnest_tokens()` and name the token-holding column `unit_token`. Store the result in `tidy_df`:

```
>>> tidy_df <- texts_df %>%
  unnest_tokens(unit_token, text)
>>> tidy_df
# A tibble: 13 x 2
      id unit_token
   <int> <chr>
1  stats
2      1 101
3      2 machine
4      2 learning
5      3 r
6      3 and
7      3 ml
8      3 workshop
9      4 r
10     4 workshop
11     4 statistics
12     4 with
13     4 r
```

Note that `unnest_tokens()` uses word-level tokenization by default; thus, the `unit_token` column contains all the word tokens extracted from respective texts, and each word occupies one row. Note that the & sign is removed from the result since `unnest_tokens()` removes all punctuation by default and converts all words into lowercase. The rest of the columns, such as `id`, are retained and duplicated for each word in the raw text string.

We can also use **n-grams** representation, where *n* refers to the number of consecutive words used to form a unique token. When *n=2*, the tokenization is called a **bigram** representation.

Convert `texts_df` into tidy data using a bigram representation by specifying the token and *n* arguments:

```
>>> tidy_df2 <- texts_df %>%
  unnest_tokens(unit_token, text, token = "ngrams", n = 2)
>>> tidy_df2
# A tibble: 9 x 2
      id unit_token
   <int> <chr>
1      1 stats 101
2      2 machine learning
3      3 r and
4      3 and ml
5      3 ml workshop
6      4 r workshop
7      4 workshop statistics
8      4 statistics with
9      4 with r
```

We can see that the resulting tokens consist of each consecutive pair of words in the original text. Again, punctuation removal and lowercase conversion are performed under the hood.

We can easily derive the word frequency distribution with the tidy data available.

2. Derive the word counts from `tidy_df`:

```
>>> tidy_df %>%
count(unit_token, sort = TRUE)
# A tibble: 10 x 2
  unit_token       n
  <chr>        <int>
1 r                3
2 workshop         2
3 101              1
4 and              1
5 learning         1
6 machine          1
7 ml               1
```

```
 8 statistics      1
 9 stats           1
10 with            1
```

Here, we used the `count ()` function to count the frequency of each unique word. We can also overlay this analysis with other `dplyr` operations, such as removing stop words (for example, `the` and `a`) from the word counts. Stop words are common words that do not convey additional meaning in text mining and are often removed from the corpus. We can inspect the list of English stop words using the `get_stopwords ()` function, as follows:

```
>>> get_stopwords()
# A tibble: 175 x 2
   word        lexicon
   <chr>       <chr>
 1 i           snowball
 2 me          snowball
 3 my          snowball
 4 myself      snowball
 5 we          snowball
 6 our         snowball
 7 ours        snowball
 8 ourselves   snowball
 9 you         snowball
10 your        snowball
# … with 165 more rows
```

3. Derive the word frequency after removing the stop words. Store the result in `tidy_df2`:

```
>>> tidy_df2 = tidy_df %>%
  filter(!(unit_token %in% get_stopwords()$word)) %>%
  count(unit_token, sort = TRUE)
>>> tidy_df2
# A tibble: 8 x 2
  unit_token      n
  <chr>       <int>
1 r               3
2 workshop        2
3 101             1
4 learning        1
5 machine         1
6 ml              1
7 statistics      1
8 stats           1
```

We can see that `and` and `with` have been removed from the result.

Next, we will work with text in the form of a document-term matrix, which is the most commonly used format when building machine learning models using textual data.

Working with a document-term matrix

We can convert the tidy DataFrame from earlier to and from a document-term matrix. Since we used unigram (single-word) representation in the previous exercise, we will continue working with unigram word frequency and look at how to transform between tidy data and a document-term matrix, as shown in the following exercise.

A commonly used package for text mining is tm. Remember to install and load this package before you continue with the following exercise.

Exercise 3.15 – converting to and from a document-term matrix

In this exercise, we will get the word frequency table in a tidy format, followed by converting the table into a sparse document-term matrix. A sparse matrix is a special data structure that contains the same amount of information but occupies much less memory space than a typical DataFrame. Lastly, we will look at converting a document-term matrix back into tidy format:

1. Derive the word frequency count for each document and word token using `tidy_df` from the previous exercise. Save the result in `count_df`:

    ```
    >>> count_df = tidy_df %>%
      group_by(id, unit_token) %>%
      summarise(count=n())
    >>> count_df
    # A tibble: 12 x 3
    # Groups:    id [4]
           id unit_token count
        <int> <chr>      <int>
    1      1 101             1
    2      1 stats           1
    3      2 learning        1
    4      2 machine         1
    5      3 and             1
    6      3 ml              1
    7      3 r               1
    8      3 workshop        1
    9      4 r               2
    10     4 statistics      1
    11     4 with            1
    12     4 workshop        1
    ```

Here, r appears twice in the fourth document, and all other words appear once. We will convert it into a document-term matrix format.

2. Convert count_df into a document-term matrix by using the cast_dtm() function from the tm package and store the result in dtm:

```
>>> dtm = count_df %>%
  cast_dtm(id, unit_token, count)
>>> dtm
<<DocumentTermMatrix (documents: 4, terms: 10)>>
Non-/sparse entries: 12/28
Sparsity            : 70%
Maximal term length: 10
Weighting           : term frequency (tf)
```

The result shows that we have a total of four documents and 10 terms. The sparsity is as high as 70% since most words appear only in their respective document. Also, the statistic used to represent a word in a document is the term frequency.

We can also look at the whole table by specifically converting it into a normal matrix:

```
>>> as.data.frame(as.matrix(dtm), stringsAsFactors=False)
  101 stats learning machine and ml r workshop statistics with
1   1     1        0       0   0  0 0        0          0    0
2   0     0        1       1   0  0 0        0          0    0
3   0     0        0       0   1  1 1        1          0    0
4   0     0        0       0   0  0 2        1          1    1
```

Now, we have the canonical document-term matrix. Note that we can use other statistics, such as tf-idf, to represent each cell in the matrix, or even use a vector of multiple numeric values to represent each word in a document. The latter is referred to as **word embedding**, which embeds each unique word into a numeric vector of a pre-specified length to achieve a more flexible and powerful representation.

3. Convert dtm back into tidy format:

```
>>> tidy_dtm = tidy(dtm)
>>> tidy_dtm
# A tibble: 12 x 3
   document term     count
   <chr>    <chr>    <dbl>
1 1        101          1
2 1        stats        1
3 2        learning     1
4 2        machine      1
5 3        and          1
6 3        ml           1
```

7	3	r	1
8	4	r	2
9	3	workshop	1
10	4	workshop	1
11	4	statistics	1
12	4	with	1

Now, we have the same tidy data as before.

Summary

In this chapter, we touched upon several intermediate data processing techniques, ranging from structured tabular data to unstructured textual data. First, we covered how to transform categorical and numeric variables, including recoding categorical variables using recode(), creating new variables using case_when(), and binning numeric variables using cut(). Next, we looked at reshaping a DataFrame, including converting a long-format DataFrame into a wide format using spread() and back again using gather(). We also delved into working with strings, including how to create, convert, and format string data.

In addition, we covered some essential knowledge regarding the stringr package, which provides many helpful utility functions to ease string processing tasks. Common functions include str_c(), str_sub(), str_subset(), str_detect(), str_split(), str_count(), and str_replace(). These functions can be combined to create a powerful and easy-to-understand string processing pipeline.

Then, we introduced regular expressions using the rebus package, which provides convenient pattern matching functionalities that work well with stringr. Its functions and keywords are easy to read, and they include START, END, ANY_CHAR, or(), one_or_more(), and others.

Lastly, we covered working with tidy text data using the tidytext package. Converting a set of textual data into a tidy format makes it easy to leverage the many utility functions from the tidyverse ecosystem. The unnest_tokens() function is often used to tidy up raw texts, and the tidy output can also be converted to and from a document-term matrix, the standard data structure used to develop machine learning models.

Text mining is a big topic, and we only covered the very basics in this chapter. Hopefully, the basic flavors presented here will encourage you to further explore the potential functionalities provided by the tidyverse ecosystem.

In the next chapter, we will switch gears and cover data visualization, taking what we have processed and converting them into visual and actionable insights.

Data Visualization with ggplot2

4

The previous chapter covered intermediate data processing techniques, focusing on dealing with string data. When the raw data has been transformed and processed into a clean and structured shape, we can take the analysis to the next level by visualizing the clean data in a graph, which we aim to accomplish in this chapter.

By the end of this chapter, you will be able to plot standard graphs using the `ggplot2` package and add customizations to present excellent visuals.

In this chapter, we will cover the following topics:

- Introducing `ggplot2`
- Understanding the grammar of graphics
- Geometries in graphics
- Controlling themes in graphics

Technical requirements

To complete the exercises in this chapter, you will need to have the latest versions of the following packages:

- The `ggplot2` package, version 3.3.6. Alternatively, install the `tidyverse` package and load `ggplot2` directly.
- The `ggthemes` package, version 4.2.4.

The versions mentioned along with the packages in the preceding list are the latest ones while I am writing this book.

All the code and data for this chapter is available at `https://github.com/PacktPublishing/The-Statistics-and-Machine-Learning-with-R-Workshop/tree/main/Chapter_4`.

Introducing ggplot2

Conveying information via graphs tends to be more effective and visually appealing than tables alone. After all, humans are much quicker at processing visual information, such as recognizing a car in an image. In building **machine learning** (**ML**) models, we are often interested in the training and test loss profile in the form of a line chart that indicates the reduction in the training and test set loss as the model gets trained for a more extended period. Observing performance metrics helps us better diagnose whether a model is **underfitting** or **overfitting**—in other words, whether the current model is too simple or overly complex. Note that the test set is used to approximate a future dataset, and minimizing the test set error helps the model generalize to new datasets, an approach known as **empirical risk minimization**. Underfitting refers to the case when the model does poorly in both training and test sets due to insufficient fitting power, while overfitting means the model does well in the training set but not in the test set due to an overly complex model. Both underfitting and overfitting lead to high error frequency on the test set and thus low generalization power.

Good visualization skills are also a signpost of a good communicator. Creating good visualizations requires carefully designing the interface while satisfying the technical constraints regarding what is achievable. When tasked with building an ML model, most of the time is often spent on data processing, model development, and fine-tuning, only leaving a disproportionately small amount of time to communicate the modeling results to stakeholders. Effective communication means that an ML model, albeit a black-box solution for people outside this field, could still be transparently and adequately explained to and understood by its internal users. Meaningful and powerful visualizations created by various offerings from ggplot2, the specific package from the tidyverse ecosystem that focuses on graphing, serve as an excellent enabler to effective communication; the outputs are generally more visually engaging and attractive than the default plotting options offered by base R. After all, creating good visualizations will be an essential skill as you climb up the corporate ladder and think more from the audience's perspective. Good presentation skills will become equally important to (if not more important than) your technical skills?, such as model development.

This section will show you how to achieve good visual communication by building simple yet powerful plots using the ggplot2 package. It will help demystify modern visualization techniques using R and prepare you for more advanced visualization techniques. We will start with a simple scatter plot example and introduce the basic plotting grammar of the ggplot2 package using the mtcars dataset, which contains a set of automobile-related observations and is automatically loaded in the working environment when loading ggplot2.

Building a scatter plot

A scatter plot is a two-dimensional plot where the value of the two variables, often numeric in type, uniquely determines each dot on the plot. It is the go-to plot when we want to assess the relationship between two numeric variables.

Let us go through an exercise to plot the relationship between the number of cylinders (the `cyl` variable) in a car and the miles per gallon (the `mpg` variable) using the `mtcars` dataset.

Exercise 4.1 – Building a scatter plot using the mtcars dataset

In this exercise, we will first examine the structure of the `mtcars` dataset and generate a bivariate scatter plot using `ggplot2`. Proceed as follows:

1. Load and examine the structure of the `mtcars` dataset, like so:

```
>>> library(ggplot2)
>>> str(mtcars)
'data.frame':   32 obs. of  11 variables:
 $ mpg : num  21 21 22.8 21.4 18.7 18.1 14.3 24.4 22.8 19.2 ...
 $ cyl : num  6 6 4 6 8 6 8 4 4 6 ...
 $ disp: num  160 160 108 258 360 ...
 $ hp  : num  110 110 93 110 175 105 245 62 95 123 ...
 $ drat: num  3.9 3.9 3.85 3.08 3.15 2.76 3.21 3.69 3.92 3.92
 ...
 $ wt  : num  2.62 2.88 2.32 3.21 3.44 ...
 $ qsec: num  16.5 17 18.6 19.4 17 ...
 $ vs  : num  0 0 1 1 0 1 0 1 1 1 ...
 $ am  : num  1 1 1 0 0 0 0 0 0 0 ...
 $ gear: num  4 4 4 3 3 3 3 4 4 4 ...
 $ carb: num  4 4 1 1 2 1 4 2 2 4 ...
```

The result shows that the `mtcars` DataFrame contains 32 rows and 11 columns, a relatively small and structured dataset that is easy to work with. Next, we will plot the relationship between `cyl` and `mpg`.

2. Use the `ggplot()` and `geom_point()` functions to generate a scatter plot based on the `cyl` and `mpg` variables. Enlarge the size of the title and text for the plot along both axes using the `theme` layer:

```
>>> ggplot(mtcars, aes(x=cyl, y=mpg)) +
   geom_point() +
   theme(axis.text=element_text(size=18),
         axis.title=element_text(size=18,face="bold"))
```

As shown in *Figure 4.1*, the generated result contains 32 dots whose positions are uniquely determined by a combination of `cyl` and `mpg`. The screenshot suggests a decreasing trend in the value of `mpg` as `cyl` increases, although the within-group variation is also pronounced across the three groups of `cyl`:

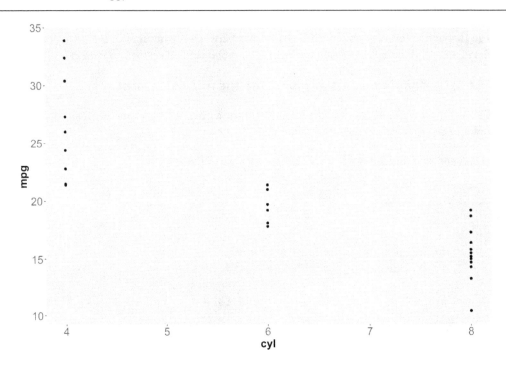

Figure 4.1 – Scatter plot between cyl and mpg

Note that the aes () function maps cyl to the *x* axis and mpg to the *y* axis. When the mapping relationship is not explicitly shown, we often assume that the first argument corresponds to the horizontal axis and the second to the vertical axis.

The script used to generate the scatter plot consists of two high-level functions: ggplot () and geom_point (). The ggplot () function specifies the dataset to be used in the first argument and the variables to be respectively plotted on the two axes in the second argument, wrapped using the aes () function (more on this later). The geom_point () function enforces the display to be in a scatter plot. These two functions are chained together via a particular + operator, indicating overlaying the second layer of operation to the first.

Also, note that the cyl variable is treated as numeric by ggplot (), as shown by the additional labels of 5 and 7 on the horizontal axis. We can verify the distinct values of cyl via the unique () function as follows:

```
>>> unique(mtcars$cyl)
 6  4  8
```

Apparently, we need to treat it as a categorical variable to avoid unwanted interpolation between different values. This can be achieved by wrapping the cyl variable via the factor () function, which converts the input argument to a categorical output:

```
>>> ggplot(mtcars, aes(factor(cyl), mpg)) +
    geom_point() +
```

```
theme(axis.text=element_text(size=18),
    axis.title=element_text(size=18,face="bold"))
```

The resulting plot is shown in *Figure 4.2*. By explicitly converting `cyl` to a categorical variable, the horizontal axis correctly indicates a distribution of dots for each unique value of `cyl`:

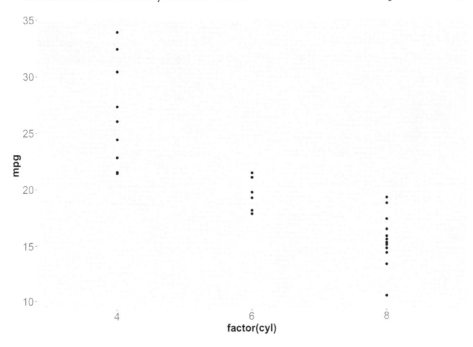

Figure 4.2 – Scatter plot after converting cyl to a categorical variable

Up until now, we have learned how to build a scatter plot by passing in the variables of interest after converting to the desired type. This works similarly to other kinds of plots, which observe a standard set of grammatical rules. Next, we will go through these fundamental rules to understand their commonalities.

Understanding the grammar of graphics

The previous example contained the three essential layers that need to be specified when plotting a graph: **data**, **aesthetics**, and **geometries**. The primary purpose of each layer is listed as follows:

- The data layer specifies the dataset to be plotted. This corresponds to the `mtcars` dataset we specified earlier.

- The aesthetics layer specifies the scale-related items that map the variables to the visual properties of the plot. Examples include the variables to be shown for the *x* axis and *y* axis, the size and color, and other plot aesthetics. This corresponds to the `cyl` and `mpg` variables we specified earlier.

- The geometry layer specifies the visual elements used for the data, such as presenting the data via points, lines, or other forms. The geom_point() command we set in the previous example tells the plot to be shown as a scatter plot.

Other layers, such as the theme layer, also help beautify the plot, which we will cover later.

The geom_point() layer from the previous example also suggests that we could easily switch to another type of plot by changing the keyword after the underscore. For example, as shown in the following code snippet, we can show the scatter plot as a boxplot for each unique value of cyl using the geom_boxplot() function:

```
>>> ggplot(mtcars, aes(factor(cyl), mpg)) +
    geom_boxplot() +
    theme(axis.text=element_text(size=18),
          axis.title=element_text(size=18,face="bold"))
```

Running this command generates the output shown in *Figure 4.3*, which visualizes a set of points as a boxplot for each distinct value of cyl. Employing a boxplot is an excellent way to detect outliers, such as the two extreme points lying outside the third boxplot:

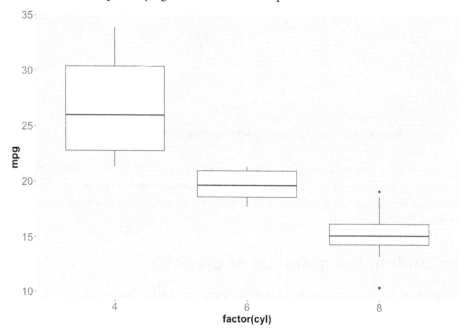

Figure 4.3 – Visualizing the same plot using a boxplot

Similarly, we could change the color and size of the points in the previous scatter plot by tweaking the aesthetics layer. Let us go through an exercise to see how this can be achieved.

Exercise 4.2 – Changing the color and size of points in a scatter plot

In this exercise, we will use the aesthetics layer to modify the color and size of the points displayed in the last scatter plot based on the disp and hp variables. The disp variable measures the engine displacement, and the hp variable indicates the gross horsepower. The points will thus vary in color and size given different values of disp and hp. Proceed as follows:

1. Change the color of the points in the scatter plot by passing disp to the color argument in the aes() function. Enlarge the size parameter of the legend as well:

```
>>> ggplot(mtcars, aes(factor(cyl), mpg, color=disp)) +
  geom_point() +
  theme(axis.text=element_text(size=18),
      axis.title=element_text(size=18,face="bold"),
        legend.text = element_text(size=20))
```

Running this command will generate the output shown in *Figure 4.4*, where the color gradient of each point changes based on the value of disp:

Figure 4.4 – Adding color to the scatter plot

2. Change the size of the points in the scatter plot by passing hp to the size argument in the aes() function, as follows:

```
>>> ggplot(mtcars, aes(factor(cyl), mpg, color=disp, size=hp)) +
    geom_point()
```

Running this command will generate the output shown in *Figure 4.5*, where the size of each point also changes based on the value of hp:

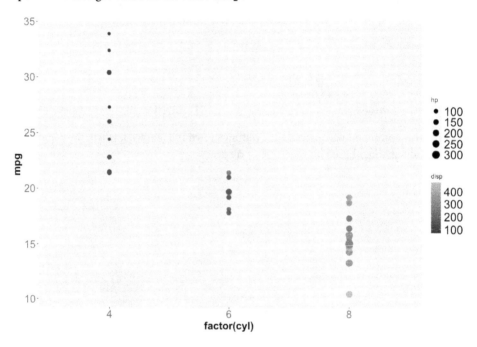

Figure 4.5 – Changing point size in the scatter plot

Although the plot now looks more enriched, be careful when adding dimensions to a single plot. In our current example, the single plot contains four dimensions of information: cyl, mpg, disp, and hp. Human brains are adept at processing two- or three-dimensional visuals but may struggle when presented with higher dimensions of graphs. The presentation style depends on what message we want to convey to our audience. Instead of lumping all dimensions together, building a separate plot with just two or three variables for illustration may be more effective. Remember—effective communication in model development lies in the quality of the message to the audience and not in the richness of the visual output.

The following exercise will let us look at the individual components of different layers in more detail.

Exercise 4.3 – Building a scatter plot with smooth curve fitting

In this exercise, we will build a scatter plot and fit a smooth curve that passes through the points. Adding a smooth curve helps us detect the overall pattern among the points and is achieved using the geom_smooth() function. Proceed as follows:

1. Build a scatter plot using hp and mpg with smooth curve fitting using geom_smooth() and coloring using disp, and adjust the opacity of the points by setting alpha=0.6 in geom_point():

```
>>> ggplot(mtcars, aes(hp, mpg, color=disp)) +
  geom_point(alpha=0.6) +
  geom_smooth() +
  theme(axis.text=element_text(size=18),
      axis.title=element_text(size=18,face="bold"),
        legend.text = element_text(size=20))
```

Running the preceding command generates the output shown in *Figure 4.6*, where the central blue curve represents a model that best fits the points and the surrounding bounds indicate the uncertainty interval. We will discuss more on the concept of a model in a later chapter:

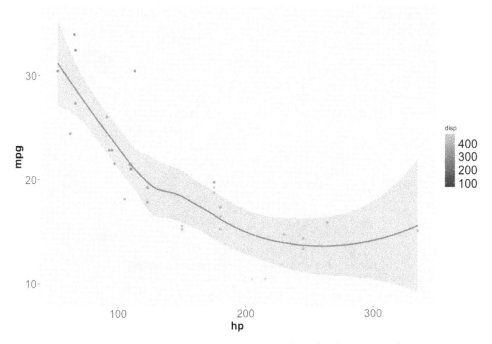

Figure 4.6 – Fitting a smooth curve among the points in a scatter plot

Since the graphics are based on the concept of additive layers, we can also generate a plot by starting with some components, storing them in a variable, and then furnishing the graph variable with additional components. Let us see how this is done in the following steps.

2. Build a scatter plot using hp and mpg with the same opacity level and store the plot in the plt variable:

```
>>> plt = ggplot(mtcars, aes(hp, mpg)) +
  geom_point(alpha=0.6) +
  theme(axis.text=element_text(size=18),
        axis.title=element_text(size=18,face="bold"))
>>> plt
```

As shown in *Figure 4.7*, directly printing out plt generates a working plot, which suggests that a plot can also be stored as an object:

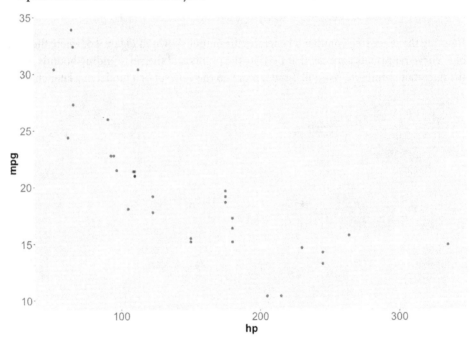

Figure 4.7 – Generating a scatter plot using hp and mpg

3. Color the points using `disp` and add a smooth curve fitting to the previous plot, like so:

```
>>> plt = plt +
  geom_point(aes(color=disp)) +
  geom_smooth()
>>> plt
```

Running these commands will generate the same plot as the one shown in *Figure 4.6*. Therefore, we can build a base plot, save it in a variable, and adjust its visual properties by adding extra layer specifications.

We can also exercise more refined control over the size, shape, and color of the points in a scatter plot, all achieved by specifying the relevant arguments. Let us see how this can be completed in the following exercise.

Exercise 4.4 – Controlling the size, shape, and color of points in a scatter plot

In this exercise, we will experiment with different input parameters to control a few visual properties of the points in the scatter plot. These controls are provided by the `geom_point()` function. Proceed as follows:

1. Generate a scatter plot between hp and mpg, and color the points using `disp`. Show the points as circles of size 4:

```
>>> ggplot(mtcars, aes(hp, mpg, color=disp)) +
  geom_point(shape=1, size=4) +
  theme(axis.text=element_text(size=18),
      axis.title=element_text(size=18,face="bold"),
        legend.text = element_text(size=20))
```

Running this command will generate the output shown in *Figure 4.8*, where we see that the points are enlarged to be circles of different colors:

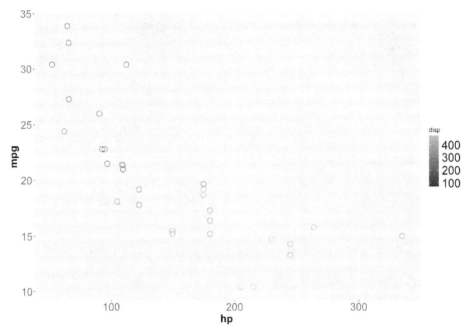

Figure 4.8 – Generating the same scatter plot with bigger-sized circles as points

Note that setting shape=1 in geom_point() presents the points as circles. We could show them in other forms by changing this argument. For example, the following command visualizes the points as triangles of a smaller size:

```
>>> ggplot(mtcars, aes(hp, mpg, color=disp)) +
    geom_point(shape=2, size=2) +
    theme(axis.text=element_text(size=18),
        axis.title=element_text(size=18,face="bold"),
        legend.text = element_text(size=20))
```

This is shown in *Figure 4.9*:

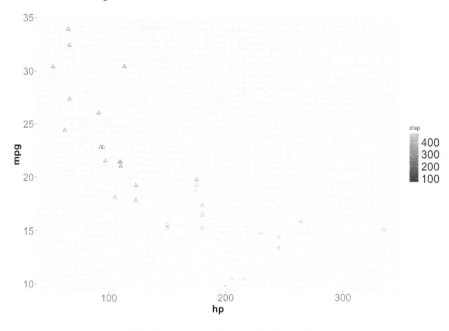

Figure 4.9 – Visualizing the points as triangles in the scatter plot

Next, we will look at how to make the scatter plot more visually appealing by filling in the inner color of the points.

2. Fill in the color of the previous scatter plot using `cyl` (after converting it to a factor type) in the `aes()` function, and set the `shape` parameter to 21, `size` to 5, and transparency (via `alpha`) to 0.6 in the `geom_point()` function:

```
>>> ggplot(mtcars, aes(wt, mpg, fill = factor(cyl))) +
  geom_point(shape = 21, size = 5, alpha = 0.6) +
  theme(axis.text=element_text(size=18),
      axis.title=element_text(size=18,face="bold"),
        legend.text = element_text(size=20))
```

Looking at the output in *Figure 4.10*, the plot now looks more visually appealing, where three groups of points are spread across different ranges of hp and mpg. A shrewd reader may wonder why we are setting shape=21 when the points are still visualized as circles. This is because 21 is a special value that allows the inner color of circles to be filled, along with their outlines or outer color:

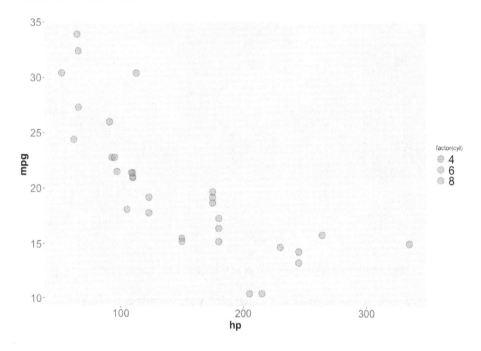

Figure 4.10 – Filling the inner color of the points in the scatter plot

Note that other than visualizing points on a graph, we can also present them as textual labels, which is more informative in a particular scenario. It could also be the case that multiple points overlap, making it difficult to tell them apart. Let's look at how to handle such situations and present the points using an alternative way via the following exercise.

Exercise 4.5 – Alternative ways of presenting points in a scatter plot

In this exercise, we will learn two different ways of presenting the points in a scatter plot: showing textual labels and jittering the overlapping points. Both techniques will add more flexibility to our plotting toolkit. Proceed as follows:

1. Visualize the brand names based on the names of each row using row.names() and plot them on the previous scatter plot of hp against mpg using geom_text():

    ```
    >>> ggplot(mtcars, aes(hp, mpg)) +
      geom_text(label=row.names(mtcars)) +
    ```

```
theme(axis.text=element_text(size=18),
      axis.title=element_text(size=18,face="bold"))
```

Running this command will generate the output shown in *Figure 4.11*, where brand names replace the points. However, some brand names overlap with each other, making it difficult to identify their specific text. Let us see how to remedy this:

Figure 4.11 – Showing brand names in a scatter plot

2. Adjust the overlapping text using the `position_jitter()` function by passing it into the `position` argument of the `geom_text()` function:

```
>>> ggplot(mtcars, aes(hp, mpg)) +
  geom_text(label=row.names(mtcars),
            fontface = "bold",
            position=position_jitter(width=20,height=20)) +
  theme(axis.text=element_text(size=18),
        axis.title=element_text(size=18,face="bold"))
```

Executing this command will generate the output shown in *Figure 4.12*, where we have additionally specified the `fontface` argument to be `bold` for better clarity. By changing the `width` and `height` parameters of the `position_jitter()` function and passing it to the `position` argument of `geom_text()`, we managed to adjust the position of the text on the graph, which is now more visually digestible:

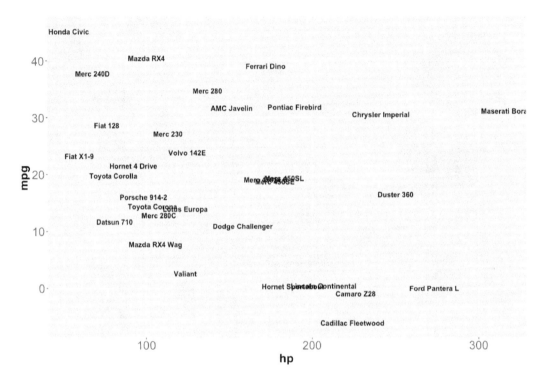

Figure 4.12 – Jittering the position of the text

Next, we will look at how to jitter overlapping points.

3. Generate a scatter plot of factored `cyl` against mpg, as follows:

```
>>> ggplot(mtcars, aes(factor(cyl), mpg)) +
    geom_point() +
    theme(axis.text=element_text(size=18),
        axis.title=element_text(size=18,face="bold"))
```

Running this command generates the output shown in *Figure 4.13*, where we intentionally used the `cyl` categorical variable to show that multiple points are overlapping on the graph:

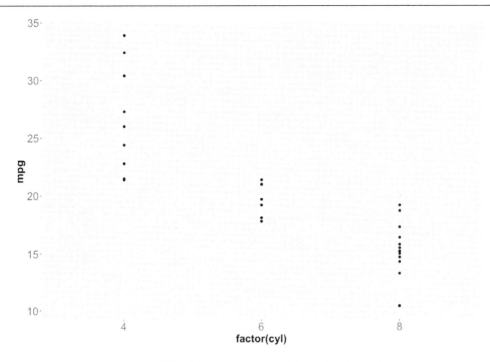

Figure 4.13 – Visualizing a scatter plot with overlapping points

Let us adjust the position of the overlapping points so that they are visually separable, giving us a sense of how many such points line up on a single spot. Note that jittering means adding random positional adjustments to the points in this case.

4. Jitter the points using geom_jitter(), like so:

```
>>> ggplot(mtcars, aes(factor(cyl), mpg)) +
    geom_jitter() +
    theme(axis.text=element_text(size=18),
        axis.title=element_text(size=18,face="bold"))
```

Running this command generates the output shown in *Figure 4.14*, where the points along each category of cyl are now separated from each other instead of aligning on the same line. Adding random jitters thus helps visually separate the overlapping points using random perturbations:

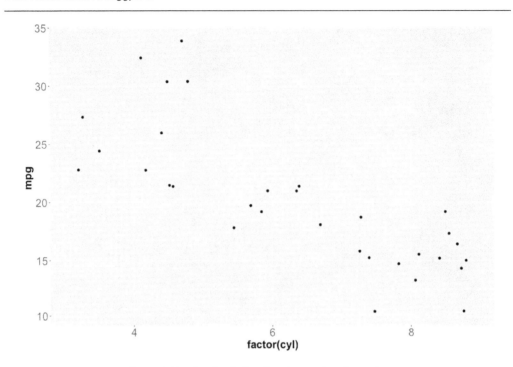

Figure 4.14 – Randomly jittering the overlapping points

Next, we will look at the geometry of graphics that determines the visual elements shown in the plot.

Geometries in graphics

The previous section mostly covered scatter plots. In this section, we will go over two additional common types of plots: bar charts and line plots. We will discuss different ways to construct these plots, focusing on the geometries that can be used to control layer-specific visual properties of the graph.

Understanding geometry in scatter plots

Let us revisit the scatter plot and zoom in on the geometry layer. The geometry layer determines how the plot actually looks, which is an essential layer in our visual communication. At the time of writing, there are over 50 geometries we can choose from, all of which start with the geom_ keyword.

Some overall guidelines apply when deciding which type of geometry to use. For example, the following list contains the possible kinds of applicable geometries for a typical scatter plot:

- **Point**, which visualizes the data as points

- **Jitter**, which adds positional jittering to a scatter plot

- **Abline**, which adds a line on the scatter plot

- **Smooth**, which smooths the plot by fitting a trend line along with the confidence bounds to help identify a particular pattern in the data

- **Count**, which counts and shows the number of observations at each location in the scatter plot

Each geometry layer is associated with its own aesthetic configurations, including both compulsory and optional settings. For example, the geom_point() function requires x and y as mandatory arguments to uniquely locate points on the plot and allows optional settings such as the alpha parameter to control the level of transparency, as well as color and fill to manage the coloring of the points, along with their shape and size parameters, and so on.

Since the geometry layer provides layer-specific control, we can set some of the visual properties in either the aesthetics layer or the geometry layer. For example, the following code generates the same plot as the one shown in *Figure 4.15*, where the coloring can be set in either the base ggplot() function or the layer-specific geom_point() function:

```
>>> ggplot(mtcars, aes(hp, mpg, color=factor(cyl))) +
  geom_point() +
  theme(axis.text=element_text(size=18),
        axis.title=element_text(size=18,face="bold"),
        legend.text = element_text(size=20))
>>> ggplot(mtcars, aes(hp, mpg)) +
  geom_point(aes(col=factor(cyl))) +
  theme(axis.text=element_text(size=18),
        axis.title=element_text(size=18,face="bold"),
        legend.text = element_text(size=20))
```

That produces the following plot:

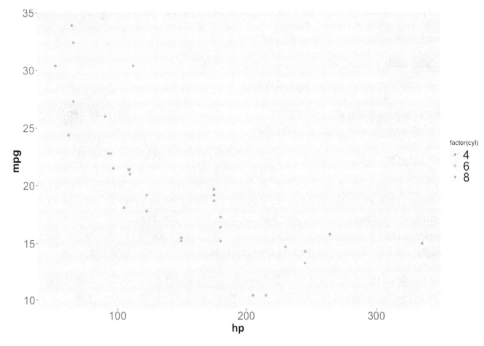

Figure 4.15 – Generating the same scatter plot using layer-specific geometry control

The flexibility from layer-specific control manifests when we have more than one layer (not necessarily a different type) to display in the plot. We will see how multiple geometry layers can be used together in the following exercise.

Exercise 4.6 – Using multiple geometry layers

In this exercise, we will display the average hp and mpg values across different groups of cyl on top of the previous scatter plot. Once obtained from the raw mtcars dataset, extra mean statistics can be added by overlaying another geometry layer, taking the same type of scatter plot. Proceed as follows:

1. Calculate the average values for all columns for each group of cyl using the dplyr library and store the result in a variable called tmp:

    ```
    >>> library(dplyr)
    >>> tmp = mtcars %>%
      group_by(factor(cyl)) %>%
      summarise_all(mean)
    ```

```
>>> tmp
# A tibble: 3 × 12
  `factor(cyl)`   mpg   cyl  disp    hp  drat    wt  qsec
  <fct>          <dbl> <dbl> <dbl> <dbl> <dbl> <dbl> <dbl>
1 4               26.7    4  105.  82.6  4.07  2.29  19.1
2 6               19.7    6  183.  122.  3.59  3.12  18.0
3 8               15.1    8  353.  209.  3.23  4.00  16.8
# … with 4 more variables: vs <dbl>, am <dbl>,
#   gear <dbl>, carb <dbl>
```

We can see that the summary statistics on the average of all columns are obtained using the summarize_all() function, a utility function that applies the input function across all columns for each group. Here, we pass the mean function to calculate the average of a column. The resulting tibble object, stored in tmp, contains the average value for all variables across three groups of cyl.

It is important to note that upon adding an extra geometry layer, the base aesthetics layer expects the same column names in each geometry layer. The base aesthetics layer in the ggplot() function applies to all geometry layers. Let us see how to add an extra geometry layer as a scatter plot to show the average hp and mpg values across different groups of cyl.

2. Add an extra layer of scatter plot to show the average hp and mpg values as big squares for each group of cyl:

```
>>> ggplot(mtcars, aes(x=hp, y=mpg, color=factor(cyl))) +
  geom_point() +
  geom_point(data=tmp, shape=15, size=6) +
  theme(axis.text=element_text(size=18),
        axis.title=element_text(size=18,face="bold"),
          legend.text = element_text(size=20))
```

Running this command will generate the output shown in *Figure 4.16*, where the big squares (obtained by setting shape=15 and size=6 in the second geom_point layer) are sourced from the tmp dataset, as specified by the data parameter in the additional geometry layer.

Note that the average hp and mpg values are automatically left-joined into the existing dataset, which shows different values of hp and mpg for each group of cyl. To ensure the two geometry layers are compatible with each other when plotted together, we need to ensure all matching coordinates (the x and y arguments) exist in the corresponding raw datasets:

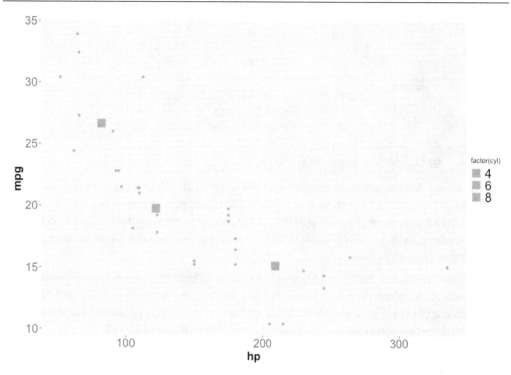

Figure 4.16 – Visualizing the average hp and mpg values for each group of cyl

This plot consists of two geometry layers, where the first layer plots each observation as small circles, and the second layer plots the average values as big boxes for hp and mpg for each group of cyl. Adding extra layers follows the same principle, as long as the data source for each layer contains the column names as specified in the base aesthetics layer.

To further illustrate the need for matching coordinates for multiple layers, let us try typing the following command in the console, where we only select the mpg and disp columns in the raw data passed to the second geometry layer. As you can see from the output, the hp column is expected, without which an error is thrown:

```
>>> ggplot(mtcars, aes(x=hp, y=mpg, color=factor(cyl))) +
  geom_point() +
  geom_point(data=tmp[,c("mpg","disp")], shape=15, size=6)
Error in FUN(X[[i]], ...) : object 'hp' not found
```

In the next section, we will look at a new type of plot: a bar chart, along with its associated geometry layer.

Introducing bar charts

A bar chart displays certain statistics (such as frequency or proportion) of a categorical or continuous variable in the form of bars. Among multiple types of bar charts, a histogram is a special type of bar chart that shows the binned distribution of a single continuous variable. Therefore, plotting a histogram is always in terms of one continuous input variable, achieved using the geom_histogram() function and only specifying the x argument. Under the hood, the function first cuts the continuous input variable into discrete bins. It then uses the internally calculated count variable to indicate the number of observations in each bin to be passed to the y argument.

Let us look at how to build a histogram in the following exercise.

Exercise 4.7 – Building a histogram

In this exercise, we will look at different ways to apply positional adjustments when displaying a histogram. Proceed as follows:

1. Build a histogram of the hp variable using the geom_histogram() layer, like so:

```
>>> ggplot(mtcars, aes(x=hp)) +
  geom_histogram() +
  theme(axis.text=element_text(size=18),
       axis.title=element_text(size=18,face="bold"))
`stat_bin()` using `bins = 30`. Pick better value with
`binwidth`.
```

Running this command will generate the output shown in *Figure 4.17*, along with the warning message on the binning. This is because the default binning value is not suitable since there are multiple gaps between the bars, making it difficult to interpret for a continuous variable. We will need to fine-tune the width of each bin using the binwidth argument:

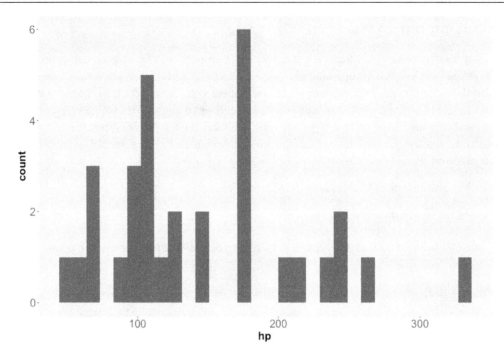

Figure 4.17 – Plotting a histogram for hp

2. Adjust the `binwidth` argument to make the histogram continuous and remove the warning message, as follows:

```
>>> ggplot(mtcars, aes(x=hp)) +
  geom_histogram(binwidth=40) +
  theme(axis.text=element_text(size=18),
        axis.title=element_text(size=18,face="bold"))
```

This produces the following output:

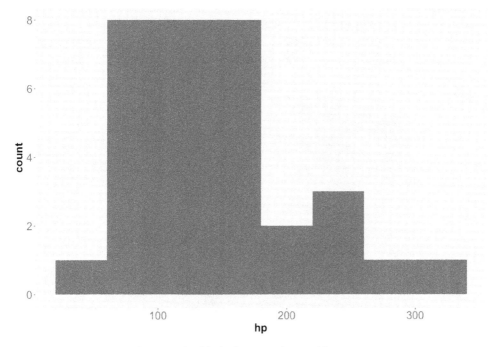

Figure 4.18 – Displaying a continuous histogram

Making a continuous-looking histogram is data-dependent and requires trial and error. In this case, setting `binwidth=40` seems to be working for us.

Next, we will introduce grouping to the previous histogram by changing the coloring of the bars.

3. Fill in the bars with different colors using the `fill` argument based on the factored `cyl`:

```
>>> ggplot(mtcars, aes(x=hp, fill=factor(cyl))) +
  geom_histogram(binwidth=40) +
  theme(axis.text=element_text(size=18),
      axis.title=element_text(size=18,face="bold"),
      legend.text = element_text(size=20))
```

Running this command generates the output shown in *Figure 4.19*, where each bar represents different groups of `cyl`. However, a shrewd reader may immediately find that, for some bars with two colors, it is difficult to discern whether they are overlapping or stacked on top of one another. Indeed, the default setting for the histogram is `position="stack"`, meaning the bars are stacked by default. To remove such confusion, we can explicitly show the bars side by side:

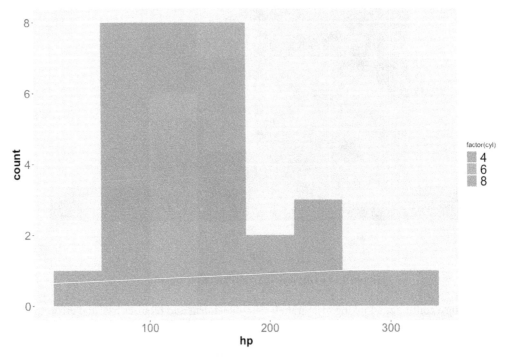

Figure 4.19 – Coloring the bars in the histogram

4. Display the bars side by side by setting `position="dodge"`, as follows:

```
>>> ggplot(mtcars, aes(x=hp, fill=factor(cyl))) +
    geom_histogram(binwidth=40, position="dodge") +
    theme(axis.text=element_text(size=18),
        axis.title=element_text(size=18,face="bold"),
        legend.text = element_text(size=20))
```

Running this command generates the output shown in *Figure 4.20*, where the bars are now shown side by side. We can further adjust the `binwidth` parameter to reduce the gaps in between:

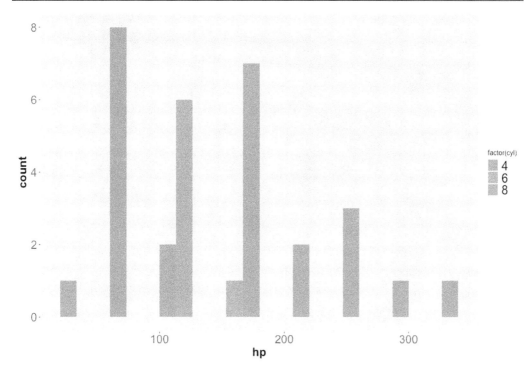

Figure 4.20 – Side-by-side bar chart

Finally, we can also show the statistics as proportions instead of counts.

5.   Show the previous histogram of hp by cyl as proportions by executing the following code:

```
>>> ggplot(mtcars, aes(x=hp, fill=factor(cyl))) +
  geom_histogram(binwidth=40, position="fill") +
  ylab("Proportion") +
  theme(axis.text=element_text(size=18),
        axis.title=element_text(size=18,face="bold"),
          legend.text = element_text(size=20))
```

Running this command generates the output shown in *Figure 4.21*, where the ylab() function is used to change the label of the *y* axis. Since the proportion of each bin needs to sum to 1, the plot contains bars of equal height, each containing one or more groups. For each bin with multiple groups, the height of each color represents the proportion of observations falling in this group of cyl within the specific bin. Such a plot is often used when we only care about the relative percentage of each group instead of the absolute count:

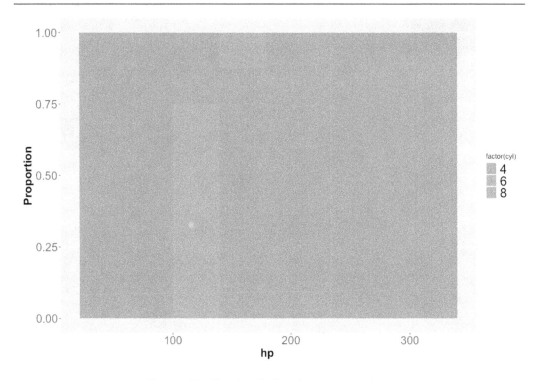

Figure 4.21 – Showing the bar chart as proportions

As mentioned earlier, a histogram is a special type of bar chart. A classical bar chart contains a categorical variable on the *x* axis, where each position represents the count of the number of observations falling into that particular category. A bar chart can be generated using the `geom_bar()` function, which allows the same positional adjustment as `geom_histogram()`. Let us go through the following exercise to learn its usage.

Exercise 4.8 – Building a bar chart

In this exercise, we will visualize the count of observations by `cyl` and `gear` as a bar chart. Proceed as follows:

1. Plot the count of observations for each unique combination of `cyl` and `gear` in a stacked bar chart, using `cyl` as the *x* axis:

    ```
    >>> ggplot(mtcars, aes(x=factor(cyl), fill=factor(gear))) +
    geom_bar(position="stack") +
    theme(axis.text=element_text(size=18),
        axis.title=element_text(size=18,face="bold"),
        legend.text = element_text(size=20))
    ```

Running this command will generate the output shown in *Figure 4.22*, where the height of the bar represents the count of observations for the particular combination of `cyl` and `gear`:

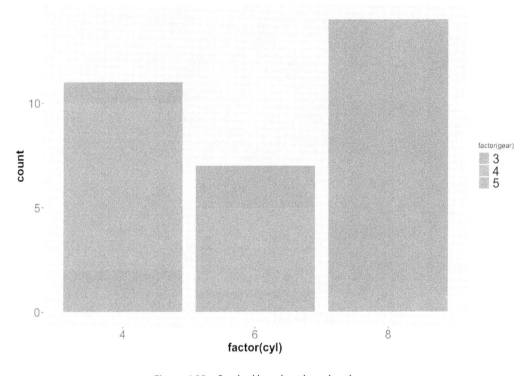

Figure 4.22 – Stacked bar chart by cyl and gear

We can also represent the bar chart using the proportion/percentage of observations in the respective group.

2. Convert the bar chart to a percentage-based plot to show the distribution of each combination, as follows:

```
>>> ggplot(mtcars, aes(x=factor(cyl), fill=factor(gear))) +
  geom_bar(position="fill") +
  theme(axis.text=element_text(size=18),
      axis.title=element_text(size=18,face="bold"),
      legend.text = element_text(size=20))
```

Running this command will generate the output shown in *Figure 4.23*:

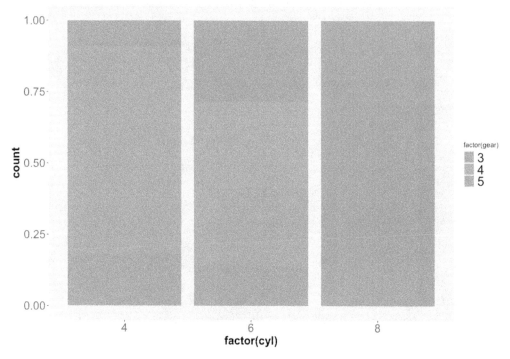

Figure 4.23 – Visualizing a bar chart as proportions

As before, we can also convert the bar chart from stacked to side by side.

3. Visualize the previous information in a side-by-side bar chart, as follows:

```
>>> ggplot(mtcars, aes(x=factor(cyl), fill=factor(gear))) +
    geom_bar(position="dodge") +
    theme(axis.text=element_text(size=18),
        axis.title=element_text(size=18,face="bold"),
        legend.text = element_text(size=20))
```

Running this command will generate the output shown in *Figure 4.24*:

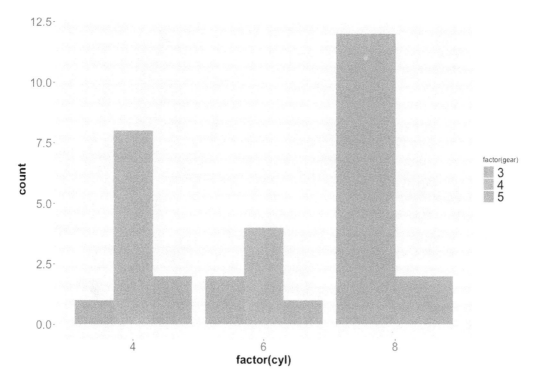

Figure 4.24 – A side-by-side bar chart

We can also customize the bar chart so that the bars partially overlap with each other. This is achieved using the `position_dodge()` function as follows, where we adjust the `width` parameter to jitter the overlapping bars to a certain extent:

```
>>> ggplot(mtcars, aes(x=factor(cyl), fill=factor(gear))) +
    geom_bar(position = position_dodge(width=0.2)) +
    theme(axis.text=element_text(size=18),
          axis.title=element_text(size=18,face="bold"),
          legend.text = element_text(size=20))
```

Running this command generates the output shown in *Figure 4.25*:

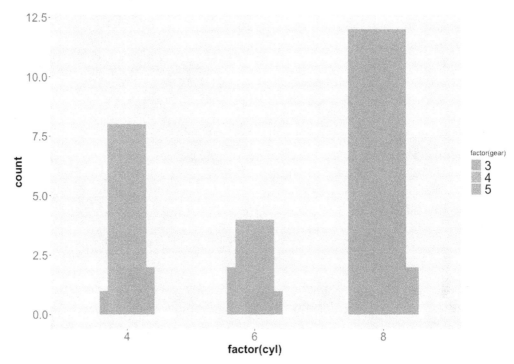

Figure 4.25 – Adjusting overlapping bars in a bar chart

Next, we will look at another popular type of plot: a line plot.

Introducing line plots

A **line plot** displays the value of one variable as the other variable changes. As with a scatter plot, a line plot can be considered scattered points connected via a line. It is mainly used to depict the relationship between two variables. For example, when two variables are positively correlated with each other, increasing one variable would lead to a seemingly proportionate increase in the other variable. Visualizing such a relationship on a line plot could result in a trend line between the two variables with a positive slope.

One of the most widely used types of line plots is the time series plot, where the value of a particular metric (such as stock price) is shown as a function of time (such as daily). In the following exercise, we will look at the quarterly earnings of Johnson & Johnson between 1960 and 1981, using the JohnsonJohnson dataset provided by base R. We will explore different ways to visualize the line chart, along with a little bit of data processing specific to time series data.

Exercise 4.9 – Building a time series plot

In this exercise, we will look at visualizing the time series data as a line plot. Proceed as follows:

1. Examine the structure of the JohnsonJohnson dataset by executing the following code:

    ```
    >>> str(JohnsonJohnson)
    Time-Series [1:84] from 1960 to 1981: 0.71 0.63 0.85 0.44 0.61
    0.69 0.92 0.55 0.72 0.77 ...
    ```

    The output suggests that the dataset is a univariate (meaning a single variable) time series ranging from 1960 to 1981. Printing out its contents (only showing the top five rows) also tells us that the frequency is quarterly, using year-quarter as a unique index for each data point in the time series:

    ```
    >>> JohnsonJohnson
          Qtr1   Qtr2   Qtr3   Qtr4
    1960  0.71   0.63   0.85   0.44
    1961  0.61   0.69   0.92   0.55
    1962  0.72   0.77   0.92   0.60
    1963  0.83   0.80   1.00   0.77
    1964  0.92   1.00   1.24   1.00
    ```

    Let us convert it to our familiar DataFrame format to ease data manipulation.

2. Convert it to a DataFrame called JohnsonJohnson2 with two columns: qtr_earning to store the quarterly time series, and date to store the approximate date:

    ```
    >>> library(zoo)
    >>> JohnsonJohnson2 = data.frame(qtr_earning=as.
    matrix(JohnsonJohnson),
              date=as.Date(as.yearmon(time(JohnsonJohnson))))
    >>> head(JohnsonJohnson2, n=3)
      qtr_earning        date
    1        0.71 1960-01-01
    2        0.63 1960-04-01
    3        0.85 1960-07-01
    ```

    The date column is obtained by extracting the time index from the JohnsonJohnson time series object, displaying as the year-month format using as.yearmon(), and then converting to a date format using as.Date().

    We will also add two extra indicator columns for plotting purposes later.

3. Add an `ind` indicator column that takes the value of TRUE if the date is equal to or beyond 1975-01-01, and FALSE otherwise. Also, extract the quarter from the `date` variable and store it in the `qtr` variable:

```
>>> JohnsonJohnson2 = JohnsonJohnson2 %>%
  mutate(ind = if_else(date >= as.Date("1975-01-01"), TRUE,
FALSE),
          qtr = quarters(date))
>>> head(JohnsonJohnson2, n=3)
  qtr_earning        date    ind qtr
1        0.71 1960-01-01 FALSE  Q1
2        0.63 1960-04-01 FALSE  Q2
3        0.85 1960-07-01 FALSE  Q3
```

In this command, we used the `quarters()` function to extract the quarter from a date-formatted field. Next, we will plot the quarterly earnings as a time series.

4. Plot `qtr_earning` as a function of `date` using a line plot, as follows:

```
>>> ggplot(JohnsonJohnson2, aes(x=date, y=qtr_earning)) +
        geom_line() +
    theme(axis.text=element_text(size=18),
        axis.title=element_text(size=18,face="bold"))
```

Running this command generates the output shown in *Figure 4.26*, where we specify the `date` column as the *x* axis and `qtr_earning` as the *y* axis, followed by the `geom_line()` layer:

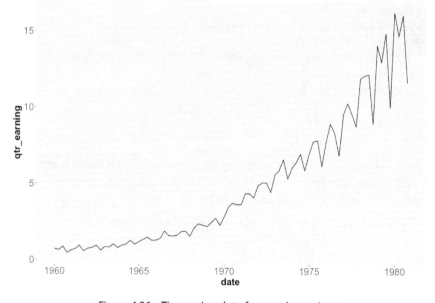

Figure 4.26 – Time series plot of quarterly earnings

The line chart for the quarterly earnings displays a long-term increasing trend and short-term fluctuations. The subject of time series forecasting focuses on using these structural components (such as trend and seasonality) to predict future values.

In addition, we can color code the time series so that different line segments display a different color according to another grouping variable.

5. Specify the color of the line plot according to the `ind` column, as follows:

```
>>> ggplot(JohnsonJohnson2, aes(x=date, y=qtr_earning,
                           color=ind)) +
    geom_line() +
    theme(axis.text=element_text(size=18),
          axis.title=element_text(size=18,face="bold"),
          legend.text = element_text(size=20))
```

Running this command generates the output shown in *Figure 4.27*, where we set `color=ind` in the base aesthetics layer to change the coloring. Note that the two line segments are disconnected since they are essentially separate time series plotted on the chart:

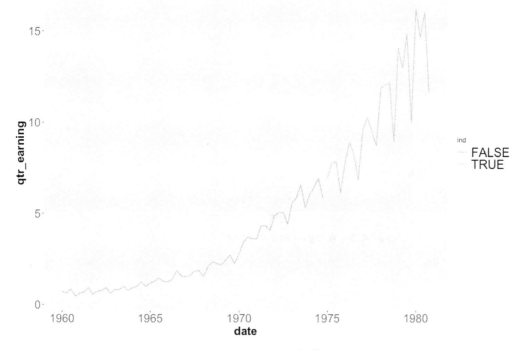

Figure 4.27 – Two line charts of different colors

We can also plot multiple lines when there are multiple categories in the grouping variable, each of which will assume a different color.

6. Plot the time series for each quarter respectively, like so:

```
>>> ggplot(JohnsonJohnson2, aes(x=date, y=qtr_earning,
                                color=qtr)) +
    geom_line() +
    theme(axis.text=element_text(size=18),
        axis.title=element_text(size=18,face="bold"),
        legend.text = element_text(size=20))
```

Running the preceding command will generate the output shown in *Figure 4.28*, where **Q1** seems to be outperforming others as it approaches 1980:

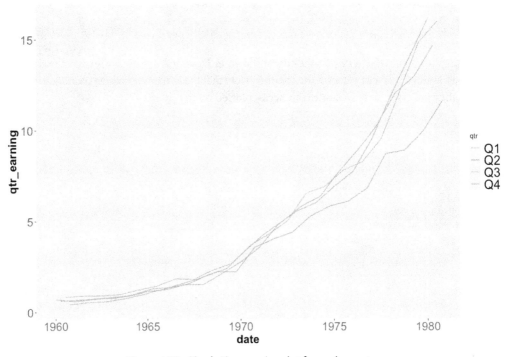

Figure 4.28 – Yearly times series plot for each quarter

In the next section, we will look at the theme layer, which controls the stylistic elements of a plot.

Controlling themes in graphics

The theme layer specifies all non-data-related properties on the plot, such as the background, legend, axis labels, and so on. Proper control of the themes in the plot could aid visual communication by highlighting critical information and directing users' attention to the intended message we would like to convey.

There are three types of visual elements controlled by the theme layer, as follows:

- **Text**, used to specify the textual display (for example, color) of the axis label
- **Line**, used to specify the visual properties of the axes such as color and line type
- **Rectangle**, used to control the borders and backgrounds of the plot

All three types are specified using functions that start with `element_`, including examples such as `element_text()` and `element_line()`. We will go over these functions in the following section.

Adjusting themes

The theme layer can be easily applied as an additional layer on the existing graph. Let us go through an exercise on how this can be achieved.

Exercise 4.10 – Applying themes

In this exercise, we will look at how to tweak the theme-related elements of the previous time series plot, including moving the legend and changing the properties of the axes. Proceed as follows:

1. Display the legend of the previous time series plot at the bottom by overlaying a theme layer with its `legend.position` argument specified as `"bottom"`. Also, enlarge the font size of the text along the axes and legend:

```
>>> ggplot(JohnsonJohnson2, aes(x=date, y=qtr_earning,
                                color=qtr)) +
  geom_line() +
  theme(legend.position="bottom",
        axis.text=element_text(size=18),
       axis.title=element_text(size=18,face="bold"),
        legend.text = element_text(size=20))
```

Running this command generates the output shown in *Figure 4.29*, where the legend is now moved to the bottom of the plot:

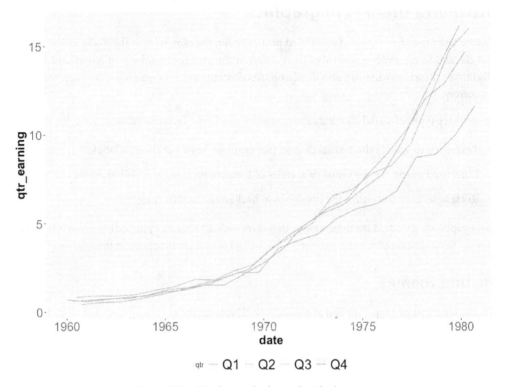

Figure 4.29 – Displaying the legend at the bottom

We can also position the legend anywhere within the plot by providing the coordinate information to the legend.position argument. The coordinates start at the lower-left corner with a value of (0,0) and span all the way to the upper-right corner, taking the value of (1,1). Since the upper-left part of the plot seems vacant, we may consider moving the legend there to save some extra space.

2. Move the legend to the upper-left corner by supplying a pair of proper coordinates:

```
>>> tmp = ggplot(JohnsonJohnson2, aes(x=date, y=qtr_earning,
                                color=qtr)) +
    geom_line() +
    theme(legend.position=c(0.1,0.8),
          axis.text=element_text(size=18),
          axis.title=element_text(size=18,face="bold"),
          legend.text = element_text(size=20))
>>> tmp
```

Here, we specified the legend's position to be (0.1, 0.8). In general, configuring a proper position using the coordinate system requires trial and error. We have also saved the result in a variable called tmp, which will be used later. The generated plot is shown in *Figure 4.30*:

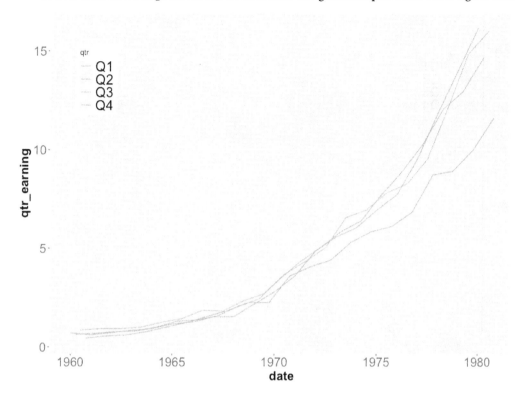

Figure 4.30 – Coordinate-based positional adjustment for the legend

Next, we will adjust the properties of the axes.

3. Based on the previous plot, change the color of the axes' titles to blue using the element_ text() function on the axis.title property. Also, make the axes' lines solid black using the element_line() function on the axis.line property:

```
>>> tmp = tmp +
  theme(
    axis.title=element_text(color="blue"),
    axis.line = element_line(color = "black", linetype =
"solid")
    )
>>> tmp
```

Running this command generates the output shown in *Figure 4.31*, where we used the `element_text()` and `element_line()` functions to adjust the visual properties (`color` and `linetype`) of the title (`axis.title`) and the lines (`axis.line`) for the axes:

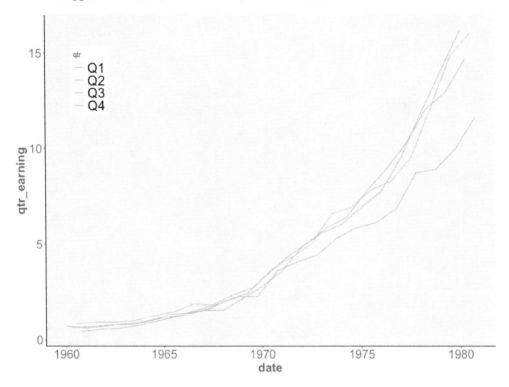

Figure 4.31 – Changing the titles and lines of the axes

Finally, we can change the default background and grids as well.

4. Remove the default grids and background in the previous plot by executing the following code:

```
>>> tmp = tmp +
    theme(
        panel.grid.major = element_blank(),
        panel.grid.minor = element_blank(),
        panel.background = element_blank()
    )
>>> tmp
```

Here, we used `panel.grid.major` and `panel.grid.minor` to access the grid properties and `panel.background` to access the background property of the plot. The `element_blank()` removes all existing configurations and is specified for all these three properties. The result is shown in *Figure 4.32*:

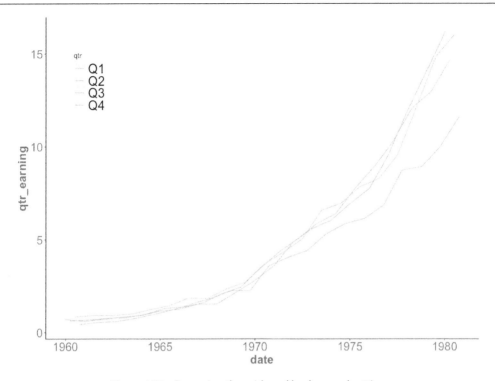

Figure 4.32 – Removing the grids and background setting

Note that we can also save the theme layer in a variable and apply it as an overlay to other plots. We treat the whole plot or a specific layer configuration as a variable, making it convenient to scale to multiple plots.

Besides creating our own themes, we can also utilize the built-in theme layers provided by ggplot2. As listed here, these built-in themes provide off-the-shelf solutions to facilitate plotting:

- theme_gray(), the default theme we used earlier
- theme_classic(), the traditional theme mostly used in scientific plotting
- theme_void(), which removes all non-data-related properties
- theme_bw(), mostly used when the transparency level is configured

For example, we can use the theme_classic() function to generate a similar plot as before, as shown in the following code snippet:

```
>>> ggplot(JohnsonJohnson2, aes(x=date, y=qtr_earning,
                                color=qtr)) +
    geom_line() +
```

```
theme_classic() +
theme(axis.text=element_text(size=18),
      axis.title=element_text(size=18,face="bold"),
      legend.text = element_text(size=20))
```

Running this command generates the output shown in *Figure 4.33*:

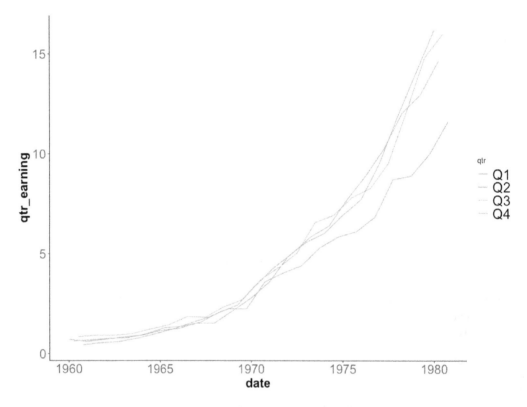

Figure 4.33 – Using out-of-the-box theme setting

Other than the built-in themes, the ggthemes package provides additional themes that further extend our choices of available themes. Let us explore this package in the next section.

Exploring ggthemes

The ggthemes package contains multiple pre-built themes. Just as using dplyr could significantly accelerate our data processing task, using the pre-built themes could also ease our graphing efforts compared with ground-up development. Let us look at a few available themes in this package.

Exercise 4.11 – Exploring themes

In this exercise, we will explore a few additional out-of-the-box themes provided by ggthemes. Remember to download and load this package before continuing with the following code examples. We will cover two theme functions. Proceed as follows:

1. Apply the theme_fivethirtyeight theme on the previous plot, as follows:

```
>>> ggplot(JohnsonJohnson2, aes(x=date, y=qtr_earning,
                                color=qtr)) +
    geom_line() +
    theme_fivethirtyeight() +
    theme(axis.text=element_text(size=18),
          axis.title=element_text(size=18,face="bold"),
          legend.text = element_text(size=20))
```

Running this command generates the output shown in *Figure 4.34*, where the legend is placed at the bottom:

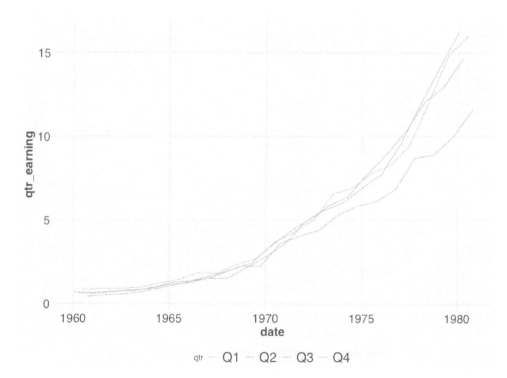

Figure 4.34 – Applying the theme_fivethirtyeight theme

2. Apply the `theme_tufte()` theme, like so:

```
>>> ggplot(JohnsonJohnson2, aes(x=date, y=qtr_earning,
                                color=qtr)) +
    geom_line() +
    theme_tufte() +
    theme(axis.text=element_text(size=18),
          axis.title=element_text(size=18,face="bold"),
          legend.text = element_text(size=20))
```

Running this command generates the output shown in *Figure 4.35*, the type of plot commonly used in scientific papers. Note that the plots in academic papers recommend only showing essential information. This means that additional configurations such as background are discouraged. Real-life plots, on the other hand, prefer a decent balance between usefulness and beauty:

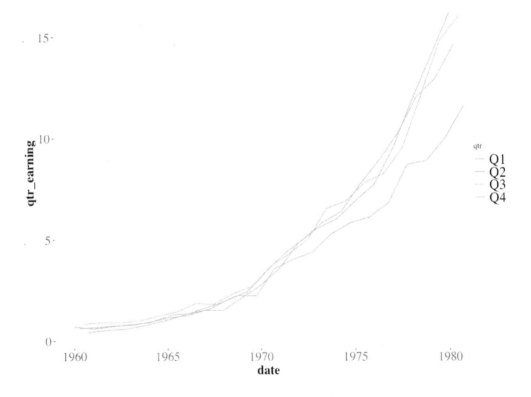

Figure 4.35 – Applying the theme_tufte theme

Throughout this section, we looked at controlling theme-related elements in a graph, which gives us great flexibility when it comes to fine-tuning and customizing a graph.

Summary

In this chapter, we introduced essential graphics techniques based on the `ggplot2` package. We started by going over the basic scatter plot and learned the grammar of developing layers in a plot. To build, edit, and improve a plot, we need to specify three essential layers: data, aesthetics, and geometries. For example, the `geom_point()` function used to build a scatter plot allows us to control the size, shape, and color of the points on a graph. We can also display them as text in addition to presenting points using the `geom_text()` function.

We also covered the layer-specific control provided by the geometry layer and showed examples using bar charts and line plots. A bar chart can help represent the frequency distribution of categorical variables and the histogram of continuous variables. A line chart supports time series data and can help identify trends and patterns if appropriately plotted.

Finally, we also covered the theme layer, which allows us to control all non-data-related visual aspects of a graph. Coupled with the built-in themes by base R and off-the-shelf themes by `ggthemes`, we have many options to choose from and accelerate the graphing effort.

In the next chapter, we will cover **exploratory data analysis** (**EDA**), a common and essential step in many data analytics and modeling tasks.

5

Exploratory Data Analysis

The previous chapter covered the basic plotting principles using `ggplot2`, including the use of various geometries and themes layers. It turns out that cleaning and massaging the raw data (covered in *Chapter 2* and *Chapter 3*) and visualizing the data (covered in *Chapter 4*) belong to the first stage of a typical data science project workflow – that is, **exploratory data analysis (EDA)**. We will cover this using a few case studies in this chapter. We will learn how to apply the coding techniques we covered earlier in this book and focus on analyzing the data through the lens of EDA.

By the end of this chapter, you will know how to uncover the structures of data using numerical and graphical techniques, discover interesting relationships among variables, and spot unusual observations.

In this chapter, we will cover the following topics:

- EDA fundamentals
- EDA in practice

Technical requirements

To complete the exercises in this chapter, you will need to have the following:

- The latest version of the `yfR` package, which is 1.0.0 at the time of writing
- The latest version of the `corrplot` package, which is 0.92 at the time of writing

The code and data for this chapter are available at `https://github.com/PacktPublishing/The-Statistics-and-Machine-Learning-with-R-Workshop/blob/main/Chapter_5/chapter_5.R`.

EDA fundamentals

When facing a new dataset in the form of a table (a DataFrame) in Excel or a dataset, EDA helps us gain insight into the underlying pattern and irregularities of variables in the dataset. This is an important first-step exercise before building any predictive model. As the saying goes, *garbage in, garbage out*. When the input variables used for model development suffer from problems, such as missing values or different scales, the resulting model will either perform poorly, converge slowly, or even hit an error in the training stage. Therefore, understanding your data and ensuring the raw materials are in check are critical steps in warrantying a good-performing model later on.

This is where EAD comes in. Instead of being a rigid statistical procedure, EAD is a set of exploratory analyses that enables you to develop a better understanding of the features and potential relationships in the data. It serves as a transitional analysis to guide modeling later on, involving both the data manipulation and visualization techniques we learned earlier. It helps summarize salient characteristics of the data through various forms of visual aids, facilitating the extraction of important features.

There are two broad types of EDA: descriptive statistics such as the mean, median, mode, and inter-quantile range, and graphical descriptions such as density plots, histograms, box plots, and so on.

A typical EAD process includes analyzing categorical and numerical variables, both standalone in univariate analysis and in combination via bivariate and multivariate analysis. Common practices include analyzing the distribution of a given set of variables and examining missing values and outliers. In the following sections, we will start by analyzing different types of data, including categorical and numerical variables. We will then go through a case study to apply and reinforce the techniques covered in previous chapters using `dplyr` and `ggplot2`.

Analyzing categorical data

In this section, we will look at how to analyze two categorical variables via graphical and numerical summaries. We will use a dataset on comic characters from the Marvel comics universe, which should not be unfamiliar to you if you are a fan of Marvel superheroes. The dataset is published by **FiveThirtyEight** and hosted on their GitHub page. We can directly read the data from the GitHub repository using the `read_csv()` function from the `readr` package, the data loading arm of the `tidyverse` universe, as shown in the following code snippet:

```
>>> library(readr)
>>> df = read_csv("https://raw.githubusercontent.com/fivethirtyeight/
data/master/comic-characters/marvel-wikia-data.csv")
>>> head(df,5)
# A tibble: 16,376 × 13
    page_id name                urlslug ID      ALIGN
EYE    HAIR  SEX    GSM    ALIVE APPEARANCES
      <dbl> <chr>              <chr>   <chr> <chr> <chr> <chr> <chr>
<chr> <chr>         <dbl>
```

```
   1    1678 "Spider-Man (Pet…  "\\/Sp…  Secr…  Good…  Haze…  Brow…  Male…
  NA    Livi…        4043
   2    7139 "Captain America…  "\\/Ca…  Publ…  Good…  Blue…  Whit…  Male…
  NA    Livi…        3360
   3   64786 "Wolverine (Jame…  "\\/Wo…  Publ…  Neut…  Blue…  Blac…  Male…
  NA    Livi…        3061
   4    1868 "Iron Man (Antho…  "\\/Ir…  Publ…  Good…  Blue…  Blac…  Male…
  NA    Livi…        2961
   5    2460 "Thor (Thor Odin…  "\\/Th…  No D…  Good…  Blue…  Blon…  Male…
  NA    Livi…        2258
   6    2458 "Benjamin Grimm …  "\\/Be…  Publ…  Good…  Blue…  No H…  Male…
  NA    Livi…        2255
   7    2166 "Reed Richards (…  "\\/Re…  Publ…  Good…  Brow…  Brow…  Male…
  NA    Livi…        2072
   8    1833 "Hulk (Robert Br…  "\\/Hu…  Publ…  Good…  Brow…  Brow…  Male…
  NA    Livi…        2017
   9   29481 "Scott Summers (…  "\\/Sc…  Publ…  Neut…  Brow…  Brow…  Male…
  NA    Livi…        1955
  10    1837 "Jonathan Storm …  "\\/Jo…  Publ…  Good…  Blue…  Blon…  Male…
  NA    Livi…        1934
# … with 16,366 more rows, and 2 more variables: `FIRST APPEARANCE`
<chr>, Year <dbl>
```

Printing out the DataFrame shows that this dataset contains $16,376$ rows and 13 columns, including the character names, IDs, and so on.

In the next section, we will look at summarizing two categorical variables using the count statistic.

Summarizing categorical variables using counts

In this section, we will cover different ways to analyze two categorical variables, including using a contingency table and a bar chart. A contingency table is a useful way to show the total counts of observations that fall into each unique combination of the two categorical variables. Let's go through an exercise on how to achieve this.

Exercise 5.1 – summarizing two categorical variables

In this exercise, we will focus on two categorical variables: ALIGN (indicating whether the character is good, neutral, or bad) and SEX (indicating the gender of the character). First, we will look at the unique values of each variable, followed by summarizing the respective total counts when combined:

1. Inspect the unique values of ALIGN and SEX:

    ```
    >>> unique(df$ALIGN)
    "Good Characters"    "Neutral Characters" "Bad
    Characters"     NA
    >>> unique(df$SEX)
    ```

```
"Male Characters"        "Female Characters"        "Genderfluid
Characters" "Agender Characters"     NA
```

The results show that both variables contain NA values. Let's remove the observations with NA values in either `ALIGN` or `SEX`.

2. Remove observations with NA values in either `ALIGN` or `SEX` in `df` using the `filter` verb function:

```
>>> df = df %>%
    filter(!is.na(ALIGN),
           !is.na(SEX))
```

We can verify whether the rows with NA values have been successfully removed by checking the dimension of the resulting DataFrame and the count of NA values in `ALIGN` and `SEX`:

```
>>> dim(df)
12942    13
>>> sum(is.na(df$ALIGN))
0
>>> sum(is.na(df$SEX))
0
```

Next, we must create a contingency table to summarize the frequency of each unique combination of values.

3. Create a contingency table between `ALIGN` and `SEX`:

```
>>> table(df$ALIGN, df$SEX)
                    Agender Characters Female Characters
Genderfluid Characters Male Characters
    Bad Characters                   20               976
                     0             5338
    Good Characters                  10              1537
                     1             2966
    Neutral Characters               13               640
                     1             1440
```

We can see that most characters are male and bad. Among all male characters, the majority are bad, whereas good or neutral characters are dominant in female characters. Let's visually present and analyze the proportions using a bar chart.

4. Create a bar chart between these two variables using `ggplot2`:

```
>>> library(ggplot2)
>>> ggplot(df, aes(x=SEX, fill=ALIGN)) +
    geom_bar() +
    theme(axis.text=element_text(size=18),
          axis.title=element_text(size=18,face="bold"),
          legend.position = c(0.2, 0.8),
```

```
legend.key.size = unit(2, 'cm'),
legend.text = element_text(size=20))
```

Running this code snippet creates *Figure 5.1*. Here, we used the properties in the theme layer to adjust the size of labels on the graph. For example, `axis.text` and `axis.title` are used to increase the size of texts and titles along the axes, `legend.position` is used to move the legend to the upper-left corner, and `legend.key.size` and `legend.text` are used to enlarge the overall display of the legend:

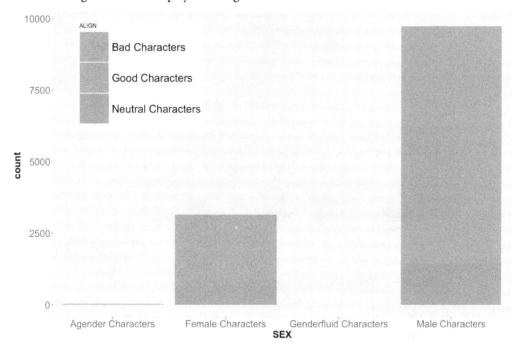

Figure 5.1 – Bar chart of ALIGN and SEX

Since the total count of `Agender Characters` and `Genderfluid Characters` is very limited, we can remove these two combinations when plotting the bar chart:

```
>>> df %>%
    filter(!(SEX %in% c("Agender Characters", "Genderfluid
Characters"))) %>%
    ggplot(aes(x=SEX, fill=ALIGN)) +
    geom_bar()
```

Running this command generates *Figure 5.2*:

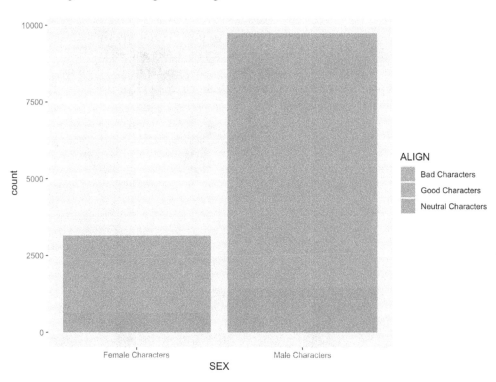

Figure 5.2 – Removing low-count combinations from the bar chart

Using counts may not be as intuitive when comparing different combinations. In this case, converting counts into proportions will help present the information on a relative scale.

Converting counts into proportions

In this section, we will go over an exercise that covers conditional proportions in a contingency table. Unlike the previous unconditional contingency table, conditioning along either dimension of a two-way contingency table results in a different distribution of proportions.

Exercise 5.2 – summarizing two categorical variables

In this exercise, we will learn how to express the previous contingency table using proportions and convert it into a conditional distribution based on a specified dimension:

1. Express the previous contingency table using proportions. Avoid scientific notation (for example, e+10) and keep three decimal places:

    ```
    >>> options(scipen=999, digits=3)
    >>> count_df = table(df$ALIGN, df$SEX)
    ```

```
>>> prop.table(count_df)
                    Agender Characters Female Characters
Genderfluid Characters Male Characters
  Bad Characters                 0.0015454         0.0754134
             0.0000000           0.4124556
  Good Characters                0.0007727         0.1187606
             0.0000773           0.2291763
  Neutral Characters             0.0010045         0.0494514
             0.0000773           0.1112656
```

The values in the contingency table are now expressed as proportions. Since the proportions are derived by dividing the previous absolute counts by the total sum, we can verify whether the total sum of proportions is equal to one by summing all the values in the table:

```
>>> sum(prop.table(count_df))
1
```

2. Obtain the contingency table as proportions after conditioning on rows (here, the ALIGN variable):

```
>>> prop.table(count_df, margin=1)
                    Agender Characters Female Characters
Genderfluid Characters Male Characters
  Bad Characters                 0.003158          0.154089
             0.000000            0.842753
  Good Characters                0.002215          0.340496
             0.000222            0.657067
  Neutral Characters             0.006208          0.305635
             0.000478            0.687679
```

We can verify the conditioning by calculating the row-wise summations:

```
>>> rowSums(prop.table(count_df, margin=1))
    Bad Characters     Good Characters Neutral Characters
                 1                   1                  1
```

In this code, setting margin=1 means row-level conditioning. We can also exercise column-level conditioning by setting margin=2.

3. Obtain the contingency table as proportions after conditioning on columns (for instance, the SEX variable):

```
>>> prop.table(count_df, margin=2)
                    Agender Characters Female Characters
Genderfluid Characters Male Characters
  Bad Characters                 0.465             0.310
             0.000               0.548
  Good Characters                0.233             0.487
             0.500               0.304
  Neutral Characters             0.302             0.203
             0.500               0.148
```

Similarly, we can verify the conditioning by calculating the column-wise summations:

```
>>> colSums(prop.table(count_df, margin=2))
     Agender Characters       Female Characters Genderfluid
 Characters          Male Characters
                 1                         1
             1                       1
```

4. Plot the unconditional proportions in a bar chart after applying the same filtering condition to SEX. Change the label of the *y* axis to `proportion`:

```
>>> df %>%
    filter(!(SEX %in% c("Agender Characters", "Genderfluid
Characters"))) %>%
    ggplot(aes(x=SEX, fill=ALIGN)) +
    geom_bar(position="fill") +
    ylab("proportion") +
    theme(axis.text=element_text(size=18),
          axis.title=element_text(size=18,face="bold"),
          legend.key.size = unit(2, 'cm'),
          legend.text = element_text(size=20))
```

Running this command generates *Figure 5.3*, where it is obvious that bad characters are predominantly male characters:

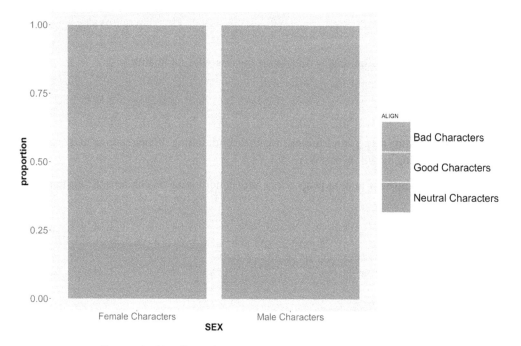

Figure 5.3 – Visualizing the unconditional proportions in a bar chart

We can also obtain a similar result from a different angle by switching the two variables in the bar chart:

```
>>> df %>%
    filter(!(SEX %in% c("Agender Characters", "Genderfluid
Characters"))) %>%
    ggplot(aes(x=ALIGN, fill=SEX)) +
    geom_bar(position="fill") +
    ylab("proportion") +
    theme(axis.text=element_text(size=18),
          axis.title=element_text(size=18,face="bold"),
          legend.key.size = unit(2, 'cm'),
          legend.text = element_text(size=20))
```

Running this command generates *Figure 5.4*, where ALIGN is used as the *x* axis and SEX is used as the grouping variable:

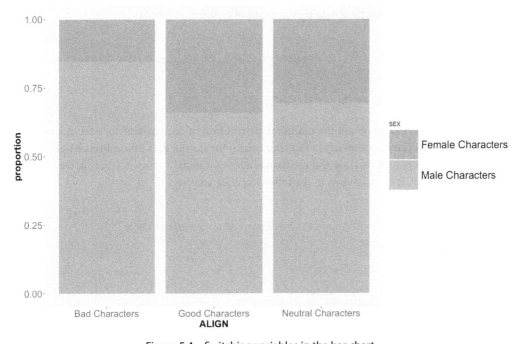

Figure 5.4 – Switching variables in the bar chart.

Next, we will look at describing one categorical variable using a marginal distribution and faceted bar chart.

Marginal distribution and faceted bar charts

Marginal distribution refers to the distribution of one variable after integrating other variables. This means that we are interested in the distribution of one specific variable, no matter how the other variables are distributed.

In the case of our previous two-way contingency table stored in `count_df`, we can derive the marginal distribution of `SEX` in the form of a frequency count by summing over all possible values of `ALIGN`. That is, we can perform column-wise summation to get the marginal count of `SEX`, as shown in the following code snippet:

```
>>> colSums(count_df)
      Agender Characters        Female Characters Genderfluid Characters
          Male Characters
                   43                     3153                        2
                 9744
```

This has the same effect as directly obtaining the count of different categories in `SEX`:

```
>>> table(df$SEX)
      Agender Characters        Female Characters Genderfluid Characters
          Male Characters
                   43                     3153                        2
                 9744
```

Now, what if we would like to obtain the marginal distribution of one variable for each category of another variable? This can be achieved via **faceting**, which breaks the data into subsets based on the unique values of a categorical variable and constructs a plot for each. To implement this, we can add a faceting layer to `ggplot2`, as shown in the following code snippet:

```
>>> df %>%
    filter(!(SEX %in% c("Agender Characters", "Genderfluid
Characters"))) %>%
    ggplot(aes(x=SEX)) +
    geom_bar() +
    facet_wrap(~ALIGN) +
    theme(axis.text=element_text(size=15),
          axis.title=element_text(size=15,face="bold"),
          strip.text.x = element_text(size = 30))
```

Running this code generates *Figure 5.5*, which contains three side-by-side bar charts for bad, good, and neutral characters, respectively. This is essentially rearranging the stacked bar chart in *Figure 5.4*. Note that faceting can be added by using the `facet_wrap` function, where `~ALIGN` indicates that the faceting is to be performed using the `ALIGN` variable. Note that we used the `strip.text.x` attribute to adjust the text size of the facet grid labels:

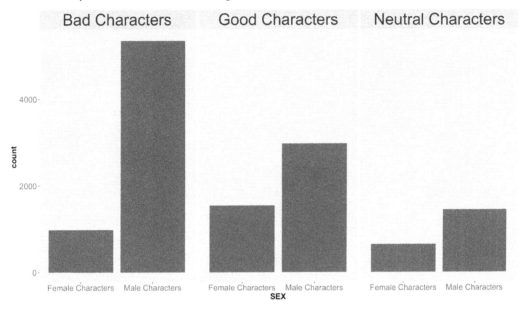

Figure 5.5 – Faceted bar chart

In addition, we can adjust the sequence of the individual bar facets by overriding the levels of `ALIGN` after converting it into a factor:

```
>>> df$ALIGN = factor(df$ALIGN, levels = c("Bad Characters", "Neutral
Characters", "Good Characters"))
```

Running the same faceting codes again will now generate *Figure 5.6*, where the sequence of facets is determined according to the levels in `ALIGN`:

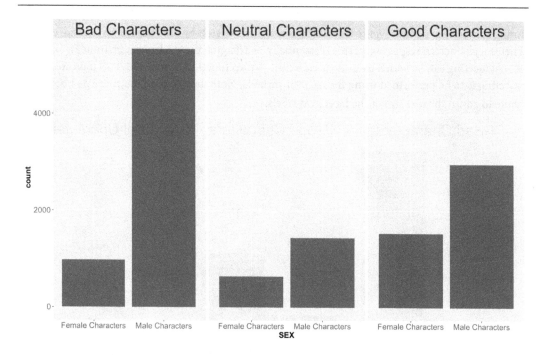

Figure 5.6 – Arranging the sequence of facets in a faceted bar chart

In the next section, we will look at different ways to explore numerical variables.

Analyzing numerical data

In this section, we will look at summarizing numerical data using different types of plots for the Marvel dataset. Since there's an infinite amount of values that a numerical/continuous variable can assume, the frequency table used earlier no longer applies. Instead, we often group the values into pre-specified bins, allowing us to work with ranges instead of single values.

Exercise 5.3 – exploring numerical variables

In this exercise, we will describe a numerical variable using a dot plot, histogram, density plot, and box plot for the `Year` variable:

1. Get a summary of the `Year` variable using the `summary()` function:

    ```
    >>> summary(df$Year)
       Min. 1st Qu.  Median    Mean 3rd Qu.    Max.    NA's
       1939    1973    1989    1984    2001    2013     641
    ```

2. Generate a dotted plot of the Year variable:

```
>>> ggplot(df, aes(x=Year)) +
  geom_dotplot(dotsize=0.2) +
  theme(axis.text=element_text(size=18),
        axis.title=element_text(size=18,face="bold"))
```

Running this command generates *Figure 5.7*, where each dot represents an observation at the corresponding location on the *x* axis. Similar observations are then stacked together on top of each top. It should be noted that using a dot plot is not the best option when the number of observations becomes large, with the *y* axis becoming meaningless due to technical limitations in ggplot2:

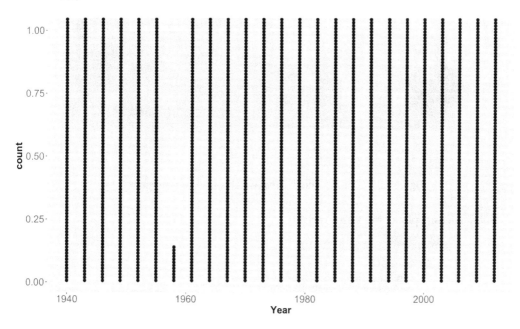

Figure 5.7 – Summarizing the Year variable using a dot plot

3. Build a histogram of the Year variable:

```
>>> ggplot(df, aes(x=Year)) +
  geom_histogram() +
  theme(axis.text=element_text(size=18),
        axis.title=element_text(size=18,face="bold"))
```

Running this command generates *Figure 5.8*, where each value of Year is grouped into bins and then the number of observations in each bin is counted to represent the height of each bin. Note that the default number of bins is 30, although this can be overwritten using the bins argument. The histogram thus presents the general shape of the distribution of the underlying variable. We can also convert it into a density plot to smooth out the steps between bins:

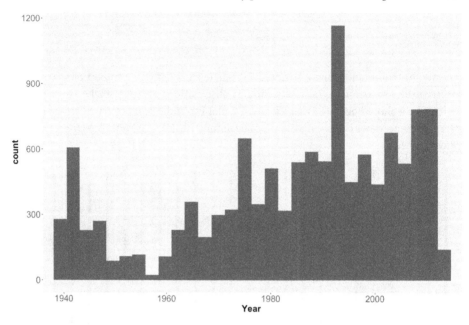

Figure 5.8 – Summarizing the Year variable using a histogram

4. Build a density plot of the `Year` variable:

```
>>> ggplot(df, aes(x=Year)) +
    geom_density() +
    theme(axis.text=element_text(size=18),
          axis.title=element_text(size=18,face="bold"))
```

Running this command generates *Figure 5.9*, where the distribution is represented as a smooth line. Note that a density plot is recommended only when there are many observations in the dataset:

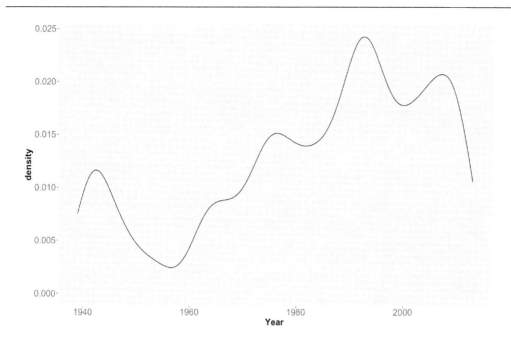

Figure 5.9 – Summarizing the Year variable using a density plot

5.   Build a box plot of the Year variable:

```
>>> ggplot(df, aes(x=Year)) +
  geom_boxplot() +
  theme(axis.text=element_text(size=18),
        axis.title=element_text(size=18,face="bold"))
```

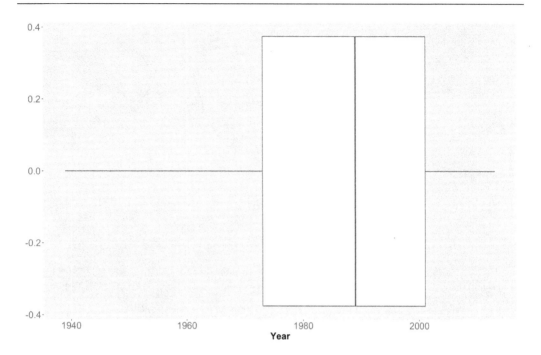

Figure 5.10 – Summarizing the Year variable using a box plot

Running this command generates *Figure 5.10*, where the central box represents the majority (25th to 75th percentile) of the observations, the middle line in the box denotes the median (50th percentile), and the outreaching whiskers include almost all "normal" observations. The outlier observation, which is none in this case, would be represented as dots outside the reach of the whiskers.

We can also add a faceting layer by SEX and observe the change in box plots across different genders.

6. Add a faceting layer to the previous box plot using the SEX variable:

```
>>> ggplot(df, aes(x=Year)) +
    geom_boxplot() +
    facet_wrap(~SEX) +
    theme(axis.text=element_text(size=18),
          axis.title=element_text(size=18,face="bold"),
          strip.text.x = element_text(size = 30))
```

Running this command generates *Figure 5.11*. As we can see, most female characters are introduced later than many of the male characters, and recent years feature more female characters than male ones:

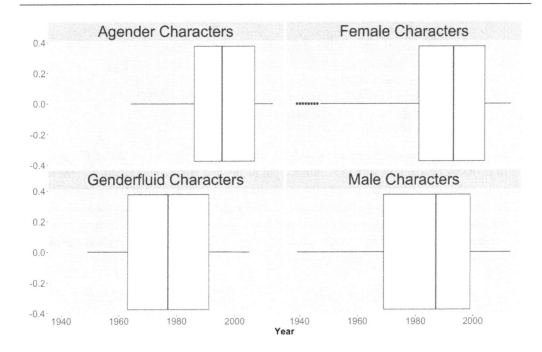

Figure 5.11 – Faceting the box plot based on SEX

In the next section, we will look at how to visualize data with higher dimensions.

Visualization in higher dimensions

The previous example used facets to present the distribution of a numerical variable in each unique value of a categorical variable. When there is more than one categorical variable, we can apply the same technique and expand the facets accordingly. This allows us to visualize the same numerical variable in higher dimensions that contain more than one categorical variable. Let's go through an exercise on visualizing the distribution of Year by ALIGN and SEX.

Exercise 5.4 – visualizing Year by ALIGN and SEX

In this exercise, we will use the facet_grid() function from ggplot2 to visualize the distribution of Year in each unique combination of ALIGN and SEX using both a density plot and a histogram:

1. Build a density plot of Year by ALIGN and SEX after applying the same filtering condition to SEX:

    ```
    >>> df %>%
      filter(!(SEX %in% c("Agender Characters", "Genderfluid
    Characters"))) %>%
      ggplot(aes(x=Year)) +
    ```

```
geom_density() +
facet_grid(ALIGN ~ SEX, labeller = label_both) +
facet_grid(ALIGN ~ SEX, labeller = label_both) +
theme(axis.text=element_text(size=18),
      axis.title=element_text(size=18,face="bold"),
      strip.text.x = element_text(size = 30),
      strip.text.y = element_text(size = 12))
```

Running this command generates *Figure 5.12*, where we used the `facet_grid()` function to create six histograms, with the columns split by the first argument, `ALIGN`, and the rows split by the second argument, `SEX`. The result shows an increasing trend (more movies were produced) for all different combinations of `ALIGN` and `SEX`. However, since the *y* axis shows the relative density only, we would need to switch to a histogram to assess the absolute frequency of occurrence. Note that we used the `strip.text.y` attribute to adjust the text size of the facet grid labels along the *y* axis:

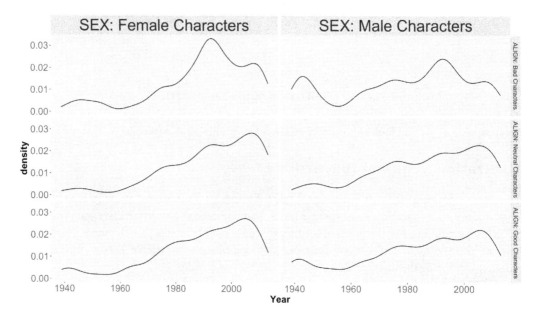

Figure 5.12 – Density plot of Year by ALIGN and SEX

2. Build the same plots using a histogram:

```
>>> df %>%
  filter(!(SEX %in% c("Agender Characters", "Genderfluid
Characters"))) %>%
  ggplot(aes(x=Year)) +
  geom_histogram() +
```

```
facet_grid(ALIGN ~ SEX, labeller = label_both) +
theme(axis.text=element_text(size=18),
        axis.title=element_text(size=18,face="bold"),
        strip.text.x = element_text(size = 30),
        strip.text.y = element_text(size = 12))
```

Running this command generates *Figure 5.13*, where we can see that good female and male characters are steadily increasing in recent years:

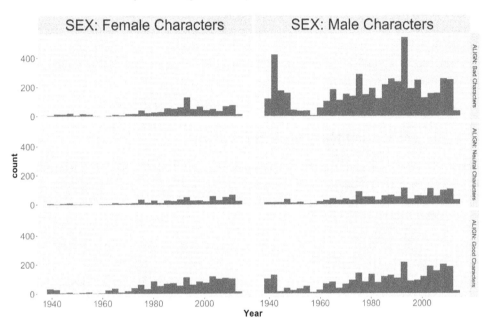

Figure 5.13 – Histogram of Year by ALIGN and SEX

In the next section, we will try different ways to measure the central concentration of a numerical variable.

Measuring the central concentration

There are different ways to measure the central concentration, or central tendency, of a numerical variable. Depending on the context and purpose, the measure of center is often used to represent a typical observation out of a numerical variable of interest.

The most popular measure of center is the mean, which is calculated as the average value of a list of numbers. In other words, we can obtain the mean value by summing all observations divided by the number of observations. This can be achieved using the mean () function in R.

Another measure of center is the median, which is the middle value after sorting the list of numbers from the smallest to the largest. This can be achieved using the median () function in R.

The third measure of center is the mode, which represents the most common observation in the list of numbers. Since there is no built-in function for calculating the mode, we must write a customized function to obtain the most frequent observation based on the count of occurrences using the `table()` function.

It is important to look at the shape of the distribution before deciding on the measure of center. For a start, note that the mean value is often drawn toward the long tail of a skewed distribution, a continuous distribution inferred from the list of numbers such as the density plot from earlier. In other words, the mean value is sensitive to the extreme values in the observations. On the other hand, the median will not suffer from such sensitivity since it is simply a measure that divides the ordered observations by half. Therefore, the median is a better and more sensible candidate measure of center when working with a skewed continuous distribution, unless additional treatment on the extreme values, often treated as outliers, is in place.

Let's look at how to obtain the three measures of center via an exercise.

Exercise 5.5 – calculating the measure of center

In this exercise, we will calculate the mean, median, and mode of APPEARANCES, which denotes the number of appearances for each character:

1. Calculate the mean of APPEARANCES:

    ```
    >>> mean(df$APPEARANCES)
    NA
    ```

    The NA result suggests that there are NA values in the observations of APPEARANCES. To verify this, we can look at the summary of this continuous variable:

    ```
    >>> summary(df$APPEARANCES)
       Min. 1st Qu.  Median    Mean 3rd Qu.    Max.    NA's
          1       1       3      20       9    4043     749
    ```

    Indeed, there are quite a few NA values. To calculate the mean value after removing these NA observations, we can enable the na.rm argument in the mean() function:

    ```
    >>> mean(df$APPEARANCES, na.rm = TRUE)
    19.8
    ```

2. Calculate the mean of APPEARANCES:

    ```
    >>> median(df$APPEARANCES, na.rm = TRUE)
    3
    ```

    When the mean and median values deviate a lot from each other, this is an obvious sign that we are working with a skewed distribution. In this case, the APPEARANCES variable is quite skewed, with the median character appearing three times and the most popular character appearing up to 4,043 times.

3. Calculate the mode of APPEARANCES:

```
>>> mode <- function(x){
  ux <- unique(x)
  ux[which.max(tabulate(match(x, ux)))]
}
>>> mode(df$APPEARANCES)
1
```

Here, we created a customized function called mode() to calculate the mode of a numerical variable, where we first extract a list of the unique values using the unique() function, then count the number of times each unique value appears using the tabulate() and match() functions, and lastly obtain the index of the maximal value using the which.max() function. The result shows that the majority of characters only appear once in the entire history of Marvel comics.

Now, let's look at a detailed breakdown of mean and median appearances by ALIGN.

4. Calculate the mean and median values of APPEARANCES by each level of ALIGN:

```
>>> df %>%
  group_by(ALIGN) %>%
  summarise(mean_appear = mean(APPEARANCES, na.rm=TRUE),
          median_appear = median(APPEARANCES, na.rm=TRUE))
  ALIGN              mean_appear median_appear
  <fct>                  <dbl>       <dbl>
1 Bad Characters          8.64          3
2 Neutral Characters     20.3           3
3 Good Characters        35.6           5
```

The result shows that good characters appear more often than bad ones.

Next, we will look at how to measure the variability of a continuous variable.

Measuring variability

As with central concentration, several metrics can be used to measure the variability or dispersion of a continuous variable. Some are sensitive to outliers, such as variance and standard deviation, while others are robust to outliers, such as **inter-quantile range (IQR)**. Let's go through an exercise on how to calculate these metrics.

Note that robust measures such as median and IQR are used in box plots, although more details are hidden compared to the full density of a given variable.

Exercise 5.6 – calculating the variability of a continuous variable

In this exercise, we will calculate different metrics on variability both manually and using built-in functions. We will start with variance, which is calculated as the average squared difference between each raw value and the mean value. Note that this is how population variance is calculated. To calculate the sample variance, we need to adjust the averaging operation by subtracting 1 from the total number of observations used in the variance calculation.

In addition, variance is a squared version of the original unit and is thus not easily interpretable. To measure the variability of the data at the same original scale, we can use standard deviation, which is calculated by taking the square root of the variance. Let's look at how to achieve this in practice:

1. Calculate the population variance of APPEARANCES after removing NA values. Keep two decimal points:

```
>>> tmp = df$APPEARANCES[!is.na(df$APPEARANCES)]
>>> pop_var = sum((tmp - mean(tmp))^2)/length(tmp)
>>> formatC(pop_var, digits = 2, format = "f")
"11534.53"
```

Here, we first remove NA values from APPEARANCES and save the result in tmp. Next, we subtract the mean value of tmp from each original value, square the result, sum all values, and then divide by the number of observations in tmp. This essentially follows the definition of variance, which measures the average variability of each observation to the central tendency – in other words, the mean value.

We can also calculate the sample variance.

2. Calculate the sample variance of APPEARANCES:

```
>>> sample_var = sum((tmp - mean(tmp))^2)/(length(tmp)-1)
>>> formatC(sample_var, digits = 2, format = "f")
"11535.48"
```

The result is now slightly different from the population variance. Note that to calculate the sample mean, we simply use one less observation in the denominator. Such adjustment is necessary, especially when we are working with limited sample data, although the difference becomes small as the sample size grows.

We can also calculate sample variance by calling the var() function.

3. Calculate sample variance using var():

```
>>> formatC(var(tmp), digits = 2, format = "f")
"11535.48"
```

The result is aligned with our previous manual calculation of the sample variance.

To obtain the measure of variability at the same unit as the original observations, we can calculate the standard deviation. This can be achieved using the sd() function.

4. Calculate the standard deviation using sd():

```
>>> sd(tmp)
107.4
```

Another measure of variability is IQR, which is the difference between the third and first quantiles and quantifies the range of the majority values.

5. Calculate the IQR using IQR():

```
>>> IQR(tmp)
8
```

We can also verify the result by calling the summary() function, which returns the different quantile values:

```
>>> summary(tmp)
   Min. 1st Qu.  Median    Mean 3rd Qu.    Max.
    1.0     1.0     3.0    19.8     9.0  4043.0
```

As discussed earlier, measures such as variance and standard deviation are sensitive to extreme values in the data, while IQR is a robust measure of outliers. We can assess the change to these measures after removing the maximal value from tmp.

6. Calculate the standard deviation and IQR after removing the maximum from tmp:

```
>>> tmp2 = tmp[tmp != max(tmp)]
>>> sd(tmp2)
101.04
>>> IQR(tmp2)
8
```

The result shows that IQR stays the same after removing the maximum, thus being a more robust measure compared to standard deviation.

We can also calculate these measures by different levels of another categorical variable.

7. Calculate the standard deviation, IQR, and count of APPEARANCES for each level of ALIGN:

```
>>> df %>%
  group_by(ALIGN) %>%
  summarise(sd_appear = sd(APPEARANCES, na.rm=TRUE),
            IQR_appear = IQR(APPEARANCES, na.rm=TRUE),
            count = n())
  ALIGN              sd_appear IQR_appear count
  <fct>                  <dbl>      <dbl> <int>
```

```
1 Bad Characters        26.4      5  6334
2 Neutral Characters    112.      8  2094
3 Good Characters       161.     14  4514
```

Next, we will dive deeper into the skewness in the distribution of a continuous variable.

Working with skewed distributions

Besides the mean and standard deviation, we can also characterize the distribution of a continuous variable using modality and skewness. Modality refers to the number of humps that exist in the continuous distribution. For example, a unimodal distribution, the most popular distribution we have seen so far in the form of a bell curve, has one peak across the whole distribution. It can grow into a bimodal distribution when there are two humps and multimodal distribution when there are three humps or more. If there is no discernable mode and the distribution appears flat across the whole support region (the range of the continuous variable), it is referred to as a uniform distribution. *Figure 5.14* summarizes the distributions of different modalities:

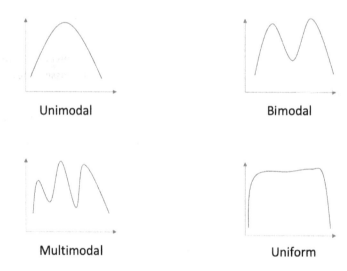

Figure 5.14 – Different types of modalities in a distribution

On the other hand, a continuous variable may be skewed toward the left or the right or appear symmetric around the central tendency. A right-skewed distribution contains more extreme values on the right tail of the distribution, while a left-skewed distribution has a long tail on the left-hand side. *Figure 5.15* illustrates the different types of skewness in a distribution:

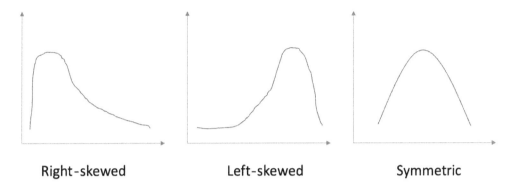

Figure 5.15 – Different types of skewness in a distribution

Distribution can also attribute its skewness to outliers in a continuous variable. When there are multiple outliers in the data, sensitive measures such as mean and variance will become distorted, causing a shift in distribution toward the outliers. Let's go through an exercise to understand how to deal with skewness and outliers in a distribution.

Exercise 5.7 – working with skewness and outliers

In this exercise, we will look at how to work with a skewed distribution that contains many extreme values, especially outliers in the data:

1. Visualize the density plot of APPEARANCES by ALIGN for observations since the year 2000. Set the transparency level to 0.2:

```
>>> tmp = df %>%
  filter(Year >= 2000)
>>> ggplot(tmp, aes(x=APPEARANCES, fill=ALIGN)) +
  geom_density(alpha=0.2) +
  theme(axis.text=element_text(size=18),
        axis.title=element_text(size=18,face="bold"),
        legend.position = c(0.8, 0.8),
        legend.key.size = unit(2, 'cm'),
        legend.text = element_text(size=20))
```

Running this command generates *Figure 5.16*, where all three distributions are quite skewed toward the right, an obvious sign of many outliers in the data:

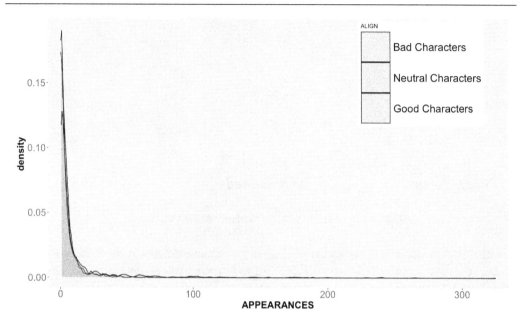

Figure 5.16 – Density plot of APPEARANCES by ALIGN

2. Remove observations whose APPEARANCES are above the 90th percentile and generate the same plot:

```
>>> tmp = tmp %>%
  filter(APPEARANCES <= quantile(APPEARANCES, 0.9, na.rm=TRUE))
>>> ggplot(tmp, aes(x=log(APPEARANCES), fill=ALIGN)) +
  geom_density(alpha=0.2) +
  theme(axis.text=element_text(size=18),
        axis.title=element_text(size=18,face="bold"),
        legend.position = c(0.8, 0.8),
        legend.key.size = unit(2, 'cm'),
        legend.text = element_text(size=20))
```

Running this command generates *Figure 5.17*, where all three distributions are much less right-skewed than before. Removing outliers is one way to work around extreme values, although the information contained in the removed observations is lost. To control the effect of outliers and retain their presence at the same time, we can transform the continuous variable using the log() function, which brings it to the logarithmic scale. Let's see how this works in practice:

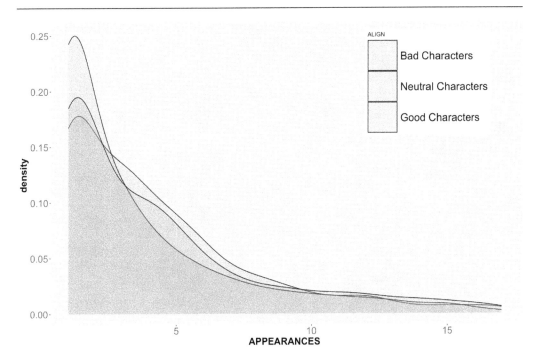

Figure 5.17 – Density plot of APPEARANCES by ALIGN after removing outliers

3. Apply log transformation to APPEARANCES and re-generate the same plot:

```
>>> ggplot(tmp, aes(x=log(APPEARANCES), fill=ALIGN)) +
    geom_density(alpha=0.2) +
    theme(axis.text=element_text(size=18),
        axis.title=element_text(size=18,face="bold"),
        legend.position = c(0.8, 0.8),
        legend.key.size = unit(2, 'cm'),
        legend.text = element_text(size=20))
```

Running this command generates *Figure 5.18*, where the three density plots appear as a bimodal distribution and not as right-skewed as before. Transforming the continuous variable using the logarithmic function could thus bring the original value to a more controlled scale:

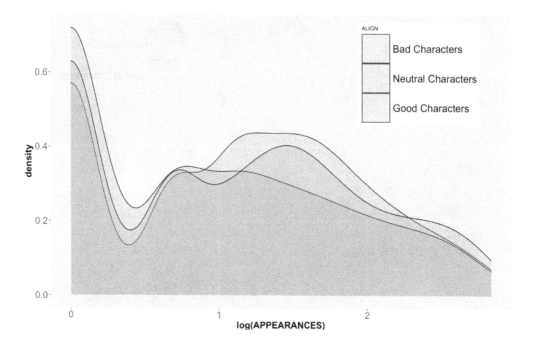

Figure 5.18 – Density plot of APPEARANCES by ALIGN after applying log transformation

In the next section, we will go through a case study to enhance our skills when conducting EDA on a new dataset.

EDA in practice

In this section, we will analyze a dataset that consists of the stock prices of the top five companies in 2021. First, we will look at how to download and process these stock indexes, followed by performing univariate analysis and bivariate analysis in terms of correlation.

Obtaining the stock price data

To obtain the daily stock prices of a particular ticker, we can use the yfR package to download the data from Yahoo! Finance, a vast repository of financial data that covers a large number of markets and assets and has been widely used in both academia and industry. The following exercise illustrates how to download the stock data using yfR.

Exercise 5.8 – downloading stock prices

In this exercise, we will look at how to specify the different parameters so that we can download stock prices from Yahoo! Finance, including the ticker name and date range:

1. Install and load the `yfR` package:

    ```
    >>> install.packages("yfR")
    >>> library(yfR)
    ```

    Note that we would need to wrap the package name inside a pair of double quotes in the `install.packages()` function.

2. Specify the starting and end date parameters, as well as the ticker names so that they cover Facebook (now META), Netflix (NFLX), Google (GOOG), Amazon (AMZN), and Microsoft (MSFT):

    ```
    >>> first_date = as.Date("2021-01-01")
    >>> last_date = as.Date("2022-01-01")
    >>> my_ticker <- c('META', 'NFLX', 'GOOG', 'AMZN', 'MSFT')
    ```

    Here, the start and end dates are formatted as the `Date` type, and the ticker names are concatenated in a vector.

3. Download the stock prices using the `yf_get()` function and store the result in `df`:

    ```
    >>> df <- yf_get(tickers = my_ticker,
                        first_date = first_date,
                        last_date = last_date)
    ```

    Running this command generates the following message, which shows that the data for all five stocks has been downloaded successfully. Each ticker has 252 rows in 2021 since there are 252 trading days in a year:

    ```
    — Running yfR for 5 stocks | 2021-01-01 --> 2022-01-01 (365
    days) —

    i Downloading data for benchmark ticker ^GSPC
    i (1/5) Fetching data for AMZN
    ✓    - found cache file (2021-01-04 --> 2021-12-31)
    ✓    - got 252 valid rows (2021-01-04 --> 2021-12-31)
    ✓    - got 100% of valid prices -- Got it!
    i (2/5) Fetching data for GOOG
    ✓    - found cache file (2021-01-04 --> 2021-12-31)
    ✓    - got 252 valid rows (2021-01-04 --> 2021-12-31)
    ✓    - got 100% of valid prices -- Good stuff!
    i (3/5) Fetching data for META
    !    - not cached
    ```

```
✓    - cache saved successfully
✓    - got 252 valid rows (2021-01-04 --> 2021-12-31)
✓    - got 100% of valid prices -- Mais contente que cusco de
cozinheira!
i (4/5) Fetching data for MSFT
✓    - found cache file (2021-01-04 --> 2021-12-31)
✓    - got 252 valid rows (2021-01-04 --> 2021-12-31)
✓    - got 100% of valid prices -- All OK!
i (5/5) Fetching data for NFLX
✓    - found cache file (2021-01-04 --> 2021-12-31)
✓    - got 252 valid rows (2021-01-04 --> 2021-12-31)
✓    - got 100% of valid prices -- Youre doing good!
i Binding price data


── Diagnostics ──────────────────────────────────────
✓ Returned dataframe with 1260 rows -- Time for some tea?
i Using 156.6 kB at /var/folders/zf/d5cczq0571n0_
x7_7rdn0r640000gn/T//Rtmp7hl9eR/yf_cache for 1 cache files
i Out of 5 requested tickers, you got 5 (100%)
```

Let's examine the structure of the dataset:

```
>>> str(df)
tibble [1,260 × 11] (S3: tbl_df/tbl/data.frame)
 $ ticker               : chr [1:1260] "AMZN" "AMZN" "AMZN"
"AMZN" ...
 $ ref_date             : Date[1:1260], format: "2021-01-04"
...
 $ price_open           : num [1:1260] 164 158 157 158 159 ...
 $ price_high           : num [1:1260] 164 161 160 160 160 ...
 $ price_low            : num [1:1260] 157 158 157 158 157 ...
 $ price_close          : num [1:1260] 159 161 157 158 159 ...
 $ volume               : num [1:1260] 88228000 53110000
87896000 70290000 70754000 ...
 $ price_adjusted       : num [1:1260] 159 161 157 158 159 ...
 $ ret_adjusted_prices  : num [1:1260] NA 0.01 -0.0249 0.00758
0.0065 ...
 $ ret_closing_prices   : num [1:1260] NA 0.01 -0.0249 0.00758
0.0065 ...
 $ cumret_adjusted_prices: num [1:1260] 1 1.01 0.985 0.992 0.999
...
 - attr(*, "df_control")= tibble [5 × 5] (S3: tbl_df/tbl/data.
frame)
```

```
    ..$ ticker              : chr [1:5] "AMZN" "GOOG" "META"
"MSFT" ...
    ..$ dl_status           : chr [1:5] „OK" „OK" „OK" „OK" ...
    ..$ n_rows              : int [1:5] 252 252 252 252 252
    ..$ perc_benchmark_dates: num [1:5] 1 1 1 1 1
    ..$ threshold_decision  : chr [1:5] "KEEP" "KEEP" "KEEP"
"KEEP" ...
```

The data that's been downloaded includes information such as the daily opening, closing, highest, and closet prices for each ticker.

In the following sections, we will use the adjusted price field, `price_adjusted`, which is adjusted for corporate events such as splits, dividends, and others. This is usually what we would use when analyzing stocks as it represents the actual financial performance of the stockholders.

Univariate analysis of individual stock prices

In this section, we will perform a graphical analysis based on the stock prices. Since the stock prices are time series data that are numerical, we will use plots such as histograms, density plots, and box plots for visualization purposes.

Exercise 5.9 – downloading stock prices

In this exercise, we will start with the time series plots for the five stocks, followed by generating other types of plots suitable for continuous variables:

1. Generate the time series plots for the five plots:

    ```
    >>> ggplot(df,
            aes(x = ref_date, y = price_adjusted,
                color = ticker)) +
      geom_line() +
      theme(axis.text=element_text(size=18),
            axis.title=element_text(size=18,face="bold"),
            legend.text = element_text(size=20))
    ```

Running this command generates *Figure 5.19*, where Netflix takes the lead in terms of stock value. However, it also suffers from a huge fluctuation, especially around November 2021:

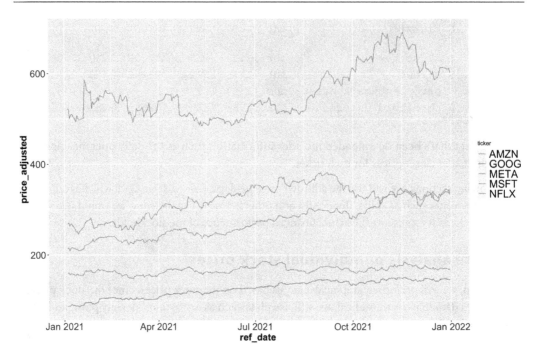

Figure 5.19 – Time series plots of the five stocks

2. Generate a histogram for each of the five stocks, with 100 bins for each histogram:

```
>>> ggplot(df, aes(x=price_adjusted, fill=ticker)) +
   geom_histogram(bins=100) +
   theme(axis.text=element_text(size=18),
        axis.title=element_text(size=18,face="bold"),
        legend.text = element_text(size=20))
```

Running this command generates *Figure 5.20*, which shows that Netflix has the biggest mean and variance in terms of stock value. Google and Amazon seem to share a similar spread, and the same goes for Facebook and Microsoft:

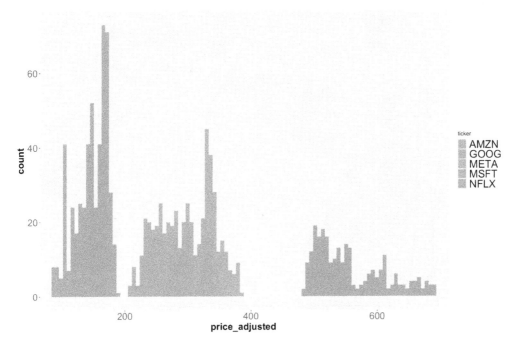

Figure 5.20 – Histograms of the five stocks

3. Generate a density plot for each of the five stocks. Set the transparency level to 0.2:

```
>>> ggplot(df, aes(x=price_adjusted, fill=ticker)) +
    geom_density(alpha=0.2) +
    theme(axis.text=element_text(size=18),
          axis.title=element_text(size=18,face="bold"),
          legend.text = element_text(size=20))
```

Running this command generates *Figure 5.21*, where the plots are now visually clearer compared to the histograms:

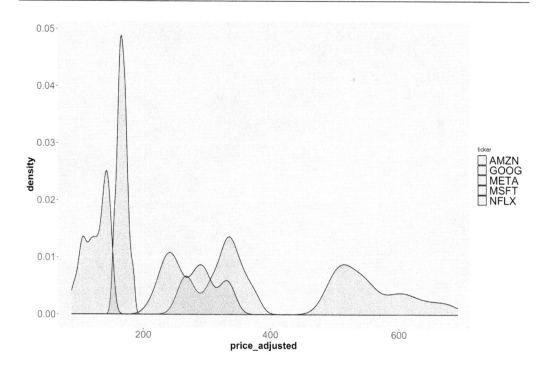

Figure 5.21 – Density plots of the five stocks

4. Generate a box plot for each of the five stocks:

```
>>> ggplot(df, aes(ticker, price_adjusted, fill=ticker)) +
    geom_boxplot() +
    theme(axis.text=element_text(size=18),
        axis.title=element_text(size=18,face="bold"),
        legend.text = element_text(size=20))
```

Running this command generates *Figure 5.22*. Box plots are good at indicating the central tendency and variation of each stock. For example, Netflix has the biggest mean and variance across all five stocks:

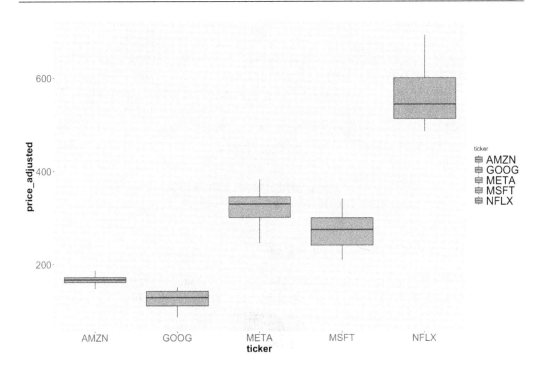

Figure 5.22 – Box plots of the five stocks

5. Obtain the mean, standard deviation, IQR, and count of each stock:

```
>>> df %>%
  group_by(ticker) %>%
  summarise(mean = mean(price_adjusted, na.rm=TRUE),
            sd = sd(price_adjusted, na.rm=TRUE),
            IQR = IQR(price_adjusted, na.rm=TRUE),
            count = n())
# A tibble: 5 × 5
  ticker  mean     sd   IQR count
  <chr>  <dbl>  <dbl> <dbl> <int>
1 AMZN    167.   8.00  10.7   252
2 GOOG    126.   18.4  31.1   252
3 META    321.   34.9  44.2   252
4 MSFT    273.   37.2  58.5   252
5 NFLX    558.   56.0  87.5   252
```

In the next section, we will look at the pairwise correlation between each pair of stocks.

Correlation analysis

Correlation measures the strength of covariation between two variables. There are several ways to calculate the specific value of correlation, with Pearson correlation being the most widely used. Pearson correlation is a value that ranges from -1 to 1, with 1 indicating two perfectly and positively correlated variables and -1 denoting perfect negative correlation. Perfect correlation means that the change in the value of one variable is always proportional to the change in the value of another variable. For example, when $y = 2x$, the correlation between variable x and y is 1 since y always changes positively in proportion to x.

Instead of manually calculating the pairwise correlation between all variables, we can use the `corrplot` package to calculate and visualize pairwise correlations automatically. Let's go through an exercise on how this can be achieved.

Exercise 5.10 – downloading stock prices

In this exercise, we will first convert the previous DataFrame from long into wide format so that each stock has a separate column indicating the adjusted price across different days/rows. The wide-format dataset will then be used to generate the pairwise correlation plot:

1. Convert the previous dataset into a wide format using the `spread()` function in the `tidyr` package. Save the result in `wide_df`:

    ```
    >>> library(tidyr)
    >>> wide_df <- df %>%
      select(ref_date, ticker, price_adjusted) %>%
      spread(ticker, price_adjusted)
    ```

    Here, we first select three variables, with `ref_date` as the row-level date index, `ticker`, whose unique values serve as the columns to be spread across the DataFrame, and `price_adjusted`, to be used to fill in the cells of the wide DataFrame. With this, we can examine the first few rows of the new dataset:

    ```
    >>> head(wide_df)
    # A tibble: 6 × 6
      ref_date    AMZN  GOOG  META  MSFT  NFLX
      <date>     <dbl> <dbl> <dbl> <dbl> <dbl>
    1 2021-01-04  159.  86.4  269.  214.  523.
    2 2021-01-05  161.  87.0  271.  215.  521.
    3 2021-01-06  157.  86.8  263.  209.  500.
    4 2021-01-07  158.  89.4  269.  215.  509.
    5 2021-01-08  159.  90.4  268.  216.  510.
    6 2021-01-11  156.  88.3  257.  214.  499.
    ```

    Now, the DataFrame has been converted from a long format into a wide format, which will facilitate the creation of correlation plots later on.

2. Generate a correlation plot using the `corrplot()` function from the `corrplot` package (to be installed if you have not done so already):

```
>>> install.packages("corrplot")
>>> library(corrplot)
>>> cor_table = cor(wide_df[,-1])
>>> corrplot(cor_table, method = "circle")
```

Running these commands generates *Figure 5.23*. Each circle represents the strength of correlation between the corresponding stocks, where a bigger and darker circle denotes a stronger correlation:

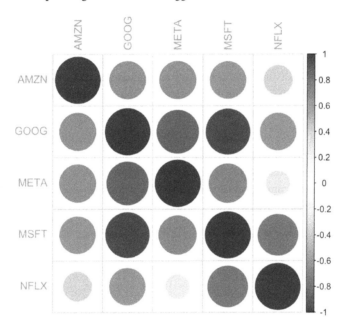

Figure 5.23 – Correlation plot between each pair of stocks

Note that the correlation plot relies on the `cor_table` variable, which stores the pairwise correlation as a table, as shown here:

```
>>> cor_table
      AMZN  GOOG  META  MSFT  NFLX
AMZN 1.000 0.655 0.655 0.635 0.402
GOOG 0.655 1.000 0.855 0.945 0.633
META 0.655 0.855 1.000 0.692 0.267
MSFT 0.635 0.945 0.692 1.000 0.782
NFLX 0.402 0.633 0.267 0.782 1.000
```

A high correlation between variables may or may not be a good thing. When the dependent variable (also called the target outcome) to be predicted is highly correlated with an independent variable (also called a predictor, feature, or covariate), we would prefer to include this feature in the prediction model due to its high covariation with the target variable. On the other hand, when two features are highly correlated, we tend to ignore one and choose the other or apply some sort of regularization and feature selection approach to downsize the impact of correlated features.

Summary

In this chapter, we introduced basic techniques to conduct EDA. We started by going over the common approaches to analyzing and summarizing categorical data, including frequency count and bar charts. We then introduced marginal distribution and faceted bar charts when working with multiple categorical variables.

Next, we switched to analyzing numerical variables and covered sensitive measures such as central tendency (mean) and variation (variance), as well as robust measures such as median and IQR. Several types of charts are available for visualizing a numerical variable, including histograms, density plots, and box plots, all of which can be combined with another categorical variable.

Finally, we went through a case study using the stock price data. We started by downloading the real data from Yahoo! Finance and applying all the EDA techniques to analyze the data, followed by creating a correlation plot to indicate the strength of covariation between each pair of variables. This allows us to develop a helpful understanding of the relationship between variables and jump-start the predictive modeling stage.

In the next chapter, we will cover r markdown, a widely used package to generate interactive reports in R.

Effective Reporting
with R Markdown

The previous chapter covered the different types of graphing techniques, all of which are static. In this chapter, we will take this one step further and discuss how to generate graphs and tables using **R Markdown** consistently.

By the end of this chapter, you will have learned the fundamentals of R Markdown reports, including how to add, fine-tune, and customize figures and tables to make interactive and effective reports. You will also know how to generate effective R Markdown reports that can add some flying colors to your presentations.

In this chapter, we will cover the following topics:

- Fundamentals of R Markdown
- Generating a financial analysis report
- Customizing R Markdown reports

Technical requirements

To complete the exercises in this chapter, you will need to have the latest versions of the following packages:

- `rmarkdown`, version 2.17
- `quantmod`, version 0.4.20
- `lubridate`, version 1.8.0

Please note that the versions of the packages mentioned in the preceding list are the latest ones while I am writing this book. All the code and data for this chapter is available at https://github.com/PacktPublishing/The-Statistics-and-Machine-Learning-with-R-Workshop/tree/main/Chapter_6.

Fundamentals of R Markdown

R Markdown is a formatting language that can help you effectively and dynamically reveal insights from the data and generate reports in the form of a PDF, an HTML file, or a web application. It allows you to tidy up your analyses via various forms of graphs and tables covered earlier in this book, and present them in a consistent, neat, and transparent manner that facilitates easy reproduction by another analyst. Either in academia or industry, demonstrating reproducibility in your analysis is an essential quality of your work. When others can easily reproduce and understand what you did in your analysis, it makes communication much easier and your work more trustworthy. Since all outputs are code-based, the ability to easily reproduce your work also makes it convenient to fine-tune the analysis when you present your initial work and come back with further modifications to be done, a common iterative process in real-life data analysis.

Using R Markdown, you can present your code along with the output (including graphs and tables) and add surrounding text as context. It is similar to Jupyter Notebook using Python, yet it has advantages backed by the tidyverse ecosystem.

R Markdown is based on the Markdown syntax, which is a simple-to-follow markup language that allows the user to create web page-like files from plain text files. Let's start by downloading the R Markdown package and creating a simple starter file.

Getting started with R Markdown

R Markdown allows us to create efficient reports to summarize our analyses and communicate the results to end users. The first thing we need to get R Markdown up and running in RStudio is to download the rmarkdown package and load it into the console, which can be done via the following commands:

```
>>> install.packages("rmarkdown")
>>> library(rmarkdown)
```

R Markdown has a dedicated type of file that ends with .Rmd. To create an R Markdown file, we can select **File** | **New File** | **R Markdown** in RStudio; this will make the window shown in *Figure 6.1* appear. The left panel contains the different formats we can choose from, where **Document** is a collection of common file types such as HTML, PDF, and Word, **Presentation** renders the R Markdown file in a presentation mode similar to PowerPoint, **Shiny** adds an interactive **Shiny** component (interactive widgets) in the R Markdown file, and **From Template** provides a list of starter templates to accelerate report generation:

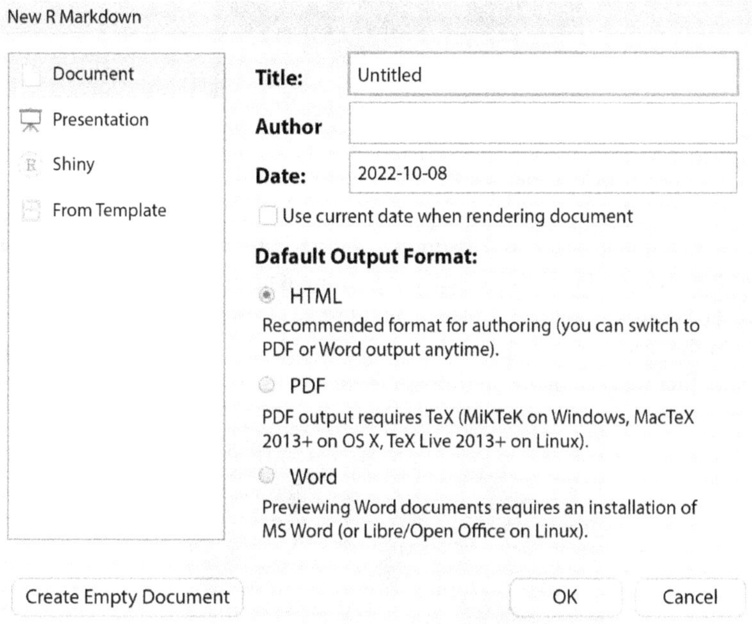

Figure 6.1 – Creating an R Markdown file

Let's start with the **Document** format and choose **HTML** as it can be easily converted into other file types later. After inputting the title (in this case, `my first rmarkdown`) and clicking **OK**, a `.Rmd` file with a basic set of instructions will be created. Not all of this information will be used, so feel free to delete unnecessary code in the script after you gain familiarity with the common components.

An R Markdown document consists of three components: the metadata for the file, the text for the report, and the code for the analysis. We'll look at each of these components in the following sections.

Getting to know the YAML header

The top of the R Markdown script, as shown in *Figure 6.2*, is a set of metadata header information wrapped by two sets of three hyphens, `- - -`, and contained in the YAML header. YAML, a human-readable data serialization language, is a syntax for hierarchical data structures commonly used for configuration files. In this case, the default information includes the title, output format, and date, represented as key-value pairs. The information contained in the header impacts the whole document. For example, to generate a PDF file, we could simply switch from `html_document` to `pdf_document` in the output configuration. This is the minimal set of information needed in a header, although you are encouraged to add the author information (via the same initial window in *Figure 6.2*) to show the authorship of your work:

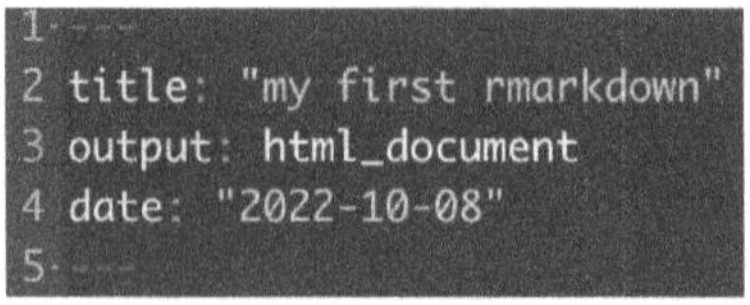

```
1 ---
2 title: "my first rmarkdown"
3 output: html_document
4 date: "2022-10-08"
5 ---
```

Figure 6.2 – YAML header of the default R Markdown script

With the header information in place, and assuming all the additional code is deleted, we can compile the R Markdown file into an HTML file by clicking on the **Knit** button in the taskbar, as shown in *Figure 6.3*. Knitting a file allows us to generate an output report from the R Markdown file. It combines the text and code in the R Markdown document into an HTML file based on the configuration in the YAML header. Note that we save the R Markdown file as `test.Rmd`:

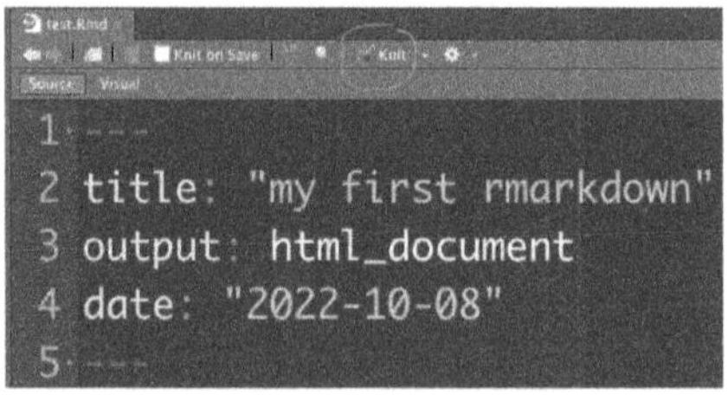

```
1 ---
2 title: "my first rmarkdown"
3 output: html_document
4 date: "2022-10-08"
5 ---
```

Figure 6.3 – Converting the R Markdown file into an HTML file using the Knit button

Knitting the R Markdown file will generate an HTML file that opens in a separate preview window. It will also save an HTML file named `test.html` in the same folder as the R Markdown file.

Next, we will learn more about the structure and syntax of the main body of an R Markdown file, including text formatting and handling code chunks.

Formatting textual information

Textual information is of equal, if not higher, importance than the code you write for analysis and modeling. Good code is often well documented, and this is even more critical when your end user is non-technical. Putting proper background information, assumption, context, and decision-making processes in place is an essential companion to your technical analysis, besides transparency and consistency of the analysis. In this section, we will review the common commands we can use to format text.

Exercise 6.1 – formatting text in R Markdown

In this exercise, we will generate the text shown in *Figure 6.4* using R Markdown:

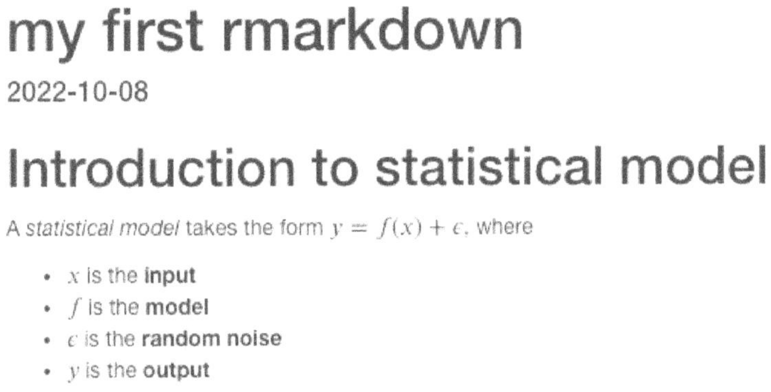

my first rmarkdown
2022-10-08

Introduction to statistical model

A *statistical model* takes the form $y = f(x) + \epsilon$, where

- x is the **input**
- f is the **model**
- ϵ is the **random noise**
- y is the **output**

Figure 6.4 – Sample text generated as an HTML file using R Markdown

The text includes a header, some words in italics or bold, a mathematical expression, and four unordered list items. Let's look at how to generate this text:

1. Write a level-one header using the # symbol:

    ```
    # Introduction to statistical model
    ```

    Note that the more hashes we use, the smaller the header will be. Remember to place the hash at the start of the line and add a single space after the hash and before the text.

2. Write the middle sentence by wrapping the text in * * for italics and $ $ for mathematical expressions:

    ```
    A *statistical model* takes the form $y=f(x)+\epsilon$, where
    ```

3. Generate the unordered list by starting each item with * and wrapping the text in ** ** for bold:

    ```
    * $x$ is the **input**
    * $f$ is the **model**
    * $\epsilon$ is the **random noise**
    * $y$ is the **output**
    ```

    Note that we can easily switch the output file from HTML to PDF, simply by changing `output: html_document` to `output: pdf_document`. The resulting output is shown in *Figure 6.5*:

my first rmarkdown

2022-10-08

Introduction to statistical model

A *statistical model* takes the form $y = f(x) + \epsilon$, where

- x is the **input**
- f is the **model**
- ϵ is the **random noise**
- y is the **output**

Figure 6.5 – Sample text generated as a PDF file using R Markdown

Knitting the R Markdown file into a PDF document may require you to install additional packages, such as LaTeX. When an error pops up complaining that the package is unavailable, simply go to the console and install this package before knitting again. We can also use the drop-down menu from the **Knit** button to choose the desired output format.

In addition, the value for the date key in the YAML header is a string. If you would like to display it as the current date automatically, you can replace the string with `"`r Sys.Date()`"`.

These are some of the common commands that we can use in a `.Rmd` file to format the texts in the resulting HTML file. Next, we will look at how to write R code in R Markdown.

Writing R code

In R Markdown, the R code is contained inside code chunks enclosed by three backticks, ` ``` `, which separate code from text in the R Markdown file. A code chunk is also accompanied by the corresponding rules and specifications on the language used and other configurations inside curly braces, `{ }`. Code chunks allow us to render code-based outputs or display the code in the report.

An example code chunk is shown in the following code snippet, where we indicate the type of language as R and perform an assignment operation:

```
```{r}
a = 1
```
```

Besides typing in the commands for a code chunk, we can also click on the code icon (starting with the letter c) in the toolbar and choose the option for the R language, as shown in *Figure 6.6*. Note that you can also use other languages such as Python, thus making R Markdown a versatile tool that allows us to use different programming languages in one working file:

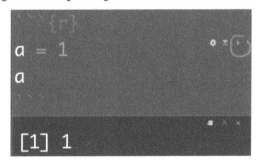

Figure 6.6 – Inserting an R code chunk

Each chunk of code can be executed by clicking on the green arrow at the right-hand side of each chunk, with the result shown just below the code chunk. For example, *Figure 6.7* shows the output after performing the assignment and printing the variable:

Figure 6.7 – Executing a code chunk

We can also specify other options in the curly braces of a code chunk. For example, we may not want to include a particular code chunk in the output of the resulting HTML file. To hide the code itself and only show the output of the code, we can add echo=FALSE to the accompanying configuration of the code chunk, as shown in the following code block:

```
```{r echo=FALSE}
a = 1
a
```
```

Figure 6.8 shows the two different types of output in the resulting HTML file:

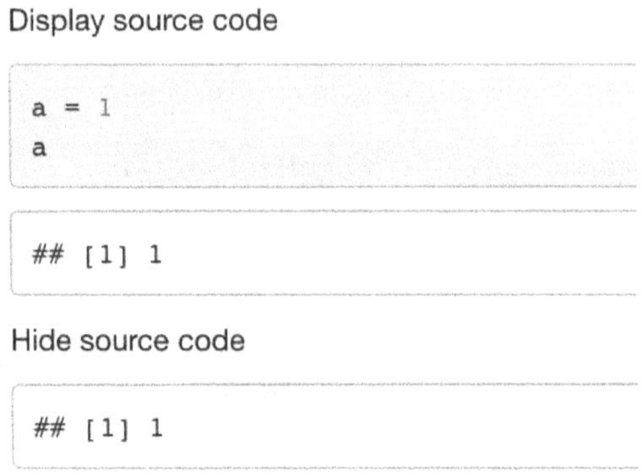

Display source code

```
a = 1
a
```

```
## [1] 1
```

Hide source code

```
## [1] 1
```

Figure 6.8 – Showing and hiding the source code in the HTML file

In addition, when we load a package in the current session, we may get a warning message in the console. In R Markdown, such a warning message will also appear in the resulting HTML. To hide the warning message, we can add warning=FALSE to the configuration. For example, in the following code snippet, we are hiding the warning message while loading the dplyr package:

```
```{r warning=FALSE}
library(dplyr)
```
```

Figure 6.9 compares the two scenarios when loading a package with or without showing the warning message:

Display warning message

```
library(dplyr)
```

```
##
## Attaching package: 'dplyr'
```

```
## The following objects are masked from 'package:stats':
##
##      filter, lag
```

```
## The following objects are masked from 'package:base':
##
##      intersect, setdiff, setequal, union
```

Hide warning message

```
library(dplyr)
```

Figure 6.9 – Hiding the warning message while loading a package

With these building blocks covered, we will conduct a case study on generating a financial analysis report using Google stock price data in the following section.

Generating a financial analysis report

In this section, we will analyze Google's stock data from Yahoo! Finance. To facilitate data download and analysis, we will use the quantmod package, which is designed to assist quantitative traders in developing, testing, and deploying statistically based trading models. Let's install the package and load it into the console:

```
>>> install.packages("quantmod")
>>> library(quantmod)
```

Next, we will generate an HTML report using R Markdown and cover the basics of data querying and analysis.

Getting and displaying the data

Let's go through an exercise to generate an initial report that automatically queries stock data from Yahoo! Finance and displays the basic information in the dataset.

Exercise 6.2 – generating the base report

In this exercise, we will set up an R Markdown file, download Google's stock price data, and display general information about the dataset:

1. Create an empty R Markdown file named `Financial analysis` and set the corresponding `output`, `date`, and `author` in the YAML file:

    ```
    ---
    title: "Financial analysis"
    output: html_document
    date: "2022-10-12"
    author: "Liu Peng"
    ---
    ```

2. Create a code chunk to load the `quantmod` package and query Google's stock price data using the `getSymbols()` function. Store the resulting data in `df`. Also, hide all messages in the resulting HTML file and add the necessary text for illustration:

    ```
    # Analyzing Google's stock data since 2007
    Getting Google's stock data
    ```{r warning=FALSE, message=FALSE}
    library(quantmod)
    df = getSymbols("GOOG", auto.assign=FALSE)
    ```
    ```

    Here, we specify `warning=FALSE` to hide the warning message when loading the package and `message=FALSE` to hide the messages that are generated when calling the `getSymbols()` function. We also specify `auto.assign=FALSE` to assign the resulting DataFrame to the `df` variable. Also, note that we can add text as comments inside the code chunks, which will be treated as typical comments that start with a hash sign, #, are.

3. Count the total number of rows and display the first and last two rows of the DataFrame via three separate code chunks. Add corresponding text to serve as documentation for the code:

    ```
    Total number of observations of `df`
    ```{r}
    nrow(df)
    ```

    Displaying the first two rows of `df`
    ```{r}
    ```
    ```

```
head(df, 2)
```
```
Displaying the last two rows of `df`
```{r}
tail(df, 2)
```
```

Note that we use ` ` to indicate inline code within the text.

At this point, we can knit the R Markdown file to observe the resulting HTML file, as shown in *Figure 6.10*. It is a good practice to check the output frequently so that any potential unexpected error can be corrected in time:

Financial analysis

Liu Peng

2022-10-12

Analyzing Google's stock data since 2007

Getting Google's stock data

```
library(quantmod)
df = getSymbols("GOOG", auto.assign=FALSE)
```

Total number of observations of df

```
nrow(df)
```

```
## [1] 3973
```

Displaying the first two rows of df

```
head(df, 2)
```

```
##            GOOG.Open GOOG.High GOOG.Low GOOG.Close GOOG.Volume GOOG.Adjusted
## 2007-01-03  11.60650  11.87200 11.48470   11.64610   309415434      11.64610
## 2007-01-04  11.68122  12.05357 11.66503   12.03638   316686586      12.03638
```

Displaying the last two rows of df

```
tail(df, 2)
```

```
##            GOOG.Open GOOG.High GOOG.Low GOOG.Close GOOG.Volume GOOG.Adjusted
## 2022-10-11    98.25   100.120    97.25      98.05    21617700         98.05
## 2022-10-12    98.27    99.648    97.67      98.30    17332800         98.30
```

Plotting the stock price data

```
chart_Series(df$GOOG.Close,name="Google Stock Price")
```

Figure 6.10 – Displaying the HTML output

4. Plot the time series of the daily closing price using the `chart_Series()` function:

```
Plotting the stock price data
```{r}
chart_Series(df$GOOG.Close,name="Google Stock Price")
```
```

Adding this code chunk to the R Markdown document and knitting it generates the same output file with an additional graph, as shown in *Figure 6.11*. The `chart_Series()` function is a utility function for plotting that's provided by `quantmod`. We can also plot it based on the `ggplot` package, as discussed in the previous chapter:

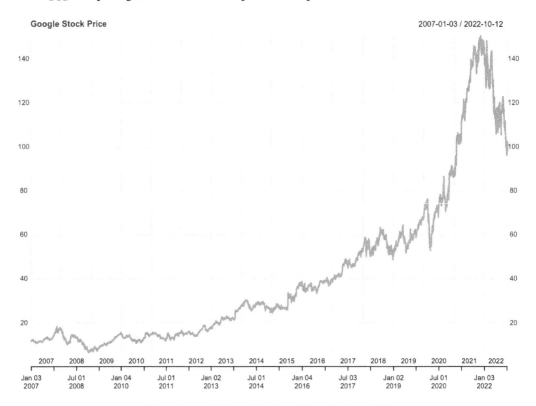

Figure 6.11 – Daily stock price of Google since 2017

In addition to generating figures from code, we can include links and images to the output. This image can be loaded either from a local drive or from the web. In the following code snippet, we are adding a line of text with a hyperlink to an example image and showing the image in the next line by directly reading it from GitHub:

```
The following image can be accessed [here](https://github.com/
PacktPublishing/The-Statistics-and-Machine-Learning-with-R-Workshop/
blob/main/Chapter_6/Image.png).
```

```
![](https://raw.githubusercontent.com/PacktPublishing/The-Statistics-
and-Machine-Learning-with-R-Workshop/main/Chapter_6/Image.png)
```

Note that we added a hyperlink to the word `here` by wrapping it inside square brackets, followed by the hyperlink in parentheses. To add an image, we can add an exclamation mark in front of the square brackets. We can also specify the size of the image by adding `{width=250px}` after the image link.

Knitting the previous code in R Markdown generates *Figure 6.12*:

Figure 6.12 – Visualizing an image from GitHub

Next, we will perform data analysis and display the result as text.

Performing data analysis

With the dataset loaded, we can perform data analysis and present insights in the resulting output document, all generated automatically and consistently. For example, we can present high-level statistics about the stock price for a particular period of interest, such as the average, maximum, and minimum prices. These statistics can be embedded into text, making the presentation style more natural and self-contained.

Exercise 6.3 – performing simple data analysis

In this exercise, we will extract the year-to-date highest, average, and lowest stock prices of Google. To achieve this, we will first transform the dataset from its original xts format into a tibble object, then summarize these statistics using dplyr. Finally, we will display this information within the text of the HTML document:

1. Load the dplyr and tibble packages and convert df from xts into tibble format. Store the resulting tibble object in df_tbl:

```
library(dplyr)
library(tibble)
df_tbl = df %>%
  as_tibble() %>%
  add_column(date = index(df), .before = 1)
```

   Here, we will use the as_tibble() function to convert the xts object into tibble format, followed by the add_column() function to insert a date column at the beginning of the DataFrame. The date information is available as an index in the original xts object.

2. Extract the year-to-date highest, average, and lowest closing prices since 2022. Store the results in max_ytd, avg_ytd, and min_ytd, respectively:

```
max_ytd = df_tbl %>%
  filter(date >= as.Date("2022-01-01")) %>%
  summarise(price = max(GOOG.Close)) %>%
  .$price

avg_ytd = df_tbl %>%
  filter(date >= as.Date("2022-01-01")) %>%
  summarise(price = mean(GOOG.Close)) %>%
  .$price

min_ytd = df_tbl %>%
  filter(date >= as.Date("2022-01-01")) %>%
  summarise(price = min(GOOG.Close)) %>%
  .$price
```

   For each statistic, we first filter by date, followed by extracting the relevant statistic based on the GOOG.Close column. Finally, we return the result as a single scalar value instead of a DataFrame.

3. Display these statistics in text:

```
Google's **highest** year-to-date stock price is `r max_ytd`.

Google's **average** year-to-date stock price is `r avg_ytd`.
```

```
Google's **lowest** year-to-date stock price is `r min_ytd`.
```

As shown in *Figure 6.13*, knitting the document outputs the statistics in the HTML file, which allows us to reference the code results in the HTML report:

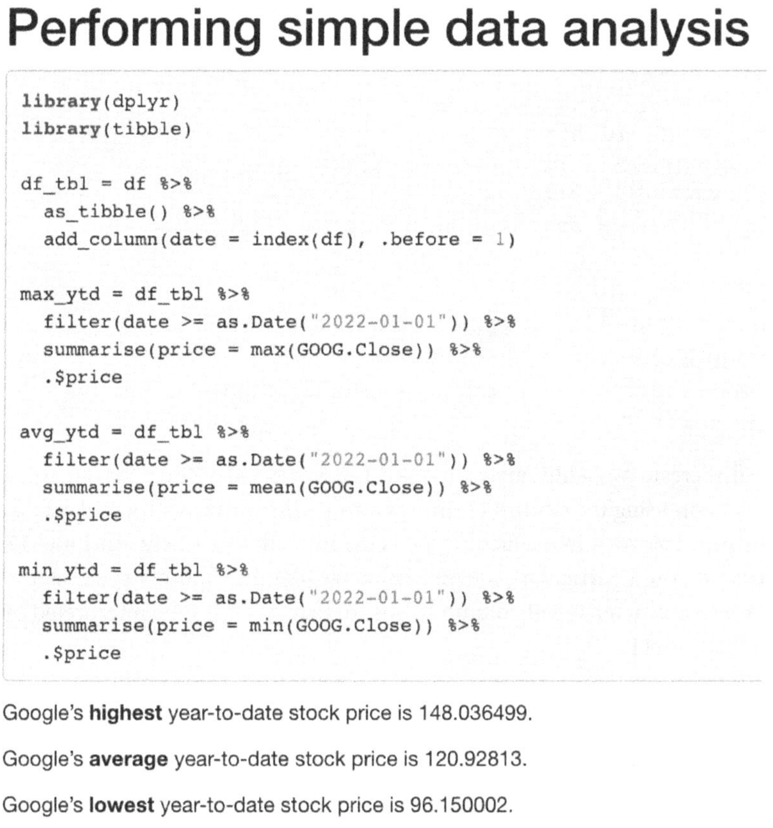

Performing simple data analysis

```
library(dplyr)
library(tibble)

df_tbl = df %>%
  as_tibble() %>%
  add_column(date = index(df), .before = 1)

max_ytd = df_tbl %>%
  filter(date >= as.Date("2022-01-01")) %>%
  summarise(price = max(GOOG.Close)) %>%
  .$price

avg_ytd = df_tbl %>%
  filter(date >= as.Date("2022-01-01")) %>%
  summarise(price = mean(GOOG.Close)) %>%
  .$price

min_ytd = df_tbl %>%
  filter(date >= as.Date("2022-01-01")) %>%
  summarise(price = min(GOOG.Close)) %>%
  .$price
```

Google's **highest** year-to-date stock price is 148.036499.

Google's **average** year-to-date stock price is 120.92813.

Google's **lowest** year-to-date stock price is 96.150002.

Figure 6.13 – Extracting simple statistics and displaying them in HTML form

In the next section, we will look at adding plots to the HTML report.

Adding plots to the report

Adding plots to the HTML report works the same way as in the RStudio console. We could simply write the plotting code in a chunk, and the graph will appear in the resulting report after we've knitted the R Markdown file. Let's go through an exercise to visualize the stock price using the `ggplot2` package we covered in the previous chapter.

Exercise 6.4 – adding plots using ggplot2

In this exercise, we will visualize the average monthly closing price for the past three years as line plots. We will also explore different configuration options for the figure in the report:

1. Create a dataset that contains the average monthly closing price between 2019 and 2021:

    ```
    library(ggplot2)
    library(lubridate)

    df_tbl = df_tbl %>%
      mutate(Month = factor(month(date), levels =
    as.character(1:12)),
             Year = as.character(year(date)))

    tmp_df = df_tbl %>%
      filter(Year %in% c(2019, 2020, 2021)) %>%
      group_by(Year, Month) %>%
      summarise(avg_close_price = mean(GOOG.Close)) %>%
      ungroup()
    ```

    Here, we first create two additional columns called Month and Year that are derived based on the date column using the month() and year() functions from the lubridate package. We also convert Month into a factor-typed column with levels between 1 and 12 so that this column can follow a particular sequence when we plot the monthly price later. Similarly, we set Year as a character-typed column to ensure that it will not be interpreted by ggplot2 as a numeric variable.

    Next, we filter the df_tbl variable by Year, group by Year and Month, and calculate the average GOOG.Close value, followed by using the ungroup() function to remove the group structure in the resulting DataFrame saved in tmp_df.

2. Plot the monthly average closing price each year as a separate line in a line chart. Change the corresponding figure label and text size:

    ```
    p = ggplot(tmp_df,
            aes(x = Month, y = avg_close_price,
                group = Year,
                color = Year)) +
      geom_line() +
      theme(axis.text=element_text(size=16),
            axis.title=element_text(size=16,face="bold"),
            legend.text=element_text(size=20)) +
      labs(titel = "Monthly average closing price between 2019 and
    2021",
            x = "Month of the year",
    ```

```
                 y = "Average closing price")

p
```

Running the previous commands in a code chunk will generate the output shown in *Figure 6.14*. Note that we have also added a header and some text to point out the purpose and context of the code. The code and the output are shown automatically after knitting the R Markdown file, thus making R Markdown a great option for producing transparent, engaging, and reproducible technical reports:

Adding plots

Show the average closing price between 2019 and 2021.

```
library(ggplot2)
library(lubridate)

df_tbl = df_tbl %>%
  mutate(Month = factor(month(date), levels = as.character(1:12)),
         Year = as.character(year(date)))

tmp_df = df_tbl %>%
  filter(Year %in% c(2019, 2020, 2021)) %>%
  group_by(Year, Month) %>%
  summarise(avg_close_price = mean(GOOG.Close)) %>%
  ungroup()

p = ggplot(tmp_df,
           aes(x = Month, y = avg_close_price,
               group = Year,
               color = Year)) +
  geom_line() +
  theme(axis.text=element_text(size=16),
        axis.title=element_text(size=16,face="bold"),
        legend.text=element_text(size=20)) +
  labs(titel = "Monthly average closing price between 2019 and 2021",
       x = "Month of the year",
       y = "Average closing price")

p
```

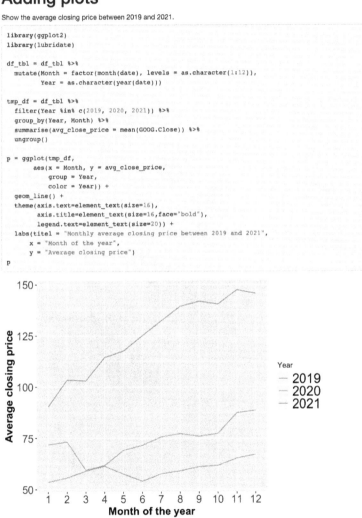

Figure 6.14 – Adding a plot to show the monthly average closing price for the past three years

We can also configure the size and position of the figure.

3. Shrink the size of the figure by setting `fig.width=5` and `fig.height=3` in the configuration section of the code chunk and show the output graph:

```
Control the figure size via the `fig.width` and `fig.height`
parameters.
```{r fig.width=5, fig.height=3}
p
```
```

Knitting the document with these added commands produces *Figure 6.15*:

Control the figure size via the `fig.width` and `fig.height` parameters.

p

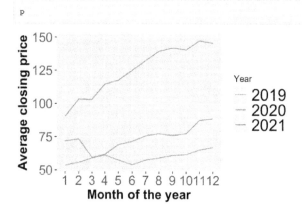

Figure 6.15 – Changing the size of the figure

4. Align the position of the figure so that it's at the center of the document:

```
Align the figure using the `fig.align` parameter.
```{r fig.width=5, fig.height=3, fig.align='center'}
p
```
```

Knitting the document with these added commands produces *Figure 6.16*:

Align the figure using the `fig.align` parameter.

p

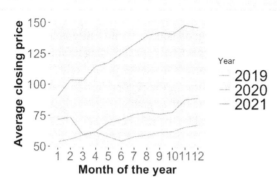

Figure 6.16 – Changing the position of the figure

5. Add a caption to the plot:

```r
Add figure caption via the `fig.cap` parameter.
```{r fig.width=5, fig.height=3, fig.align='center', fig.cap='
Figure 1.1 Monthly average closing price between 2019 and 2021'}
p
```
```

Knitting the document with these added commands produces *Figure 6.17*:

Add figure caption via the `fig.cap` parameter.

p

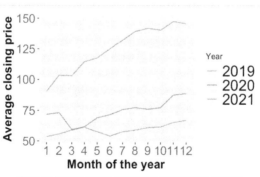

Figure 1.1 Monthly average closing price between 2019 and 2021

Figure 6.17 – Adding a caption to the figure

Besides graphs, tables are also commonly used to present and summarize information in a report. We'll look at how to generate tables in the next section.

Adding tables to the report

Representing information in a table is a good complement to its graph counterpart when the report user is interested in delving further into the details or even using it for further analysis. For end users, being able to access and use the data in the report plays a pivotal role as this gives them more control over the already preprocessed information in the report. In other words, an R Markdown-based HTML report not only summarizes information as graphs for easy digestion but also provides detailed information on specific data sources as tables to facilitate ad hoc analysis.

We can add tables using the `kable()` function from the `knitr` package, the backbone engine that supports executing code in each code chunk, and then perform dynamic report generation upon knitting the R Markdown document. Note that it is a good practice to preprocess and clean the data before presenting it as a table via `kable()`; the task should only be visualizing a clean and organized table.

Let's go through an exercise to look at how to add clean tables to the report.

Exercise 6.5 – adding tables using kable()

In this exercise, we will show the top five rows of the `tmp_df` variable in a table, then demonstrate different configuration options for displaying the table:

1. Show the top five rows of `tmp_df` using the `kable()` function from the `knitr` package:

    ```
    # Adding tables
    Printing `tmp_df` as a static summary table via the `kable()`
    function.
    ```{r}
    library(knitr)

    kable(tmp_df[1:5,])
    ```
    ```

Knitting the document with these added commands produces *Figure 6.18*:

Adding tables

Printing `tmp_df` as a static summary table via the `kable()` function.

```
library(knitr)

kable(tmp_df[1:5,])
```

Year	Month	avg_close_price
2019	1	53.61321
2019	2	55.71211
2019	3	58.95988
2019	4	61.31131
2019	5	57.53816

Figure 6.18 – Adding a table to the report

2. Change the column names of the table using the `col.names` argument:

```
Changing column names via the `col.names` parameter.
```{r}
kable(tmp_df[1:5,], col.names=c("Year", "Month", "Average
closing price"))
```
```

Knitting the document with these added commands produces *Figure 6.19*:

Changing column names via the `col.names` parameter.

```
kable(tmp_df[1:5,], col.names=c("Year", "Month", "Average closing price"))
```

Year	Month	Average closing price
2019	1	53.61321
2019	2	55.71211
2019	3	58.95988
2019	4	61.31131
2019	5	57.53816

Figure 6.19 – Changing column names in the table

We can also modify the column alignment within the table using the `align` argument. The default column alignment is on the right for the numeric column and the left for all other types of columns. As shown in *Figure 6.19*, the `Year` (character typed) and `Month` (factor typed) columns are aligned on the left, and `Average closing price` (numeric) is aligned on the right. The alignment is specified column-wise using a single letter, with `"l"` denoting aligning to the left, `"c"` to the center, and `"r"` to the right.

3. Align all columns to the center using the `align` argument:

```
Align the table via the `align` argument.
```{r}
kable(tmp_df[1:5,], col.names=c("Year", "Month", "Average
closing price"), align="ccc")
```
```

Here, we specify `align="ccc"` to align all columns to the center. Knitting the document with these added commands produces *Figure 6.20*:

Align the table via the `align` argument.

```
kable(tmp_df[1:5,], col.names=c("Year", "Month", "Average closing price"), align="ccc")
```

Year	Month	Average closing price
2019	1	53.61321
2019	2	55.71211
2019	3	58.95988
2019	4	61.31131
2019	5	57.53816

Figure 6.20 – Centering all the columns in the table

Lastly, we can also add a caption to the table.

4. Add a caption to the table using the `caption` argument:

```
Add table caption via the `caption` parameter.
```{r}
kable(tmp_df[1:5,], col.names=c("Year", "Month", "Average
closing price"), align="ccc", caption="Table 1.1 Average monthly
closing price")
```
```

Knitting the document with these added commands produces *Figure 6.21*:

Add table caption via the `caption` parameter.

```
kable(tmp_df[1:5,], col.names=c("Year", "Month", "Average closing price"), align="ccc", caption="Table 1.1 Averag
e monthly closing price")
```

Table 1.1 Average monthly closing price

Year	Month	Average closing price
2019	1	53.61321
2019	2	55.71211
2019	3	58.95988
2019	4	61.31131
2019	5	57.53816

Figure 6.21 – Adding a caption to the table

In the next section, we will discuss some common options we can use to modify the code chunk outputs after knitting the R Markdown document.

Configuring code chunks

We have seen several options from previous exercises that we can use to control the output style of a code chunk. For example, by setting `warning=FALSE` and `message=FALSE`, we could hide potential warnings and messages in the resulting output document.

There are other commonly used options. For example, we can use the `include` option to decide whether the code and results appear in the output report or not. In other words, setting `include=FALSE` will hide the code and results of the specific code chunk in the report, although the code will still be executed upon knitting the R Markdown document. By default, we have `include=TRUE` and all the code and execution results will appear in the report.

Another related option is `echo`, where setting `echo=FALSE` hides the code and only shows the execution outputs in the report. We can consider this option when we're generating plots in the report since most users are more interested in the graphical analysis compared to the process that generates the graph. Again, by default, we have `echo=TRUE`, which displays the code in the report before the plots.

Besides this, we may only be interested in showing some code instead of executing all of it. In this case, we can set `eval=FALSE` to make sure that the code in the code chunk does not impact the overall execution and result of the report. This is in contrast to setting `include=FALSE`, which hides the code but still executes it in the backend, thus bearing an effect on the subsequent code. By default, we have `eval=FALSE`, which evaluates all the code in the code chunk. *Figure 6.22* summarizes these three options:

	Code execution	Code appearance	Result appearance
include=FALSE	Yes	No	No
echo=FALSE	Yes	No	Yes
eval=FALSE	No	Yes	No

Figure 6.22 – Common options for configuring code chunks

Next, we will go over an exercise to practice different options.

Exercise 6.6 – configuring code chunks

In this exercise, we will go through a few ways to configure the code chunks we covered previously:

1. Display the maximum closing price for the past five years in a table. Show both the code and the result in the report:

```
# Code chunk options
Display both code and result be default.
```{r}
tmp_df = df_tbl %>%
   mutate(Year = as.integer(Year)) %>%
   filter(Year >= max(Year)-5,
          Year < max(Year)) %>%
   group_by(Year) %>%
   summarise(max_closing = max(GOOG.Close))
kable(tmp_df)
```
```

Here, we first convert Year into an integer-typed variable, then subset the DataFrame to keep only the last five years of data, followed by extracting the maximum closing price for each year. The result is then shown via the kable() function.

Knitting the document with these added commands produces *Figure 6.23*. The result shows that Google has been making new highs over the years:

Code chunk options

Display both code and result be default.

```
tmp_df = df_tbl %>%
  mutate(Year = as.integer(Year)) %>%
  filter(Year >= max(Year)-5,
         Year < max(Year)) %>%
  group_by(Year) %>%
  summarise(max_closing = max(GOOG.Close))
kable(tmp_df)
```

Year	max_closing
2017	53.8570
2018	63.4165
2019	68.0585
2020	91.3995
2021	150.7090

Figure 6.23 – Displaying the maximum closing price for the past five years

2. Obtain the highest closing price in a code chunk with the code and result hidden in the report by setting include=FALSE. Display the result in a new code chunk:

```
Execute the code chunk but hide both code and result in the
output by setting `include=FALSE`.
```{r include=FALSE}
total_max_price = max(df_tbl$GOOG.Close)
```

Display the code and result.
```{r}
total_max_price
```
```

Knitting the document with these commands produces *Figure 6.24*:

Execute the code chunk but hide both code and result in the output by setting `include=FALSE`.

Display the code and result.

```
total_max_price
```

```
## [1] 150.709
```

Figure 6.24 – Hiding the code and result in one code chunk and displaying the result separately

3. For the running table, hide the code chunk and only display the result in the report by setting `echo=FALSE`:

    ```
    Execute the code chunk and only display the result in the output
    by setting `echo=FALSE`.
    ```{r echo=FALSE}
    kable(tmp_df)
    ```
    ```

    Knitting the document with these commands produces *Figure 6.25*:

    Execute the code chunk and only display the result in the output by setting `echo=FALSE` .

Year	max_closing
2017	53.8570
2018	63.4165
2019	68.0585
2020	91.3995
2021	150.7090

    Figure 6.25 – Hiding the code chunk and displaying the result in the report

4. Only display the code on table generation in the code chunk and do not execute it in the report by setting `eval=FALSE`:

    ```
    Do not execute the code chunk and only display the code in the
    output by setting `eval=FALSE`.
    ```{r eval=FALSE}
    kable(tmp_df)
    ```
    ```

    Knitting the document with these commands produces *Figure 6.26*:

    Do not execute the code chunk and only display the code in the output by setting `eval=FALSE` .

    ```
    kable(tmp_df)
    ```

    Figure 6.26 – Displaying the code chunk without executing it in the report

5. Print a test message and a warning message in separate blocks. Then, put the same contents in a single block by setting `collapse=TRUE`:

    ```
    All results are in separate blocks by default.
    ```{r}
    print("This is a test message")
    warning("This is a test message")
    ```
    ```

```
Collapsing all results in one block by setting `collapse=TRUE`.
```{r collapse=TRUE}
print("This is a test message")
warning("This is a test message")
```
```

Knitting the document with these commands produces *Figure 6.27*, which shows that both the printed and warning messages are shown in a single block together with the code:

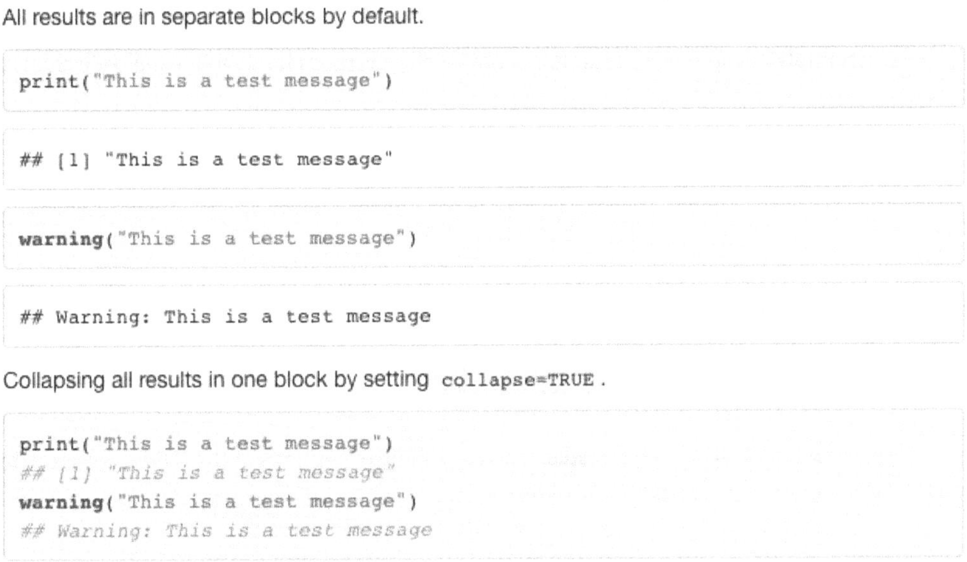

All results are in separate blocks by default.

```
print("This is a test message")
```

```
## [1] "This is a test message"
```

```
warning("This is a test message")
```

```
## Warning: This is a test message
```

Collapsing all results in one block by setting collapse=TRUE.

```
print("This is a test message")
## [1] "This is a test message"
warning("This is a test message")
## Warning: This is a test message
```

Figure 6.27 – Displaying the code and results in one block

In addition, we can hide the warning by configuring the warning attribute in the code chunk.

6.   Hide the warning by setting warning=FALSE:

```
Hide warning by setting `warning=FALSE`.
```{r collapse=TRUE, warning=FALSE}
print("This is a test message")
warning("This is a test message")
```
```

Knitting the document with these commands produces *Figure 6.28*, where the warning has now been removed from the report:

Hide warning by setting `warning=FALSE`.

```
print("This is a test message")
## [1] "This is a test message"
warning("This is a test message")
```

Figure 6.28 – Hiding the warning in the report

Setting the display parameters for each code chunk becomes troublesome when we need to repeat the same operation for many chunks. Instead, we can make a global configuration that applies to all chunks in the R Markdown document by using the `knitr::opts_chunk()` function at the beginning of the document. For example, the following code snippet hides the warnings for all following code chunks:

```
Set global options.
```{r include=FALSE}
knitr::opts_chunk$set(warning=FALSE)
```
```

In the next section, we will look at how to customize R Markdown reports, such as by adding a table of contents and changing the report style.

Customizing R Markdown reports

In this section, we will look at adding metadata such as a table of contents to the report, followed by introducing more options for changing the report style.

Adding a table of contents

When reading a report for the first time, a table of contents provides an overview of the report and thus helps readers quickly navigate the different sections of the report.

To add a table of contents, we can append a colon to the `html_document` field in the YAML header and set `toc: true` as a separate line with one more indentation than the `html_document` field. This is shown in the following code snippet:

```
---
title: "Financial analysis"
output:
    html_document:
        toc: true
date: "2022-10-12"
author: "Liu Peng"
---
```

Knitting the document with these commands produces *Figure 6.29*, where a table of contents is now displayed at the top of the report. Note that when you click on a header in the table of contents, the report will directly jump to that section, which is a nice and user-friendly feature:

Financial analysis

Liu Peng

2022-10-12

- Analyzing Google's stock data since 2007
- Performing simple data analysis
- Adding plots
- Adding tables
- Code chunk options

Figure 6.29 – Adding a table of contents to the report

We can also set `toc_float=true` to make the table of contents float. With this property specified, the table of contents will remain visible as the user scrolls through the report. The following code snippet includes this property in the YAML header:

```
---
title: "Financial analysis"
output:
    html_document:
        toc: true
        toc_float: true
date: "2022-10-12"
author: "Liu Peng"
---
```

Knitting the document with these commands produces *Figure 6.30*, where the table of contents appears on the left-hand side and remains visible as the user navigates different sections:

Figure 6.30 – Setting up a floating table of contents in the report

Next, we will look at creating a report with parameters in the YAML header.

Creating a report with parameters

Recall that our running dataset contains the daily stock prices of Google since 2007. Imagine that we need to create a separate annual report for each year; we may need to manually edit the `year` parameter for each report, which would be a repetitive process. Instead, we can set an input parameter in the YAML header as a global variable that's accessible to all code chunks. When generating other similar reports, we could simply change this parameter and rerun the same R Markdown file.

We can set a parameter input by adding the `params` field, followed by a colon in the YAML header. Then, we must add another line, indent it, and add the key and value of the parameter setting, which are separated by a colon. Note that the value of the parameter is not wrapped in quotations.

Let's go through an exercise to illustrate this.

Exercise 6.7 – generating reports using parameters

In this exercise, we will configure parameters to generate reports that are similar and only differ in the parameter setting:

1. Add a `year` parameter to the YAML header and set its value to `2020`:

    ```
    ---
    title: "Financial analysis"
    output:
        html_document:
            toc: true
            toc_float: true
    date: "2022-10-12"
    author: "Liu Peng"
    params:
        year: 2020
    ---
    ```

    Here, we use the `params` field to initiate the parameter setting and add `year: 2020` as a key-value pair.

2. Extract the summary of the closing price using the `summary()` function for 2020:

    ```
    # Generating report using parameters
    Summary statistics of year `r params$year`
    ```{r}
    df_tbl %>%
      filter(Year == params$year) %>%
      select(GOOG.Close) %>%
    ```
    ```

```
    summary()
    ```
```

Knitting the document with these commands produces *Figure 6.31*, where we use `Year ==` `params$year` as a filtering condition in the `filter()` function:

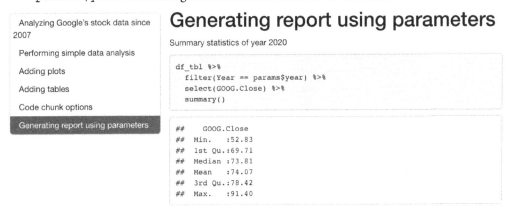

Figure 6.31 – Generating summary statistics of the closing price for 2020 using parameters

3.  Change the parameter setting and generate the same statistics for 2021:

    ```
    ---
    title: "Financial analysis"
    output:
        html_document:
            toc: true
            toc_float: true
    date: "2022-10-12"
    author: "Liu Peng"
    params:
        year: 2021

    ---
    ```

Knitting the document with these commands produces *Figure 6.32*. With a simple change of value in the parameters, we can generate a report for a different year without editing the contents following the YAML header:
    ```

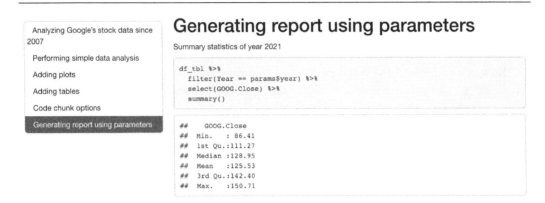

Generating report using parameters

Summary statistics of year 2021

```
df_tbl %>%
  filter(Year == params$year) %>%
  select(GOOG.Close) %>%
  summary()
```

```
##    GOOG.Close
##  Min.   : 86.41
##  1st Qu.:111.27
##  Median :128.95
##  Mean   :125.53
##  3rd Qu.:142.40
##  Max.   :150.71
```

Figure 6.32 – Generating summary statistics of the closing price for 2021 using parameters

We can also create a report based on multiple parameters, which can be appended as key-value pairs in the YAML header.

4. Generate the same statistics for the closing price for Q1 2021:

```
Summary statistics of year `r params$year` and quarter `r params$quarter`
```{r}
df_tbl %>%
  mutate(Qter = quarters(date)) %>%
  filter(Year == params$year,
         Qter == params$quarter) %>%
  select(GOOG.Close) %>%
  summary()
```
```

Here, we create a new column to represent the quarter using the quarters() function based on the date, followed by filtering using the year and the quarter parameters set in the YAML header. Knitting the document with these commands produces *Figure 6.33*:

Summary statistics of year 2021 and quarter Q1

```
df_tbl %>%
  mutate(Qter = quarters(date)) %>%
  filter(Year == params$year,
         Qter == params$quarter) %>%
  select(GOOG.Close) %>%
  summary()
```

```
##       GOOG.Close
##   Min.    : 86.41
##   1st Qu.: 94.56
##   Median :102.25
##   Mean    : 99.21
##   3rd Qu.:104.08
##   Max.    :106.42
```

Figure 6.33 – Generating summary statistics of the closing price for 2021 Q1 using multiple parameters

In the following section, we will look at the style of the report using **Cascading Style Sheets** (**CSS**), a commonly used web programming language to adjust the style of web pages.

Customizing the report style

The report style includes details such as the color and font of text in the report. Like any web programming framework, R Markdown offers controls that allow attentive users to make granular adjustments to the report's details. The adjustable components include most HTML elements in the report, such as the title, body text, code, and more. Let's go through an exercise to learn about different types of style control.

Exercise 6.8 – customizing the report style

In this exercise, we will customize the report style by adding relevant configurations within the `<style>` and `</style>` flags. The specification starts by choosing the element(s) to be configured, such as the main body (using the `body` identifier) or code chunk (using the `pre` identifier). Each property should start with a new line, have the same level of indentation, and have one more level of indentation than the preceding HTML element.

In addition, all contents to be specified are key-value pairs that end with a semicolon and are wrapped within curly braces, { }. The style configuration can also exist anywhere after the YAML header. Let's look at a few examples of specifying the report style:

1. Change the color of the text in the main body to red and the background color to #F5F5F5, the hex code that corresponds to gray:

    ```
    # Customizing report style
    <style>
    body {
      color: blue;
      background-color: #F5F5F5;
    }
    </style>
    ```

    Here, we directly use the word blue to set the color attribute of the text in the body and the hex code to set its background color; these two approaches are equivalent. Knitting the document with these commands produces *Figure 6.34*:

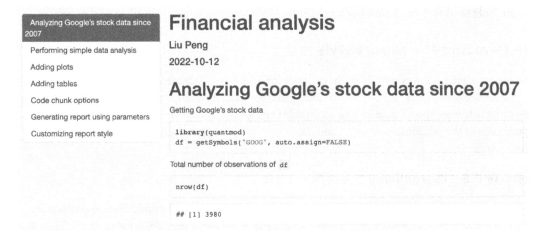

Figure 6.34 – Changing the color of the text in the body of the report

2. Change the color of the code in the code chunks to red by specifying color: red in the pre attribute:

    ```
    # Customizing report style
    <style>
    body {
      color: blue;
      background-color: #F5F5F5;
    }
    ```

```
pre {
  color: red;
}
</style>
```

Knitting the document with these commands produces *Figure 6.35*:

Figure 6.35 – Changing the color of the code in the report

3. For the table of contents, change the color of the text and border to green, and set the font size to 16px:

```
# Customizing report style
<style>
body {
  color: blue;
  background-color: #F5F5F5;
}

pre {
  color: red;
}

#TOC {
  color: green;
  font-size: 16px;
  border-color: green;
}
</style>
```

Note that the style for the table of contents is specified using #TOC without any space in between. Knitting the document with these commands produces *Figure 6.36*:

Figure 6.36 – Changing the color and font size of the table of contents in the report

4.   For the header, change the color to orange, the opacity to 0.9, and the font size to 20px:

```
# Customizing report style
<style>
body {
  color: blue;
  background-color: #F5F5F5;
}

pre {
  color: red;
}

#TOC {
  color: green;
  font-size: 16px;
  border-color: green;
}

#header {
  color: orange;
  opacity: 0.8;
  font-size: 20px;
}
</style>
```

Note that the style for the header is specified using #header without any space in between. Knitting the document with these commands produces *Figure 6.37*:

Figure 6.37 – Changing the color, font size, and opacity of the header in the report

That concludes this exercise. Now, let's wrap up this chapter.

Summary

In this chapter, we introduced R Markdown, a flexible, transparent, and consistent report-generation tool. We started by going over the fundamentals of R Markdown, including the basic building blocks such as YAML headers and code chunks, followed by text formatting techniques.

Next, we covered a case study using Google's stock data. After downloading the stock data from the web, we generated a report to summarize the statistics of the daily closing price, added plots and tables to the report, performed data processing, and displayed the results with different styling options. We also explored a few different ways to configure a code chunk.

Lastly, we discussed how to customize the R Markdown reports. The topics we covered included adding a table of contents to the report, creating repetitive reports using parameters in the YAML header, and changing the visual style of the report by editing the visual properties of different components using CSS.

With the next chapter, we will begin *Part 2* of this book and introduce the fundamentals of linear algebra and calculus using R.

Part 2: Fundamentals of Linear Algebra and Calculus in R

As we venture into the second part of this book, we aim to deepen your understanding of the mathematical foundations underpinning the data science field. While the first part laid the groundwork with statistics essentials and R programming basics, this section elevates your skill set by delving into the more intricate aspects of linear algebra and calculus, all illustrated through R implementations.

By the end of this part, you'll have a well-rounded understanding of the mathematical foundations crucial for advanced data science studies. This part of the book will give you the advanced mathematical and computational tools you need to engage with further advanced topics in statistics and machine learning.

This part has the following chapters:

- *Chapter 7, Linear Algebra in R*

- *Chapter 8, Intermediate Linear Algebra in R*

- *Chapter 9, Calculus in R*

Linear Algebra in R

The previous chapter introduced an efficient and effective reporting approach using R Markdown. *Part 1* of this book essentially covered the practical aspects of getting things done using R. *Part 2* of this book goes back to the fundamentals, covering two essential pillars of mathematics: linear algebra and calculus. Understanding these basics will better prepare us to appreciate and work with common mathematical operations to the point that these operations feel natural to us. *Part 2* aims to develop that level of literacy, starting with a fundamental review of linear algebra with R in this chapter.

By the end of this chapter, you will have learned about the basic concepts of linear algebra, including vectors, matrices, and the system of equations. You will also be able to interpret basic notations in linear algebra and work with common matrices using R.

In this chapter, we will cover the following topics:

- Introducing linear algebra
- Common matrix operations and properties
- Solving system of linear equations

Technical requirements

All the code and data for this chapter is available at `https://github.com/PacktPublishing/The-Statistics-and-Machine-Learning-with-R-Workshop/blob/main/Chapter_7/working.R`.

Introducing linear algebra

This chapter delves into one of the most important branches of mathematics: **linear algebra**. Linear algebra deals with linear operations of mathematical objects, including vectors, matrices, and tensors (high-dimensional matrices), the most common forms of data. For example, the typical table we use to store data in Excel consists of a series of columns. Each column is called a vector, which stores a specific number of elements and takes the form of a column instead of a row by default. A collection of these column vectors forms a matrix, a two-dimensional Excel table, or DataFrame, as we used to call it in the previous chapters. We can also view the same table as a collection of row vectors, where each vector lives in the form of a row.

Let's put these in context. The following code snippet loads the sleep dataset and prints out the first six rows and three columns. We use **A** to denote this 6x3 matrix in the following exposition:

```
>>> data(sleep)
>>> head(sleep)
  extra group ID
1   0.7     1  1
2  -1.6     1  2
3  -0.2     1  3
4  -1.2     1  4
5  -0.1     1  5
6   3.4     1  6
```

Let's introduce some notation here. We use a bold lowercase letter, a, to denote a column vector, say the extra variable, which consists of six elements in the vector. In other words, a is a six-dimensional vector. The column vector, a, can be *transposed* into a row vector, denoting a^T. Since a^T is a row vector, we can write $a^T = [0.7, -1.6, -0.2, -1.2, -0.1, 3.4]$, which is also the typical way of writing row vectors (or transposed column vectors) in many books. Similarly, a row vector can be transposed into the (original) column vector, giving us $(a^T)^T = a$.

A graphical illustration will help here. *Figure 7.1* depicts the process of transposing the column vector, a, into a row vector, a^T, then transposing it again into the original column vector, $(a^T)^T = a$. Therefore, we can view the matrix, A, as a collection of column vectors concatenated horizontally, or a collection of row vectors concatenated vertically:

$$a = \begin{bmatrix} 0.7 \\ -1.6 \\ -0.2 \\ -1.2 \\ -0.1 \\ 3.4 \end{bmatrix} \xrightarrow{\text{Transpose}} a^T = [0.7, -1.6, -0.2, -1.2, -0.1, 3.4]$$

$$(a^T)^T = \begin{bmatrix} 0.7 \\ -1.6 \\ -0.2 \\ -1.2 \\ -0.1 \\ 3.4 \end{bmatrix} = a$$

Figure 7.1 – Transposing the column vector into a row vector, and then transposing it again into the original column vector

Note that the rows are also referred to as observations and that the columns are referred to as features or attributes.

We will start by working with vectors in the following section.

Working with vectors

A vector in R chains together data elements of the same type. Multiple vectors join hands, often side by side, to form a matrix. Therefore, it is the building block in linear algebra, and we must start from there.

There are multiple ways to create a vector. We will explore a few such options in the following exercise.

Exercise 7.1 – creating a vector

This exercise will introduce three common ways of creating a vector. Let's get started:

1. Create a vector of integers from 1 to 6 using the c () function, where c stands for concatenation:

    ```
    x = c(1, 2, 3, 4, 5, 6)
    >>> x
    [1] 1 2 3 4 5 6
    ```

    Note that [1] indicates that the result is a one-dimensional vector.

    Since this list is sequential and equally spaced by 1, we can use the colon notation and write 1 : 6 to create the same list, as shown in the following code snippet:

    ```
    x = c(1:6)
    ```

    One more function we can use to create a sequence of integers is the seq () function, which generates a sequence of numbers equally spaced based on the by argument.

2. Create the same vector using seq ():

    ```
    x = seq(1, 6, by=1)
    ```

    These three implementations create the same vector.

    In addition, we can use the rep () function to create a list of repeated numbers. The following code snippet produces a 6-element vector whose data is all ones:

    ```
    y = rep(1, 6)
    >>> y
    [1] 1 1 1 1 1 1
    ```

A vector in R is mutable, meaning we can change the value in the vector's specific position(s). Let's look at how to achieve this.

Exercise 7.2 – modifying a vector

Modifying the contents of a vector is a common operation, especially in the context of processing a DataFrame using column-wise filtering conditions. The algebra of vectors follows mostly the same principle as scalars. To see this, we must first index the desired elements in the vector, followed by using an assignment operation to override the value of the specific element. Let's take a look:

1. Change the value of the second element in the vector, x, to 20:

    ```
    x[2] = 20
    >>> x
    [1]  1 20  3  4  5  6
    ```

    So, we can access an element in a vector by wrapping the position in a squared bracket. We can also perform bulk operations that exert the same effect on all elements in a vector.

2. Double all elements in the vector, x:

    ```
    >>> x*2
    [1]  2 40  6  8 10 12
    ```

    Here, the multiplier, 2, is broadcast to each element in the vector to perform the respective multiplication. The same broadcasting mechanism applies to the addition operation and more.

3. Increment all elements in x by one:

    ```
    >>> x + 1
    [1]  2 21  4  5  6  7
    ```

    Of course, we can obtain the same result by adding x to y, a vector of ones with the same length:

    ```
    >>> x + y
    [1]  2 21  4  5  6  7
    ```

Now that we've reviewed the basics of working with vectors, we can enter the realm of matrices.

Working with matrices

A matrix is a stack of vectors superimposed together. An m x n matrix can be considered n m-dimensional column vectors stacked horizontally, or m n-dimensional row vectors stacked vertically. For each of the DataFrames that we have worked with, each row is an observation, and each column represents a feature.

Let's look at how to create matrices by completing an exercise.

Exercise 7.3 – creating a matrix

We can use the matrix() function to create a matrix. This function accepts three arguments: the data to be passed into the matrix via data, the number of rows via nrow, and the number of columns via ncol:

1. Create a 3x2 matrix, X, filled with the number 2:

```
X = matrix(data=2, nrow=3, ncol=2)
>>> X
     [,1] [,2]
[1,]    2    2
[2,]    2    2
[3,]    2    2
```

Note the use of the broadcasting mechanism here. As we only pass in a single number, it is replicated across all the cells in the matrix. Also, observe the indexing pattern in the result. The position on the left indexes each row before the comma and each column by the position on the right after the comma. We can verify this observation by indexing the element located in the second row and the first column:

```
>>> X[1,2]
2
```

Let's verify the class of this object:

```
>>> class(X)
[1] "matrix" "array"
```

The result shows that X is both a matrix and (multi-dimensional) array.

We can also create a matrix based on a vector.

2. Create a 3x2 matrix using the x variable. Fill the matrix by rows:

```
>>> matrix(x, nrow=3, ncol=2, byrow=TRUE)
     [,1] [,2]
[1,]    1   20
[2,]    3    4
[3,]    5    6
```

Here, filling by rows means sequentially placing each element in the original vector, x, row-wise, beginning from the first row from left to right and jumping to the next row when hitting the end of the current row. We can also design the filling by columns.

3. Fill in the matrix by columns instead:

```
>>> matrix(x, nrow=3, ncol=2, byrow=FALSE)
     [,1] [,2]
```

```
[1,]     1     4
[2,]    20     5
[3,]     3     6
```

As with the vector, we can change a specific element in a matrix by first locating the element via indexing and assigning the new value, as shown here:

```
X[1,1]  =  10
>>> X
        [,1]  [,2]
[1,]     10     2
[2,]      2     2
[3,]      2     2
```

A common operation in linear algebra is to multiply a matrix by a vector. Such an operation gives rise to many interesting and important interpretations regarding matrix manipulation. We'll look at this operation in the following section.

Matrix vector multiplication

An overarching rule when multiplying a matrix by a vector is the equality of the inner dimension. In other words, the column dimension (meaning the number of columns) of the matrix, when multiplied by a vector on its right, has to be equal to the row dimension of the multiplying vector. For example, given an m x n matrix, it can only multiply a column vector of size n x 1, and such multiplication results in another vector of size m x 1.

Let's look at a concrete example here. Recall that **X** is a 3x2 matrix. Multiplying it by a 2x1 vector, *y*, should produce a 3x1 vector, *z*. The transition from the original vector, *y*, to the new vector, *z*, has an extra meaning here: *y* has changed space and now lives in a three-dimensional world instead of two! We can also say the old vector, *y*, is projected and stretched to the new vector, *z*, because of the projection matrix, *X*. Such a projection, or transformation, constitutes the majority of operations in modern neural networks. By projecting matrices (also called representations) across different layers, a neural network can *learn* different levels of abstraction.

Let's assume *y* is a vector of ones. The rule of matrix-vector multiplication says that each entry in the resulting vector, *z*, is the inner product of the corresponding vectors. The inner product between two vectors, also called the dot product, is the sum of the products of the corresponding elements in each vector. For example, the first entry in *z* is 12. Its positional index, which is [1,1], says that it requires the first row in *X*, which is [10, 2], and the first column in *y*, which is [1, 1], to enter the inner product operation. Multiplying the corresponding elements in each vector and summing the results gives us 10*1+2*1=12. Similarly, the second entry is calculated as 2*1+2*1=4. *Figure 7.2* summarizes this matrix-vector multiplication:

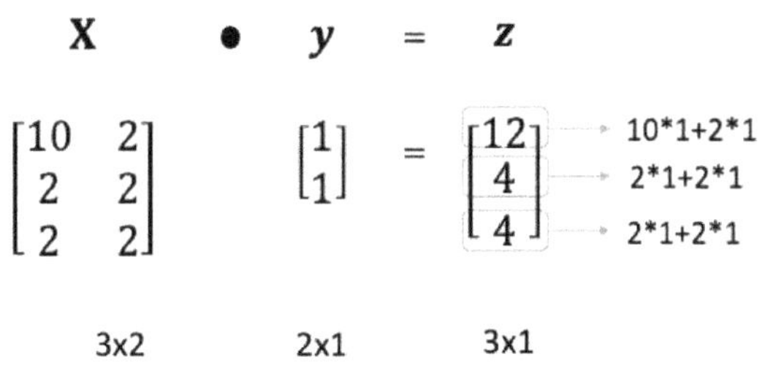

$$\mathbf{X} \quad \bullet \quad y \quad = \quad \mathbf{z}$$

$$\begin{bmatrix} 10 & 2 \\ 2 & 2 \\ 2 & 2 \end{bmatrix} \qquad \begin{bmatrix} 1 \\ 1 \end{bmatrix} = \begin{bmatrix} 12 \\ 4 \\ 4 \end{bmatrix} \begin{array}{l} \text{— } 10*1+2*1 \\ \text{— } 2*1+2*1 \\ \text{— } 2*1+2*1 \end{array}$$

$$\quad 3\text{x}2 \qquad\qquad 2\text{x}1 \qquad 3\text{x}1$$

Figure 7.2 – Illustrating the matrix-vector multiplication process

Now, let's look at how to perform matrix-vector multiplication in R.

Exercise 7.4 – applying matrix-vector multiplication

The matrix-vector multiplication operation comes via the %*% symbol in R. The %*% notation is very different from the * notation alone, which stands for element-wise multiplication. The following exercise illustrates such a difference:

1. Multiply the previous matrix, *X*, by a 2x1 vector of ones:

    ```
    >>> X %*% c(1,1)
          [,1]
    [1,]    12
    [2,]     4
    [3,]     4
    ```

    The result shows that the generated vector is 3x1. Note that the multiplication will not go through if the dimensions do not check:

    ```
    >>> X %*% c(1,1,1)
    Error in X %*% c(1, 1, 1) : non-conformable arguments
    ```

    The error says that the dimensions of the vector, c(1,1,1), do not match those in the X matrix.

    We can verify the calculation of each cell in the resulting vector.

2. Calculate the first entry in the resulting vector separately:

    ```
    >>> X[1,] %*% c(1,1)
         [,1]
    [1,]   12
    ```

The result shows that the first entry is the dot product of the first row, X[1,], and the first (and only) column, c(1,1). We can also re-express the dot product as explicit element-wise multiplication and summation.

3. Re-express the previous dot product as explicit element-wise multiplication and summation:

    ```
    >>> sum(X[1,] * c(1,1))
    [1] 12
    ```

    Note that both results return the same value but assume a different data structure: the dot product operation returns a matrix, while the re-expressed operation returns a vector.

4. Let's verify the same calculation using the second entry in the resulting vector:

    ```
    >>> X[2,] %*% c(1,1)
         [,1]
    [1,]    4
    >>> sum(X[2,] * c(1,1))
    [1] 4
    ```

    We can also examine the behavior of multiplying a matrix with a scalar.

5. Double each element in the matrix, X:

    ```
    >>> X * 2
         [,1] [,2]
    [1,]   20    4
    [2,]    4    4
    [3,]    4    4
    ```

    Note that when switching to the element-wise multiplication symbol, *, the dimension of the resulting matrix remains unchanged. In addition, the broadcasting mechanism is in play here, where a scalar of 2 is multiplied by each element in the matrix, X. The effect is as if we're multiplying X by another matrix of the same dimensions:

    ```
    >>> X * matrix(2, 3, 2)
         [,1] [,2]
    [1,]   20    4
    [2,]    4    4
    [3,]    4    4
    ```

Now, let's cover matrix-matrix multiplication. We will refer to it as matrix multiplication for short.

Matrix multiplication

Matrix multiplication is the most widely used form of operation in many domains. Take the neural network model, for example. *Figure 7.3* shows a simple network architecture called a fully connected neural network, where all neurons are fully connected. The input data, represented by the input layer with 10 rows and 3 columns, will enter a series of matrix multiplications (plus nonlinear transformations, which are ignored here) to learn useful representations (Z_1) and, therefore, accurate predictions (Z_2). There are two matrix multiplications here. The first matrix multiplication happens between the 10x3 input data, X, and the weight matrix W_1, which produces a 10x4 hidden representation, Z_1, in the hidden layer. The second matrix multiplication transforms Z_1 into the final 10x2 output, Z_2, using another 4x2 weight matrix, W_2. We can also interpret a series of matrix multiplications as transforming/projecting the input data, X, to the hidden representation, Z_1, and then to the output, Z_2:

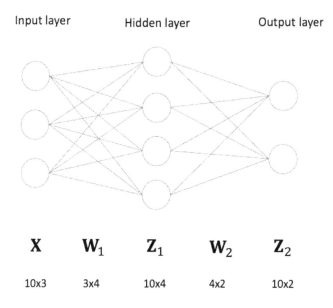

Figure 7.3 – Schematic of a simple two-layer fully connected neural network

As you can see from *Figure 7.3*, the input data, represented by the input layer with 10 rows and 3 columns, will enter a series of matrix multiplications (plus nonlinear transformations, which are ignored here) to learn useful representations (Z_1) and, therefore, accurate predictions (Z_2). The first matrix multiplication happens between the 10x3 input data, X, and the weight matrix, W_1, resulting in a 10x4 hidden representation, Z_1. The second matrix multiplication transforms Z_1 into the final 10x2 output, Z_2, using another 4x2 weight matrix, W_2.

There is more to these matrix multiplications. Let's assume the input data has many features. By applying these matrix multiplications with the automatically learned weight matrices, these features can be weighed correspondingly to produce accurate predictions in the output layer. Automatic feature learning, including those nodes in the hidden layer, is a distinguishing characteristic of modern neural networks compared to traditional manual feature learning.

Let's look at a single matrix multiplication in more detail. *Figure 7.4* illustrates the calculation process of multiplying a 2x2 matrix, $\begin{bmatrix} 1 & 3 \\ 2 & 4 \end{bmatrix}$, by another 2x2 matrix, $\begin{bmatrix} 2 & 2 \\ 2 & 2 \end{bmatrix}$. We can also view each column in the resulting matrix as a result of matrix-vector multiplication, as covered earlier; these columns are then concatenated to form the new output matrix:

$$\begin{bmatrix} 1 & 3 \\ 2 & 4 \end{bmatrix} \begin{bmatrix} 2 & 2 \\ 2 & 2 \end{bmatrix}$$

$$= \begin{bmatrix} 1*2+3*2 & 1*2+3*2 \\ 2*2+4*2 & 2*2+4*2 \end{bmatrix}$$

$$= \begin{bmatrix} 8 & 8 \\ 12 & 12 \end{bmatrix}$$

Figure 7.4 – Breaking down the matrix multiplication process

Let's go through an exercise to put the practical part of things into context.

Exercise 7.5 – working with matrix multiplication

Matrix multiplication still relies on the %*% sign in R. In this exercise, we will reproduce the example in *Figure 7.4*, with a bit of stretch in terms of the order of multiplication:

1. Reproduce the previous matrix multiplication example:

    ```
    >>> matrix(1:4, 2, 2) %*% matrix(2, 2, 2)
          [,1] [,2]
    [1,]    8    8
    [2,]   12   12
    ```

    Here, we use the matrix() function to create two matrices, the first generated by converting a vector and the second by duplicating a scalar value via broadcasting.

    The order of multiplication is of key importance in matrix algebra; shuffling the order will produce a different result in most cases. Let me switch these two matrices.

2. Switch the order of multiplication for these two matrices:

    ```
    >>> matrix(2, 2, 2) %*% matrix(1:4, 2, 2)
          [,1] [,2]
    [1,]    6   14
    [2,]    6   14
    ```

    Please verify the result of the matrix multiplication and appreciate the importance of order in multiplying two matrices: matrix multiplication is *not* commutative.

    Additionally, element-wise multiplication happens using the * operator.

3. Double every element in the previous matrix on the left:

    ```
    >>> matrix(1:4, 2, 2) * matrix(2, 2, 2)
          [,1] [,2]
    ```

```
[1,]    2    6
[2,]    4    8
```

Several special matrices are of particular interest. We'll review them in the next section.

The identity matrix

There are many special types of matrices. The first type of special matrix is the **identity matrix**, which assumes a value of 1 at the on-diagonal positions and 0 elsewhere. The biggest characteristic of an identity matrix is the identity preservation of the original multiplying matrix – that is, any matrix that can successfully multiply by an identity matrix obtains the same result as itself. This sounds like multiplying by one, and we are indeed doing so when the identity matrix is 1x1.

Exercise 7.6 – working with the identity matrix

Let's go through an exercise to see how identity matrices work:

1. Create a 2x2 identity matrix using the `diag()` function:

```
>>> diag(2)
     [,1] [,2]
[1,]    1    0
[2,]    0    1
```

   To create an identity matrix, we simply need to pass the number of ones in the diagonal of the matrix. A diagonal matrix can also be created from a vector:

```
>>> diag(c(1,2,3), nrow=3, ncol=3)
     [,1] [,2] [,3]
[1,]    1    0    0
[2,]    0    2    0
[3,]    0    0    3
```

   Here, the elements in the vector are used to fill in the diagonal positions of the diagonal matrix while leaving all the rest of the cells as 0.

   Let's multiply the 2x2 identity matrix by the previous running matrix.

2. Multiply this identity matrix by the previous 2x2 matrix numbered 1 to 4:

```
>>> matrix(1:4, 2, 2) %*% diag(2)
     [,1] [,2]
[1,]    1    3
[2,]    2    4
```

   We can see that there is no change in the resulting matrix. Let's verify this again:

```
>>> matrix(1:4, 2, 2)
     [,1] [,2]
```

```
[1,]    1    3
[2,]    2    4
```

Lastly, let's verify that the result is still the same after switching the order of multiplication.

3. Move the multiplying identity matrix to the left and perform matrix multiplication:

```
>>> diag(2) %*% matrix(1:4, 2, 2)
       [,1] [,2]
[1,]    1    3
[2,]    2    4
```

The result shows no change to the original matrix. If we use X to denote the original matrix and I to denote the identity matrix, we will get $XI = IX = X$.

Other operations can derive new matrices based on the original one, such as transposing or inversing a matrix.

Transposing a matrix

We already have some experience transposing a vector, which is a special case of transposing a matrix. Transposing a matrix means flipping the original matrix, X, to generate a new one, X^T, whose columns and rows are now switched. Transposing a transposed matrix, X^T, gives back the original matrix, $(X^T)^T = X$.

A special type of matrix is called the **symmetric matrix**, such that $X^T = X$. A symmetric matrix is also a **square matrix**; otherwise, the transposed dimensions will not check.

Let's see this in practice.

Exercise 7.7 – transposing a matrix

In this exercise, we will transpose a matrix using the t () function and transpose it again to see whether it matches the original matrix:

1. Create a square matrix, X, with elements one to four filled by column:

```
X = matrix(1:4, 2, 2)
>>> X
       [,1] [,2]
[1,]    1    3
[2,]    2    4
```

2. Transpose the matrix using t ():

```
>>> t(X)
       [,1] [,2]
[1,]    1    2
[2,]    3    4
```

Here, we can see that the transposed matrix has its rows and columns flipped from the original matrix. The diagonal elements stay the same, though.

3. Transpose the transposed matrix again and verify whether it is equal to the original matrix:

```
>>> t(t(X))
     [,1] [,2]
[1,]    1    3
[2,]    2    4
```

A visual inspection shows that it is indeed the same original matrix. However, we can also use the built-in `all.equal()` function to perform a systematic check:

```
>>> all.equal(X, t(t(X)))
[1] TRUE
```

The result shows that these two matrices are equal to each other, thus being the same matrix.

Let's review the last type of operation: inversing a matrix.

Inverting a matrix

Inversing a scalar number is intuitive. For any number, x, its inverse (or reciprocal) is $\frac{1}{x}$ if $x \neq 0$. This condition ensures that x is invertible. Similarly, not all matrices are invertible.

Formally, we say that a matrix, X, is invertible if it can, when multiplied by its inverse, X^{-1}, produce an identity matrix, I. In other words, $XX^{-1} = X^{-1}X = I$.

There are some interesting properties of an invertible matrix as well. For example, inverting an inverse matrix gives us the original matrix: $(X^{-1})^{-1} = X$. Also, since the product of an identity matrix with itself is an identity matrix, the inverse of an identity matrix is thus the identity matrix itself, a special case compared to others.

The inverse of a matrix can be obtained using the `solve()` function in R, which will give us an error if the matrix is not invertible. Let's practice this via the following exercise.

Exercise 7.8 – inverting a matrix

In this exercise, we will invert both an identity matrix and a standard square matrix using the `solve()` function:

1. Invert a two-dimensional identity matrix:

```
>>> solve(diag(2))
     [,1] [,2]
[1,]    1    0
[2,]    0    1
```

The result shows that the inverse of an identity matrix is itself.

2. Invert the running matrix, X, from the previous example:

```
Xinv = solve(X)
>>> Xinv
        [,1]  [,2]
[1,]     -2   1.5
[2,]      1  -0.5
```

We can verify the validity of this inverse matrix by multiplying it by the original matrix and checking whether it gives us an identity matrix.

3. Multiply by the original matrix to verify it based on the definition of matrix inversion:

```
>>> Xinv %*% X
        [,1]  [,2]
[1,]      1     0
[2,]      0     1
>>> X %*% Xinv
        [,1]  [,2]
[1,]      1     0
[2,]      0     1
```

The result shows that `Xinv` is indeed the inverse matrix of X.

In the following section, we will look at matrix-vector multiplication from the perspective of solving a system of linear equations, which is an essential task in many machine learning algorithms.

Solving a system of linear equations

The matrix-vector multiplication operation gives rise to a system of equations. In a typical machine learning algorithm, data comes in the form of a matrix, X, and the target outcome is a vector, y. When the model that's used is a straightforward linear model, we assume the input-output relationship as $Xw = y$, where w represents the vector of features/coefficients. An n x p matrix of input data multiplies a p x 1 vector, w, of features to produce, as expected, an n x 1 output vector, y. The essence of linear regress is thus to solve for the exact values in w such that the system of linear equations in $Xw = y$ are satisfied.

The equivalence between matrix-vector multiplication and the system of linear equations may take some time to become noticeable. Let's pause and look at this equivalence.

System of linear equations

We are already familiar with the process of calculating a matrix-vector multiplication operation. A 2x2 matrix, X, when multiplying a 2x1 vector, w, will result in a 2x1 vector, y. The first element in y, positioned as (1, 1), comes from the dot product (weighted sum) between the first row in X and the first column in y. Similarly, the second element in y, positioned as (2, 1), comes from the dot product between the second row in X and the first column in y. The positional index of each element in y determines the ingredients used in the respective dot product operation.

Such an interpretation, however, is a raw and low-level one. A more advanced interpretation lies in the column space of X and the associated linear combination of the column vectors in X weighted by the weights in w in the same matrix-vector multiplication. A concrete example will help here.

Suppose we have a simple matrix, $X = \begin{bmatrix} 1 & 3 \\ 2 & 4 \end{bmatrix}$, and a vector, $w^T = [1,1]$. The output vector is $y = \begin{bmatrix} 1*1 + 3*1 \\ 2*1 + 4*1 \end{bmatrix} = \begin{bmatrix} 4 \\ 6 \end{bmatrix}$ based on the usual method of calculation. The column view gives another process of calculation: $y = 1*\begin{bmatrix} 1 \\ 2 \end{bmatrix} + 1*\begin{bmatrix} 3 \\ 4 \end{bmatrix} = \begin{bmatrix} 1 \\ 2 \end{bmatrix} + \begin{bmatrix} 3 \\ 4 \end{bmatrix} = \begin{bmatrix} 4 \\ 6 \end{bmatrix}$. Here, the two columns, $\begin{bmatrix} 1 \\ 2 \end{bmatrix}$ and $\begin{bmatrix} 3 \\ 4 \end{bmatrix}$, are weighted by each element in w, respectively. In other words, these two columns are linearly combined to produce the output column vector, y. This forms the basis of the system of linear equations. *Figure 7.5* summarizes these two different perspectives of matrix-vector multiplication for this example:

$$\begin{bmatrix} 1 & 3 \\ 2 & 4 \end{bmatrix} \times \begin{bmatrix} 1 \\ 1 \end{bmatrix} = \begin{bmatrix} 1*1 + 3*1 \\ 2*1 + 4*1 \end{bmatrix} = \begin{bmatrix} 4 \\ 6 \end{bmatrix}$$

The usual way of calculating matrix-vector multiplication

$$\begin{bmatrix} 1 & 3 \\ 2 & 4 \end{bmatrix} \times \begin{bmatrix} 1 \\ 1 \end{bmatrix} = 1*\begin{bmatrix} 1 \\ 2 \end{bmatrix} + 1*\begin{bmatrix} 3 \\ 4 \end{bmatrix} = \begin{bmatrix} 1 \\ 2 \end{bmatrix} + \begin{bmatrix} 3 \\ 4 \end{bmatrix} = \begin{bmatrix} 4 \\ 6 \end{bmatrix}$$

The more advanced column view of matrix-vector multiplication

Figure 7.5 – Illustrating the two different ways of calculating matrix-vector multiplication

Now, assume the weight vector, w, is unknown, with two unknown elements, w_1 and w_2 – in other words, $w^T = [w_1, w_2]$. Also, assume that the input matrix, X, and the output vector, y, are known. This is a common situation where we are given the input-output data pairs and asked to estimate the coefficients of the linear model such that $Xw = y$. Now comes the system of linear equations. Plugging in the previous weighted combination of column vectors gives us

$w_1*\begin{bmatrix} 1 \\ 2 \end{bmatrix} + w_2*\begin{bmatrix} 3 \\ 4 \end{bmatrix} = \begin{bmatrix} w_1 \\ 2w_1 \end{bmatrix} + \begin{bmatrix} 3w_2 \\ 4w_2 \end{bmatrix} = \begin{bmatrix} w_1 + 3w_2 \\ 2w_1 + 4w_2 \end{bmatrix} = \begin{bmatrix} 4 \\ 6 \end{bmatrix}$. We have a system of equations, as follows:

$$\begin{cases} w_1 + 3w_2 = 4 \\ 2w_1 + 4w_2 = 6 \end{cases}$$

Solving this system of linear equations gives us $w_1 = w_2 = 1$. The equivalence of matrix-vector multiplication and the system of linear equations is manifested here.

Let's go through an exercise to appreciate the equivalence in terms of code.

Exercise 7.9 – understanding the system of linear equations

In this exercise, we will first use the usual matrix-vector multiplication procedure to calculate the result, followed by providing a detailed and manual column-view implementation of the same operation. We will see that both give the same result:

1. Create a 2x2 matrix, X, that consists of four numbers, 1 to 4, and is filled by column:

    ```
    X = matrix(c(1:4), nrow=2, ncol=2)
    >>> X
          [,1] [,2]
    [1,]    1    3
    [2,]    2    4
    ```

2. Create a vector, w, of length two, both filled with a value of 1:

    ```
    w = c(1,1)
    >>> w
    [1] 1 1
    ```

    Note that a vector is a column vector by default. When the 2x2 matrix, X, multiplies the 2x1 vector, w, as shown in the following code, the vector, w, is expressed as a column vector before entering the multiplication operation.

3. Multiply X and w and save the result in y:

    ```
    y = X %*% w
    >>> y
          [,1]
    [1,]    4
    [2,]    6
    ```

    Now, the y vector is displayed as an explicit column vector. Lastly, let's verify the calculation using the column view.

4. Perform the same matrix-vector multiplication using the column view, meaning a weighted sum of column vectors:

    ```
    >>> w[1] * X[,1] + w[2] * X[,2]
    [1] 4 6
    ```

    Here, we use X[,1] and X[,2] to access the first and second columns, respectively. The result agrees with the previous one, although following a different format as it is displayed as a row vector now.

Matrix-vector multiplication gives rise to a system of linear equations. However, this system of linear equations may or may not have a solution. Even if there is a solution, it may not be unique. In the following section, we will examine the potential solution to matrix-vector equations.

The solution to matrix-vector equations

Let's continue with the example where a 2x2 matrix, X, multiplies a 2x1 weight vector, w, to produce a 2x1 output vector, y. We will discuss three cases here: without a solution, with a unique solution, and with infinitely many solutions.

We will start with the second case. Recall our previous values in X, w, and y. In $\begin{bmatrix} 1 & 3 \\ 2 & 4 \end{bmatrix} \text{x} \begin{bmatrix} 1 \\ 1 \end{bmatrix} = \begin{bmatrix} 4 \\ 6 \end{bmatrix}$, we multiplied the former two (X and w) to produce the third one (y) and showed how to derive the values of w given X and y by solving a system of linear equations in $\begin{bmatrix} 1 & 3 \\ 2 & 4 \end{bmatrix} \text{x} \begin{bmatrix} w_1 \\ w_2 \end{bmatrix} = \begin{bmatrix} 4 \\ 6 \end{bmatrix}$. In other words, the solution is unique, and we can solve it analytically by writing out the explicit system of linear equations and solving the unknowns. It will be more involved when we have more unknowns to solve, but the process stays the same. In a typical linear regression setting, when the number of unknowns (the length of the weight vector, w) is the same as the number of equations (the number of rows in X), and provided that the input matrix, X, is a nice one (for example, if its columns are uncorrelated), we can find a unique solution to the system of linear equations.

Now, let's tweak the input matrix, X, so that no solution can be found. One way to do this is to set the second row vector of X to zero, giving us $X = \begin{bmatrix} 1 & 3 \\ 0 & 0 \end{bmatrix}$. Thus, we have the following:

$$\begin{bmatrix} 1 & 3 \\ 0 & 0 \end{bmatrix} \text{x} \begin{bmatrix} w_1 \\ w_2 \end{bmatrix} = \begin{bmatrix} 4 \\ 6 \end{bmatrix}$$

$$\begin{cases} w_1 + 3w_2 = 4 \\ \quad\quad 0 = 6 \end{cases}$$

It is obvious that the second equation fails, and therefore this system of linear equations cannot have a solution. This is called an *inconsistent* system since 0 cannot be 6. Note that in the first equation, there are infinitely many pairs of (w_1, w_2) that satisfy the equation. This also leads to our third case.

Now, suppose we change the output vector slightly and make the second entry of y zero, giving us $y = \begin{bmatrix} 4 \\ 0 \end{bmatrix}$. Thus, we have the following:

$$\begin{bmatrix} 1 & 3 \\ 0 & 0 \end{bmatrix} \text{x} \begin{bmatrix} w_1 \\ w_2 \end{bmatrix} = \begin{bmatrix} 4 \\ 0 \end{bmatrix}$$

$$\begin{cases} w_1 + 3w_2 = 4 \\ \quad\quad 0 = 0 \end{cases}$$

Now, this system of linear equations has at least one solution and is *consistent*, but with infinitely many solutions. There is not much we can take away from solving this system of equations; since there are infinitely many solutions, we are unable to evaluate which one is the best to report. Real-world optimization problems are typically concerned with finding *the* single most optimal solution or the best sub-optimal solution that is closest to the empirically unattainable optimal solution.

Figure 7.6 summarizes these three cases:

$$\begin{bmatrix} 1 & 3 \\ 2 & 4 \end{bmatrix} \times \begin{bmatrix} w_1 \\ w_2 \end{bmatrix} = \begin{bmatrix} 4 \\ 6 \end{bmatrix} \implies \begin{cases} w_1 + 3w_2 = 4 \\ 2w_1 + 4w_2 = 6 \end{cases}$$ Consistent and has one unique solution

$$\begin{bmatrix} 1 & 3 \\ 0 & 0 \end{bmatrix} \times \begin{bmatrix} w_1 \\ w_2 \end{bmatrix} = \begin{bmatrix} 4 \\ 6 \end{bmatrix} \implies \begin{cases} w_1 + 3w_2 = 4 \\ 0 = 6 \end{cases}$$ Inconsistent and has no solution

$$\begin{bmatrix} 1 & 3 \\ 0 & 0 \end{bmatrix} \times \begin{bmatrix} w_1 \\ w_2 \end{bmatrix} = \begin{bmatrix} 4 \\ 0 \end{bmatrix} \implies \begin{cases} w_1 + 3w_2 = 4 \\ 0 = 0 \end{cases}$$ Consistent and has infinitely many solutions

Figure 7.6 – Three different cases when solving a system of linear equations

It turns out that we can also view the process of solving a system of linear equations from a geometric perspective. Let's dive in.

Geometric interpretation of solving a system of linear equations

Once again, let's start with the case where we have a unique solution for the system of linear equations with two unknowns (w_1 and w_2) and two equations:

$$\begin{bmatrix} 1 & 3 \\ 2 & 4 \end{bmatrix} \times \begin{bmatrix} w_1 \\ w_2 \end{bmatrix} = \begin{bmatrix} 4 \\ 6 \end{bmatrix}$$

$$\begin{cases} w_1 + 3w_2 = 4 \\ 2w_1 + 4w_2 = 6 \end{cases}$$

If we introduce a two-dimensional coordinate system and put w_1 and w_2 on the x axis and y axis, respectively, we will see two lines on the coordinate system, each representing a linear equation in the system. Now, we can re-express the previous system of equations as follows:

$$\begin{cases} w_2 = -\dfrac{w_1}{3} + \dfrac{4}{3} \\ w_2 = -\dfrac{w_1}{2} + \dfrac{3}{2} \end{cases}$$

Therefore, solving the system of linear equations corresponds with finding the intersection point of these two lines, since it is only at the intersection point that both equations are satisfied. The following codes help us plot the two lines:

```
plot(x=1, y=1, xlab="w1", ylab="w2", xlim=c(0, 5), ylim=c(0, 5))
abline(a=4/3, b=-1/3)
abline(a=3/2, b=-1/2)
```

Here, we use the `plot()` function to draw a two-dimensional coordinate system with a circle representing the point (1,1). We also add two lines using the `abline()` function, which accepts the intercept and slope of the line as input arguments.

Running this code generates *Figure 7.7*. We can see that these two lines happen to meet at point (1,1), which is not a coincidence!

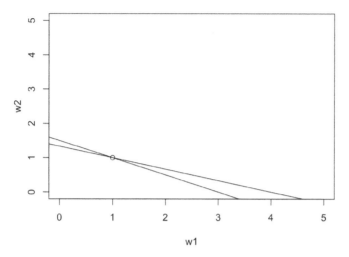

Figure 7.7 – Plotting the system of linear equations as intersecting lines

Now, we will look at the case with no solution for the system of equations. The previous example shows an inconsistent system of equations, where the second equation simply does not stand. With the geometric interpretation in the coordinate system, we can make another line parallel to the first one, such as by shifting the line up by one unit, to produce a case with no solution. Specifically, we can have the following system of equations:

$$\begin{cases} w_1 + 3w_2 = 4 \\ w_1 + 3(w_2 - 1) = 4 \end{cases}$$

Here, we minus 1 from w_2 to shift the line upward by one unit. Again, we can express these equations as a function of w_1:

$$\begin{cases} w_2 = -\dfrac{w_1}{3} + \dfrac{4}{3} \\ w_2 = -\dfrac{w_1}{3} + \dfrac{7}{3} \end{cases}$$

Let's plot these two lines on the coordinate system via the following code. The only change is the bigger intercept in the second line compared to the first one:

```
plot(x=1, y=1, xlab="w1", ylab="w2", xlim=c(0, 5), ylim=c(0, 5))
abline(a=4/3, b=-1/3)
abline(a=7/3, b=-1/3)
```

Running this code generates *Figure 7.8*. Since these two lines are parallel, there will be no intersection between the two lines, so we'll end up with no solution to the system of linear equations:

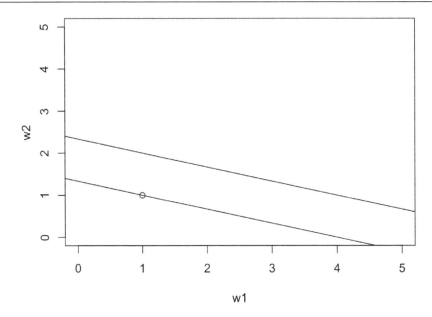

Figure 7.8 – Plotting two parallel lines

Now, let's move on to the case with infinitely many solutions. As you may have guessed, we just need to make the second equation the same as the first one, thus creating two overlapping lines in the coordinate system. Any point on these two overlapping lines is a valid solution, and there is an infinite number of such points on the line(s).

Working with a system of linear equations may not guarantee a solution. Even if there are as many unknown variables as there are rows in the system, this isn't guaranteed to give us a solution. In the following section, we'll look at under what conditions we are guaranteed to get a unique solution.

Obtaining a unique solution to a system of linear equations

Recall our analytical framework: $Xw = y$, where we are given input-output pairs (X, y) and would like to solve the unknown vector, w. If things were simple and all these variables were scalars, we would simply divide both sides by X and obtain the solution, $w = X^{-1}y$, provided that X is invertible. Note that we would multiply both sides by X^{-1} on the left to perform the division. This brings us to the first condition: the matrix, X, needs to have a corresponding inverse matrix; that is, it is invertible.

1. *Figure 7.9* illustrates the correspondence between simple scalar calculations in regular algebra and matrix manipulations in matrix algebra. Note that by definition of matrix inverse, we have $X^{-1}X = I$. We used this result in the derivation. The identity matrix is special; any matrix multiplying an identity matrix will stay unchanged. Thus, we have $Iw = w$. Again, we assume the matrix, X, is invertible:

Scalar manipulation in regular algebra	Matrix manipulation in matrix algebra

$$3x = 2$$
$$\frac{3x}{3} = \frac{2}{3}$$
$$x = \frac{2}{3}$$

$$Xw = y$$
$$X^{-1}Xw = X^{-1}y$$
$$Iw = X^{-1}y$$
$$w = X^{-1}y$$

Figure 7.9 – Comparing scalar and matrix manipulations when solving the system of linear equations

The second condition is that the *determinant* of the matrix, X, cannot be zero. The determinant is a summary measure of the size of the matrix; we will discuss this in more detail in the next chapter. The third condition is what we alluded to earlier: the rows and columns of the matrix, X, can form a *basis* for the corresponding row and column space. There should be no correlated rows or columns in the matrix. A basis of a vector space is a set of vectors that can uniquely generate all the other vectors in the same vector space via linear combinations.

In addition, we would need a somewhat trivial condition: when $Xw = 0$, we have $w = 0$.

Computing the inverse and determinant of a matrix is straightforward in R. In the following code, we are calculating the inverse of the matrix, X, using the `solve()` function and its determinant using the `det()` function:

```
>>> solve(X)
       [,1] [,2]
[1,]    -2  1.5
[2,]     1 -0.5
>>> det(X)
[1] -2
```

The next chapter will touch on the matrix determinant, norm, trace, and other special properties in more detail.

Now that we can calculate the matrix inverse, let's learn how to obtain the solution to the system of linear equations using the inverse of the input matrix, X, and output, vector y.

Exercise 7.10 – obtaining the unique solution to a system of linear equations

In this exercise, we will make use of the matrix inverse to obtain the solution to a system of linear equations:

1. Construct the input matrix, X, and output vector, y, based on the previous example:

```
X = matrix(c(1:4), nrow=2, ncol=2)
>>> X
```

```
          [,1] [,2]
    [1,]    1    3
    [2,]    2    4
    y = c(4, 6)
    >>> y
    [1] 4 6
```

Note that the data we chose here is based on the previous running example, where the solution to the unknown variables is $w = [1,1]$.

2. Calculate the solution to the system of linear equations:

```
    w_hat = solve(X) %*% y
    >>> w_hat
          [,1]
    [1,]    1
    [2,]    1
```

As we can see, the solution, which can be called the coefficient vector, matches the true result exactly. We can also multiply the input matrix by this solution to obtain an estimated output and check whether it is the same as the given output vector.

3. Calculate the output via matrix-vector multiplication between the input matrix and the estimated coefficient vector:

```
    >>> X %*% w_hat
          [,1]
    [1,]    4
    [2,]    6
```

The result matches the given output vector.

Solving a system of linear equations with a square input matrix is a simple setting. When working with real data, the input matrix will likely be non-square, with more rows or columns. When there are more rows than columns, we have a limited number of unknown variables that need to satisfy more equations. This is an **overdetermined system**. On the other hand, when there are more unknown variables to satisfy fewer equations, we have an **underdetermined system**. We will discuss these two situations in the following section.

Overdetermined and underdetermined systems of linear equations

The example we covered earlier, which consisted of two equations and two unknown variables to solve, gives a solution that passes through both lines on the coordinate system. This is called **interpolation**, where the solution perfectly satisfies both equations without any errors. In the language of machine learning, interpolation means the trained model can score 100% on the training dataset. This is not necessarily a good thing since the model will likely run into the risk of overfitting – that is, it will do pretty well on the training set but not so well on the test set.

Complex models such as neural networks and other nonlinear models tend to have very low or even zero training errors – that is, they are likely to interpolate the training data. The zero training cost occurs when the model becomes sufficiently complex to interpolate the training data. Perfect prediction in the training data is likely to happen when the number of coefficients, p, used by the model is equal to or larger than the number of observations, n, in the training input. In the case of linear regression, when $p = n$, we solve a system of linear equations where the number of free variables, whose optimal values we are solving for, is equal to the number of linear equations in the system.

A simple example is a two-dimensional training set with two observation points on a coordinate system, as we saw earlier. When only one parameter is available, the model is reflected as a straight horizontal line in the coordinate system. This univariate model is too simple to fit these two points unless they happen to live along the fitted line. The resulting system of equations is said to be overdetermined and under-parameterized since we have more linear equations than the unknown variables in the system.

When we have two parameters to fit the model, the problem can be solved with an exact solution: two linear equations and two unknowns, a straightforward exercise. We can fit a line anywhere in the coordinate system by adjusting its intercept and slope. So, the problem becomes fitting a line that passes through the two points. We can train a model that passes through the two points and produces zero training error, thus interpolating between these two points.

When there are more than two unknown variables – say we have three weights in the model – the problem becomes underdetermined and over-parameterized. The solution, when solvable, will be infinite. Having more than two parameters corresponds to fitting two points with a curved line. Any curve that passes through the two points produces zero training error, yet the curve can be arbitrarily wiggly and complex, depending on the number of parameters used. *Figure 7.10* summarizes these three scenarios:

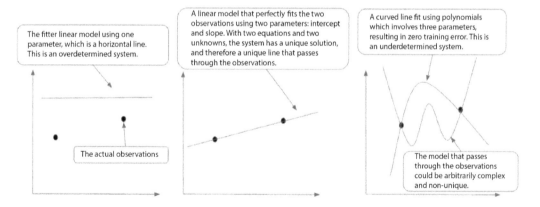

Figure 7.10 – Three scenarios of model complexity

Figure 7.10 shows fitting two two-dimensional observations using different model complexities. The left plot contains a 0th degree polynomial model that contains only one parameter, thus being a horizontal line. The middle plot has an equal number of parameters to the number of observations, making the solution unique and exact. The right plot contains models with more than two parameters, where the solution is non-unique due to being an underdetermined system.

Perfect interpolation occurs when $p = n$ – that is, the number of features is equal to the number of observations. The perfect interpolation continues with $p > n$ when the system of equations becomes underdetermined and over-parameterized, leading to infinitely many solutions that correspond to models of an arbitrary shape. All these models pass through the observed data points and thus give zero training error.

Summary

In this chapter, we introduced the basics of linear algebra, including working with vectors and matrices and performing matrix-vector multiplication. We highlighted a few special matrices, such as the identity matrix, and common operations, such as transposing and inverting a matrix.

Next, we used matrix-vector multiplication to solve a system of linear equations under different settings. We introduced the geometric interpretation that corresponds to the system of linear equations, along with how to obtain the solution using matrix inverse and multiplication operations.

Lastly, we touched upon common settings of the input matrix in the machine learning context, covering both underdetermined and overdetermined systems. Developing such an understanding will be crucial when we delve into statistical modeling and machine learning in the third part of this book.

In the next chapter, we will discuss slightly more advanced concepts in matrix algebra and implementations in R.

8

Intermediate Linear Algebra in R

The previous chapter covered the basics of linear algebra and its calculations in R. This chapter will go a step further by extending to intermediate linear algebra and cover topics such as the determinant, rank, and trace of a matrix, eigenvalues and eigenvectors, and **principal component analysis (PCA)**. Besides providing an intuitive understanding of these abstract yet important mathematical concepts, we'll cover the practical implementations of calculating these quantities in R.

By the end of this chapter, you will have grasped important matrix properties, such as determinant and rank, and gained hands-on experience in calculating these quantities.

In this chapter, we will cover the following topics:

- Introducing the matrix determinant
- Introducing the matrix trace
- Understanding the matrix norm
- Getting to know eigenvalues and eigenvectors
- Introducing principal component analysis

Technical requirements

To run the code in this chapter, you will need to have the following:

- The latest version of the `Matrix` package, which is 1.5.1 at the time of writing
- The latest version of the `factoextra` package, which is 1.0.7 at the time of writing

All the code and data for this chapter is available at `https://github.com/PacktPublishing/The-Statistics-and-Machine-Learning-with-R-Workshop/blob/main/Chapter_8/working.R`.

Introducing the matrix determinant

The **determinant** of a matrix is a special scalar value that can be calculated from a matrix. Here, the matrix needs to be square, meaning it has an equal number of rows and columns. For a 2x2 square matrix, the determinant is simply calculated as the difference between the product of the diagonal elements and the off-diagonal elements.

Mathematically, suppose our 2x2 matrix is $A = \begin{bmatrix} a & b \\ c & d \end{bmatrix}$. Its determinant, $|A|$, is thus calculated as follows:

$$\det(A) = |A| = ad - bc$$

Please do not confuse these vertical lines with the absolute operation sign. They represent the determinant in the context of a matrix, and the determinant of a matrix can be negative as well.

Let's say our 2x2 matrix is $A = \begin{bmatrix} 2 & 6 \\ 1 & 8 \end{bmatrix}$. We can find its determinant like so:

$$|A| = 2*8 - 6*1 = 10$$

Calculating the determinant of a matrix is the easy part, but understanding its use is of equal importance. Before we cover its properties, first, we'll review the calculation in R to get a straightforward understanding of the scalar output value.

In the following code snippet, we are creating a matrix, A, from a vector with proper configurations (two rows, filling by row). As usual, we verify the content in the matrix by printing it out to the console. We then call the det () function to calculate its determinant:

```
A <- matrix(c(2, 6, 1, 8), nrow=2, byrow=TRUE)
>>> A
     [,1] [,2]
[1,]    2    6
[2,]    1    8
>>> det(A)
[1] 10
```

There is also a corresponding formula for calculating the determinant of a 3x3 matrix or even higher dimension. We will not entertain these cases here as understanding the properties of the determinant is more important at this stage.

Interpreting the determinant

Recall that any matrix can be thought of as a transformation or projection that changes the input from one space to another. There are two things to note for such a change: *quantity* and *direction*. Quantity measures the percentage change in the magnitude of the original size of the matrix, while the direction indicates the sign of the transformation, taking either a positive or a negative value. Here, the matrix size can be considered as the area of a 2x2 matrix or the volume of a 3x3 matrix.

The columns in the matrix represent the collection of linear transformation that either stretches or squishes the space of the original input, thus changing the size of the matrix. So, the determinant measures how much the collection of linear transformations stretches or squishes the input. It gives a factor by which the area or volume of a region increases or decreases. Also, since directionality matters, the change may flip the input, as indicated by a negative determinant.

Let's look at an example. Suppose we have a 2x2 input matrix, $\begin{bmatrix} 1 & 0 \\ 0 & 1 \end{bmatrix}$, and we would like to transform it via another 2x2 matrix, $\begin{bmatrix} 3 & 0 \\ 0 & 2 \end{bmatrix}$. A direct multiplication gives an output of $\begin{bmatrix} 3 & 0 \\ 0 & 2 \end{bmatrix}$, which can be verified by carrying out the matrix multiplication rule:

$$\begin{bmatrix} 1 & 0 \\ 0 & 1 \end{bmatrix}\begin{bmatrix} 3 & 0 \\ 0 & 2 \end{bmatrix} = \begin{bmatrix} 3 & 0 \\ 0 & 2 \end{bmatrix}$$

There is no change to the output since the input matrix is essentially an identity matrix, and we know that any matrix multiplied by an identity matrix remains unchanged. No surprise here. However, when viewing $\begin{bmatrix} 1 & 0 \\ 0 & 1 \end{bmatrix}$ as the input matrix, with $\begin{bmatrix} 3 & 0 \\ 0 & 2 \end{bmatrix}$ on the left as the transformation matrix and $\begin{bmatrix} 3 & 0 \\ 0 & 2 \end{bmatrix}$ on the right as the output matrix, we can see that the transformation matrix increases the area of the input matrix by a factor of six.

To see this, imagine the input matrix, $\begin{bmatrix} 1 & 0 \\ 0 & 1 \end{bmatrix}$, on a two-dimensional coordinate system. The area of the input matrix is $1 * 1 = 1$, while the area of the output matrix is $3 * 2 = 6$, which happens to be the determinant of the transformation matrix.

This is not a coincidence. The net effect of the transformation matrix is thus to magnify the area of the input matrix by 6, maintaining the same direction. And it is not difficult to obtain the same increase in area in a negative direction when changing the transformation matrix to $\begin{bmatrix} -3 & 0 \\ 0 & 2 \end{bmatrix}$ or $\begin{bmatrix} 3 & 0 \\ 0 & -2 \end{bmatrix}$:

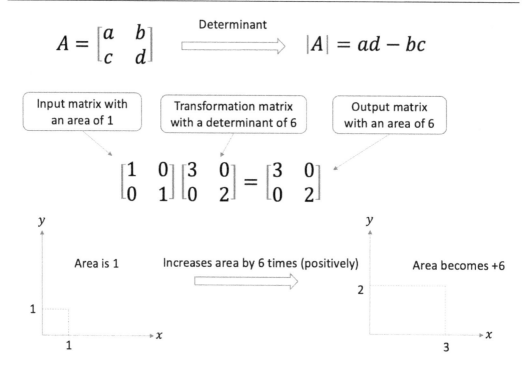

$$A = \begin{bmatrix} a & b \\ c & d \end{bmatrix} \quad \xrightarrow{\text{Determinant}} \quad |A| = ad - bc$$

Input matrix with an area of 1

Transformation matrix with a determinant of 6

Output matrix with an area of 6

$$\begin{bmatrix} 1 & 0 \\ 0 & 1 \end{bmatrix} \begin{bmatrix} 3 & 0 \\ 0 & 2 \end{bmatrix} = \begin{bmatrix} 3 & 0 \\ 0 & 2 \end{bmatrix}$$

Area is 1

Increases area by 6 times (positively)

Area becomes +6

Figure 8.1 – Illustrating the effect of the matrix determinant in determining the change in the area of the input matrix

Figure 8.1 summarizes this important property.

Connection to the matrix rank

The rank of a matrix, A, is the maximal number of linearly independent columns in the matrix. This number has a connection to the determinant of a matrix. Specifically, the rank is the number of rows (or columns) of the largest square submatrix of A such that its determinant is nonzero.

Let's look at an example. Suppose A is a 2x3 matrix, $\begin{bmatrix} 1 & 2 & 3 \\ 3 & 2 & 4 \end{bmatrix}$. First, we find the largest square submatrix, which is $\begin{bmatrix} 1 & 2 \\ 3 & 2 \end{bmatrix}$ or $\begin{bmatrix} 2 & 3 \\ 2 & 4 \end{bmatrix}$. Both matrices have a nonzero determinant. Thus, the rank of A is. This means we can use this technique to find the rank of a matrix. *Figure 8.2* summarizes this approach:

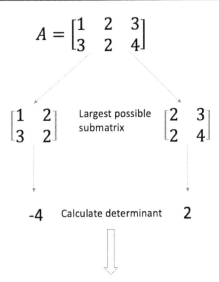

$$A = \begin{bmatrix} 1 & 2 & 3 \\ 3 & 2 & 4 \end{bmatrix}$$

$\begin{bmatrix} 1 & 2 \\ 3 & 2 \end{bmatrix}$ Largest possible submatrix $\begin{bmatrix} 2 & 3 \\ 2 & 4 \end{bmatrix}$

-4   Calculate determinant   **2**

Since the determinant is nonzero, the rank of A is two

Figure 8.2 – Deriving the rank of a matrix using the determinant

Let's look at how to obtain the rank of a matrix computationally:

1. For this, we need to load the `Matrix` package in R and call the `rankMatrix()` function.

2. As shown in the following code snippet, first, we create the 3x2 matrix, A, and print it out. When designing this matrix, we simply fill in a vector that consists of the row-wise concatenation of A:

```
library(Matrix)
A = matrix(c(1,2,3,3,2,4), nrow=2, byrow=TRUE)
>>> A
     [,1] [,2] [,3]
[1,]    1    2    3
[2,]    3    2    4
```

3. Next, we call the `rankMatrix()` function to obtain its rank:

```
>>> rankMatrix(A)
[1] 2
attr(,"method")
[1] "tolNorm2"
attr(,"useGrad")
[1] FALSE
attr(,"tol")
[1] 6.661338e-16
```

4. Multiple attributes are returned. We can access the first attribute as follows:

```
>>> rankMatrix(A)[1]
[1] 2
```

In the next section, we will look at another important property of a matrix: the trace.

Introducing the matrix trace

The **trace** is a quantity that only applies to square matrices, such as the covariance matrix often encountered in ML. It is denoted as tr(A) for a square matrix, A, and is calculated as the sum of the diagonal elements in a square matrix. Let's take a look:

1. In the following code snippet, we are creating a 3x3 matrix, A, and using the `diag()` function to extract the diagonal elements and sum them up to obtain the trace of the matrix. Note that we first create a DataFrame consisting of three columns, each having three elements, and then convert it into a matrix format to store in A:

```
A = as.matrix(data.frame("c1"=c(1,2,3),"c2"=c(2,5,2),"c3
"=c(-1,8,3)))
>>> A
      c1 c2 c3
[1,]   1  2 -1
[2,]   2  5  8
[3,]   3  2  3
>>> diag(A)
[1] 1 5 3
>>> sum(diag(A))
[1] 9
```

2. Since there is no built-in function that can calculate the trace in one shot, we can build a customized one to perform this task. As shown in the following code snippet, the customized function named `trace()` essentially loops through all the diagonal elements of the input square matrix and adds them up as the return value:

```
trace <- function(A) {
  # get matrix dimension
  n = dim(A)[1]
  # track trace value
  tr = 0
  # add diagonal elements to trace
  for(k in 1:n) {
    l = A[k,k]
    tr = tr + l
  }
```

```
        return(tr[[1]])
    }
```

3.   Testing out this function gives us the same trace as before:

```
>>> trace(A)
9
```

There are some interesting properties regarding the trace of the matrix, as we'll see in the next section.

Special properties of the matrix trace

To illustrate these properties, we'll first create another matrix, B:

```
B = as.matrix(data.frame("c1"=c(1,0,1),"c2"=c(1,1,2),"c3"=c(-1,2,0)))
>>> B
      c1 c2 c3
[1,]   1  1 -1
[2,]   0  1  2
[3,]   1  2  0
```

We will introduce five properties that are commonly used in statistical modeling. All of these properties will be verified in our example with matrices A and B:

- **Property 1**: The trace of the sum of two square matrices is the sum of the traces of the two matrices:

$$tr(A + B) = tr(A) + tr(B)$$

```
>>> trace(A + B) == trace(A) + trace(B)
[1] TRUE
```

Here, we use the equality sign to check whether the left-hand side is equal to the right-hand side. A return of TRUE means that the property has been verified.

- **Property 2**: The trace of a matrix is equal to the trace of the matrix's transpose:

$$tr(A) = tr(A^T)$$

```
>>> trace(A) == trace(t(A))
[1] TRUE
```

Note that we used the t () function to obtain the transpose of a matrix:

```
>>> t(A)
     [,1] [,2] [,3]
c1     1    2    3
c2     2    5    2
c3    -1    8    3
```

- **Property 3**: For a matrix multiplied by a scalar value, its trace is the same as the original trace multiplied by the same scalar:

$$\text{tr}(\alpha A) = \alpha \text{tr}(A)$$

```
>>> trace(2*A)  ==  2*trace(A)
[1]  TRUE
```

Here, we chose a scalar coefficient of 2. Feel free to change this value and verify whether the property still holds.

- **Property 4**: The trace is cyclical:

$$\text{tr}(AB) = \text{tr}(BA)$$

```
>>> trace(A %*% B)  ==  trace(B %*% A)
[1]  TRUE
```

This property says that when multiplying *A* by *B*, the trace of the resulting matrix is the same as the one from multiplying *B* by *A*.

Note the use of the %*% notation when performing matrix multiplication.

- **Property 5**: The trace is invariant:

$$\text{tr}(A) = \text{tr}(BA\,B^{-1})$$

```
>>> trace(A)  ==  trace(crossprod(crossprod(B,A),solve(B)))
[1]  TRUE
```

This property says that if we multiply *B* by *A*, and then multiply the inverse, B^{-1}, the trace of the resulting matrix is the same as the trace of matrix *A*. *Figure 8.3* summarizes these five properties:

Property 1: The trace of the sum of two square matrices is the sum of the traces of the two matrices:
$$\text{tr}(A + B) = \text{tr}(A) + \text{tr}(B)$$

Property 2: The trace of a matrix is equal to the trace of the matrix's transpose:
$$\text{tr}(A) = \text{tr}(A^T)$$

Property 3: For a matrix multiplied by a scalar value, its trace is the same as the original trace multiplied by the same scalar:
$$\text{tr}(\alpha A) = \alpha \text{tr}(A)$$

Property 4: The trace is cyclical:
$$\text{tr}(AB) = \text{tr}(BA)$$

Property 5: The trace is invariant:
$$\text{tr}(A) = \text{tr}(BAB^{-1})$$

Figure 8.3 – Five properties of matrix trace

In the next section, we will cover another important summary measure of a matrix: the matrix norm.

Understanding the matrix norm

The **norm** of a matrix is a scalar value that measures the magnitude of the matrix. Therefore, the norm is a way to measure the size or length of a vector or a matrix. For example, the weights of a deep neural network are stored in matrices, and we would typically constrain the norm of the weights to be small to prevent overfitting. This allows us to quantify the magnitude, which is useful when comparing different vectors or matrices, which often consist of multiple elements. As it generalizes from the vector norm, we will first go through the basics of the vector norm.

Understanding the vector norm

Suppose we have a vector, $a = [1,0,-1]$, and another vector, $b = [1,2,0]$. To assess the similarity between these two vectors, we can argue that they are the same in the first element only and different for the remaining two elements. To compare these two vectors holistically, we need a single metric – one that summarizes the whole vector. The norm is one way to go forward.

There are different norms for an arbitrary vector of length n. All forms come from the following generalized form of L^r-norm:

$$\|x\|_p = \left(\sum_{i=1}^{n} |x_i|^p \right)^{1/p}$$

Here, the double vertical bars denote the norm. This is called the L^p-norm because p is used as a placeholder to represent the specific type of norm. Common values of p include 1, 2, and ∞, although theoretically, it could take on any positive integer value. For example, to calculate the L^1-norm of the vector, we can simply plug $p = 1$ into the formula.

We'll go through these common norms in the following sections.

Calculating the L^1-norm of a vector

Substituting $p = 1$ in the previous equation gives us the L^1-norm:

$$\|x\|_1 = \sum_{i=1}^{n} |x_i|$$

This can be considered as the total length of the vector, x, which is calculated as the sum of the absolute values of all entries in the vector. When we have a two-element vector, $x = [x_1, x_2]$, the L^1-norm will be $\|x\|_1 = |x_1| + |x_2|$.

First, let's create a 3x1 matrix to represent a column vector:

```
a = as.matrix(c(1,2,3))
>>> a
      [,1]
[1,]    1
```

```
[2,]     2
[3,]     3
```

We can call the norm() function to calculate the L^1-norm in R, as shown in the following code snippet:

```
>>> norm(a)
[1] 6
```

Note that the norm() function calculates the L^1-norm by default. To set the type of norm explicitly, we can pass in the type="1" argument, as follows:

```
>>> norm(a, type="1")
[1] 6
```

Next, we will move on to the L^2-norm.

Calculating the L^2-norm of a vector

The L^2-norm is the most common type of norm we usually work with. Also called the Euclidean norm, the L^2-norm measures the usual distance between two points. Plugging $p = 2$ into the previous formula gives us the following definition of the L^2-norm:

$$\|\mathbf{x}\|_2 = \sqrt{\sum_{i=1}^{n} |x_i|^2} = \sqrt{\sum_{i=1}^{n} x_i^2}$$

The calculation process involves squaring each entry, adding them up, and then taking the square root. Similarly, when we have a two-element vector, $x = [x_1, x_2]$, the L^2-norm will be $\|x\|_2 = \sqrt{x_1^2 + x_2^2}$.

We can calculate the L^2-norm of a vector by specifying type="2", as follows:

```
>>> norm(a, type="2")
[1] 3.741657
```

Next comes the max norm.

Calculating the L^∞-norm of a vector

The L^∞-norm, or **max norm**, finds the largest absolute value of all the elements within the vector. This norm is often used in worse-case scenarios – for example, representing the maximum noise injected into a signal. Its definition is as follows:

$$\|x\|_\infty = \max_{1 \leq i \leq n} |x_i|$$

So, the calculation process involves pairwise comparisons of absolute values in search of the maximum.

To calculate the L^∞-norm, we can specify type="2", as follows:

```
>>> norm(a, type="I")
3
```

Now, we'll move on to the matrix norm.

Understanding the matrix norm

The **matrix norm** is similar to the vector norm, although the calculations are slightly different. We can still use the norm() function to do the job. We will use X to denote a general matrix, $m \times n$, where X_{ij} denotes the element located at the ith row and j^{th} column. All forms of matrix norm come from the following generalized form of L^p-norm:

$$\|X\|_p = \left(\sum_{i=1}^{m} \sum_{j=1}^{n} |X_{ij}|^p \right)^{1/p}$$

First, let's create a 3x3 matrix:

```
X = as.matrix(data.frame("c1"=c(1,2,3),"c2"=c(2,5,2),"c3"=c(-1,8,3)))
>>> X
     c1 c2 c3
[1,]  1  2 -1
[2,]  2  5  8
[3,]  3  2  3
```

Next, we will look at the L^1-norm of the X matrix.

Calculating the L^1-norm of a matrix

The L^1-norm for a matrix is similar to its vector form but slightly different. As shown here, to calculate the L^1-norm for a matrix, we must first sum the absolute values of each column, then take the largest summation as the L^1-norm:

$$\|X\|_1 = \max_{1 \leq j \leq n} \sum_{i=1}^{m} |X_{ij}|$$

Since the summation is performed column-wise, the L^1-norm for a matrix is also called the column-sum norm.

We can calculate the L^1-norm using the same command that we used earlier:

```
>>> norm(X, type="1")
[1] 12
```

A visual inspection shows that the third column gives the maximum summation of absolute values of 12. We can also quickly check the column-wise summation of the absolute values of the matrix, as follows:

```
>>> colSums(abs(X))
c1 c2 c3
 6  9 12
```

Calculating the Frobenius norm of a matrix

The L^2-norm of a matrix is more involved at this stage, so we will focus on a similar kin that is widely used in practice: the **Frobenius norm**. The Frobenius norm is calculated by summing all squared entries of the matrix and taking the square root:

$$\|X\|_F = \sqrt{\sum_{i=1}^{m}\sum_{j=1}^{n}|X_{ij}|^2}$$

We can calculate the Frobenius norm by setting `type="f"`, as follows:

```
>>> norm(X, type="f")
[1] 11
```

Let's verify the calculations by carrying out the manual process of squaring all entries, summing them up, and taking the square root:

```
>>> sqrt(sum(X^2))
[1] 11
```

Now, we will look at the infinity norm of a matrix.

Calculating the infinity norm of a matrix

The **infinity norm** of a matrix works similarly to the L^1-norm of a matrix, although the order of sequence is different. In particular, we would sum up the absolute values for each row and return the largest summation:

$$\|X\|_\infty = \max_{1 \le j \le m}\sum_{i=1}^{n}|X_{ij}|$$

Thus, the infinity norm is also referred to as the **row-sum norm**. We can calculate the L^∞-norm by specifying `type="I"` in the `norm()` function:

```
>>> norm(X, type="I")
[1] 15
```

Again, it is a good habit to verify the result by manually carrying out the calculation process:

```
>>> max(rowSums(abs(X)))
[1] 15
```

Figure 8.4 summarizes these three norms for both vectors and matrices:

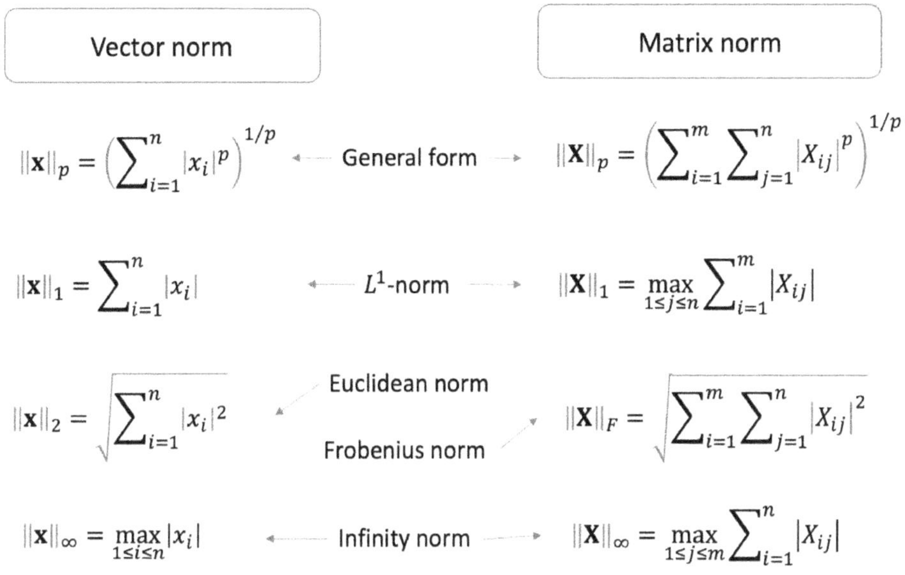

$$\boxed{\text{Vector norm}} \qquad\qquad \boxed{\text{Matrix norm}}$$

$$\|\mathbf{x}\|_p = \left(\sum_{i=1}^{n} |x_i|^p\right)^{1/p} \quad \longleftarrow \text{ General form } \longrightarrow \quad \|\mathbf{X}\|_p = \left(\sum_{i=1}^{m}\sum_{j=1}^{n} |X_{ij}|^p\right)^{1/p}$$

$$\|\mathbf{x}\|_1 = \sum_{i=1}^{n} |x_i| \quad \longleftarrow \text{ } L^1\text{-norm} \text{ } \longrightarrow \quad \|\mathbf{X}\|_1 = \max_{1 \le j \le n} \sum_{i=1}^{m} |X_{ij}|$$

$$\|\mathbf{x}\|_2 = \sqrt{\sum_{i=1}^{n} |x_i|^2} \quad \overset{\text{Euclidean norm}}{\underset{\text{Frobenius norm}}{\nearrow}} \quad \|\mathbf{X}\|_F = \sqrt{\sum_{i=1}^{m}\sum_{j=1}^{n} |X_{ij}|^2}$$

$$\|\mathbf{x}\|_\infty = \max_{1 \le i \le n} |x_i| \quad \longleftarrow \text{ Infinity norm } \longrightarrow \quad \|\mathbf{X}\|_\infty = \max_{1 \le j \le m} \sum_{i=1}^{n} |X_{ij}|$$

Figure 8.4 – Common norms for vectors and matrices

Having covered these fundamentals, let's move on to the next important topic: eigenvalues and eigenvectors.

Getting to know eigenvalues and eigenvectors

The **eigenvalue**, often denoted by a scalar value of λ, and the **eigenvector**, often denoted by $\mathbf{v}$, are essential properties of a square matrix, A. Two central ideas are required to understand the purpose of eigenvalues and eigenvectors. The first is that the matrix, A, is a transformation that maps one input vector to another output vector, which possibly changes the direction. The second is that the eigenvector is a special vector that does not change direction after going through the transformation induced by A. Instead, the eigenvector gets scaled along the same original direction by a multiple of the corresponding scalar eigenvalue. The following equation sums this up:

$$\mathbf{A}\mathbf{v} = \lambda\mathbf{v}$$

These two points capture the essence of **eigendecomposition**, which represents the original matrix, A, in terms of its eigenvalues and eigenvectors and thus allows easier matrix operations in many cases. Let's start by understanding a simple case: **scalar-vector multiplication**.

Understanding scalar-vector multiplication

Matrix-vector multiplication can result in many forms of transformations, such as rotation, reflection, dilation, contraction, projection, and a combination of these operations. With eigenvalues and eigenvectors, we can decompose these operations into a series of simpler ones. For the case of scalar-vector multiplication, when the scalar value is in the range of 0 and 1, this will make the elements of the vector smaller, thus *contracting* the vector.

Suppose we want to multiply a vector, $\mathbf{v}$, by a scalar, λ, giving us $\lambda\mathbf{v}$. Since $\mathbf{v}$ contains one or more elements, multiplication essentially applies to each element in the vector. The following code snippet shows the result of multiplying a scalar by a vector, where each of the elements gets doubled:

```
v = c(1,2,3)
lambda = 2
>>> lambda * v
[1] 2 4 6
```

Now comes a key technique: introducing the identity matrix, $\mathbf{I}$. Since the vector has three elements, we can introduce a 3x3 identity matrix into the equation. In the previous chapter, we learned that multiplying an identity matrix will not change the result, so this is something we can proceed with:

$$\lambda\mathbf{v} = \lambda\mathbf{I}\mathbf{v}$$

Here, we first create a 3x3 identity matrix, $\mathbf{I}$, using the diag() function, as follows:

```
I = diag(3)
>>> I
     [,1] [,2] [,3]
[1,]    1    0    0
[2,]    0    1    0
[3,]    0    0    1
```

Multiplying it by the scalar, λ, changes all diagonal elements to 2 in $\lambda\mathbf{I}$:

```
>>> lambda * I
     [,1] [,2] [,3]
[1,]    2    0    0
[2,]    0    2    0
[3,]    0    0    2
```

Since $\lambda\mathbf{I}$ is a 3x3 matrix, the previous scalar-vector multiplication becomes matrix-vector multiplication now. This means that we need to switch to the %*% sign to perform the inner multiplication:

```
>>> (lambda * I) %*% v
     [,1]
[1,]    2
```

```
[2,]     4
[3,]     6
```

The result is the same as the previous scalar-vector product, although it is now expressed as a column vector instead of a row vector. From the perspective of matrix-vector multiplication, the matrix *transforms* the vector by doubling every element in the vector.

Next, we will formally define the notion of eigenvalues and eigenvectors.

Defining eigenvalues and eigenvectors

By introducing the identity matrix, we managed to convert a scalar-vector multiplication, $\lambda \mathbf{v}$, into a matrix-vector multiplication, $\lambda \mathbf{I} \mathbf{v}$. This makes all the difference in understanding the key equation: $\mathbf{A}\mathbf{v} = \lambda\mathbf{v}$, where the left is a matrix-vector multiplication and the right is a scalar-vector multiplication. By writing $\mathbf{A}\mathbf{v} = \lambda\mathbf{I}\mathbf{v}$, this equation suddenly makes more sense as it has the same type of operation on both sides.

Now, we must define λ and $\mathbf{v}$ to give them proper names. For a square matrix, $\mathbf{A}$, we say that the scalar, λ, is an eigenvalue of $\mathbf{A}$, together with an associated eigenvector, $\mathbf{v} \neq 0$, if $\mathbf{A}\mathbf{v} = \lambda\mathbf{v}$ is true. This equation says that the matrix-vector multiplication in $\mathbf{A}\mathbf{v}$ produces the same vector as the scalar-vector multiplication. λ and $\mathbf{v}$ together are called an eigenpair.

We can quickly verify the equality mentioned in $\mathbf{A}\mathbf{v} = \lambda\mathbf{v}$. Suppose $\mathbf{A} = \begin{bmatrix} 2 & 3 \\ 0 & 1 \end{bmatrix}$, $\lambda = 2$, and $\mathbf{v}^T = [1,0]$. The calculation on the left-hand side gives us the following:

$$\mathbf{A}\mathbf{v} = \begin{bmatrix} 2 & 3 \\ 0 & 1 \end{bmatrix} \begin{bmatrix} 1 \\ 0 \end{bmatrix} = \begin{bmatrix} 2 \\ 0 \end{bmatrix}$$

The calculation on the right-hand side gives us the following:

$$\lambda\mathbf{v} = 2 \begin{bmatrix} 1 \\ 0 \end{bmatrix} = \begin{bmatrix} 2 \\ 0 \end{bmatrix}$$

Therefore, the equality checks.

Geometrically, the eigenvector is a vector that stays fixed in its original direction when we apply a matrix transformation by $\mathbf{A}$. It stays on the same line and remains invariant upon multiplication. Also, there is often a collection of such eigenvectors (with the corresponding eigenvalues) for a square matrix.

The following code snippet verifies the same:

```
A = matrix(c(2,3,0,1), byrow=TRUE, nrow=2)
lambda = 2
v = c(1,0)
>>> A%*%v
        [,1]
[1,]     2
[2,]     0
```

```
>>> lambda*v
[1] 2 0
```

Note that the eigenvector focuses entirely on the *direction* of the transformation-invariant vector, rather than the magnitude. To see this, we can double the eigenvector and find that the equality still checks:

$$\mathbf{Av} = \begin{bmatrix} 2 & 3 \\ 0 & 1 \end{bmatrix} \begin{bmatrix} 2 \\ 0 \end{bmatrix} = \begin{bmatrix} 4 \\ 0 \end{bmatrix}$$

$$\lambda\mathbf{v} = 2\begin{bmatrix} 2 \\ 0 \end{bmatrix} = \begin{bmatrix} 4 \\ 0 \end{bmatrix}$$

We can verify the equality by taking their difference:

```
>>> A%*%v - lambda*v
        [,1]
[1,]      0
[2,]      0
```

Figure 8.5 summarizes our understanding so far:

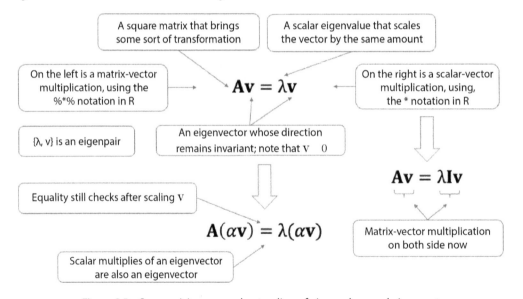

Figure 8.5 – Summarizing our understanding of eigenvalues and eigenvectors

Next, we'll look at how to compute the eigenvalues and eigenvectors of a square matrix.

Computing eigenvalues and eigenvectors

The previous examples assume that we have access to the eigenvalues and eigenvectors. In practice, these need to be computed from the original square matrix. This section focuses on how to obtain the solutions to the eigenvalues and eigenvectors.

Let's start from where we left off. By introducing the identity matrix, we managed to obtain $\mathbf{Av} = \lambda \mathbf{Iv}$. Moving things around, we have $\mathbf{Av} - \lambda \mathbf{Iv} = 0$. Combining similar terms gives us $(\mathbf{A} - \lambda \mathbf{I})\mathbf{v} = 0$. We know that $\mathbf{v} \neq 0$ by definition. Based on the invertibility of a matrix, if there is a non-zero vector in the null space (the set of all vectors that end up as zero) of a matrix, then this matrix is not invertible. Therefore, $\mathbf{A} - \lambda \mathbf{I}$ is not invertible.

There is one convenient property that connects the determinant of a matrix with the invertibility – that is, when a matrix is not invertible, its determinant has to be zero. Similarly, if a matrix is invertible, its determinant cannot be zero. Thus, we have the following:

$$\det(\mathbf{A} - \lambda \mathbf{I}) = 0$$

This gives us a system of linear equations, which we can use to solve the values of λ, as shown in the previous chapter. Let's look at a concrete example.

Suppose we have a 2x2 square matrix, $\mathbf{A} = \begin{bmatrix} 0 & 1 \\ -2 & -3 \end{bmatrix}$. Plugging this into the previous equation gives us the following:

$$\det\left(\begin{bmatrix} 0 & 1 \\ -2 & -3 \end{bmatrix} - \lambda \begin{bmatrix} 1 & 0 \\ 0 & 1 \end{bmatrix} \right) = 0$$

By multiplying the scalar, λ, into the identity matrix, we get the following:

$$\det\left(\begin{bmatrix} 0 & 1 \\ -2 & -3 \end{bmatrix} - \begin{bmatrix} \lambda & 0 \\ 0 & \lambda \end{bmatrix} \right) = 0$$

Combining the two matrices gives us this:

$$\det\left(\begin{bmatrix} -\lambda & 1 \\ -2 & -3 - \lambda \end{bmatrix} \right) = 0$$

By applying the definition of the matrix determinant, we get the following:

$$-\lambda(-3 - \lambda) - 1*(-2) = 0$$

$$\lambda^2 + 3\lambda + 2 = 0$$

$$(\lambda + 1)(\lambda + 2) = 0$$

$$\lambda = -1, -2$$

Thus, we have two solutions: $\lambda = -1$ and $\lambda = -2$. The next step is to find the corresponding eigenvectors for both; we will focus on $\lambda = -1$ in the following exposition.

Since the square matrix, $\mathbf{A}$, is 2x2, we know that the eigenvector, v, needs to be two-dimensional. By denoting $v^T = \begin{bmatrix} v_1, v_2 \end{bmatrix}$ and plugging $\lambda = -1$ and $\mathbf{A} = \begin{bmatrix} 0 & 1 \\ -2 & -3 \end{bmatrix}$ into $\mathbf{A}v = \lambda \mathbf{I}v$, we get the following:

$$\begin{bmatrix} 0 & 1 \\ -2 & -3 \end{bmatrix} \begin{bmatrix} v_1 \\ v_2 \end{bmatrix} = \begin{bmatrix} -v_1 \\ -v_2 \end{bmatrix}$$

This gives us the following system of equations:

$$\begin{cases} v_2 = -v_1 \\ -2v_1 - 3v_2 = -v_2 \end{cases}$$

Solving this system of equations gives us $v_2 = -v_1$, which corresponds to an infinite number of solutions. This makes sense as the eigenvector focuses more on direction instead of absolute magnitude. In this case, the direction is represented by a line, $y = -x$, in a two-dimensional coordinate system. *Figure 8.6* summarizes the process of finding the eigenvalues and eigenvectors of a square matrix:

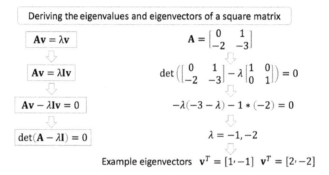

Figure 8.6 – The process of deriving the eigenvalues and eigenvectors of a square matrix

Let's say we take $v_1 = 1$. The resulting eigenvector becomes $v^T = [1, -1]$. When $v_1 = 2$, we get $v^T = [2, -2]$. To calculate the eigenvalues and eigenvectors via eigen-decomposition, we can simply call the `eigen()` function in R, as shown in the following code snippet:

```
>>> eigen(A)
eigen() decomposition
$values
[1] 2 1

$vectors
      [,1]        [,2]
[1,]    1 -0.9486833
[2,]    0  0.3162278
```

There are two entries in the `values` attribute and two column vectors in the `vectors` attribute, indicating a total of two eigenpairs in the matrix. We can access the first eigenvalue as follows:

```
>>> eigen(A)$values[1]
[1] 2
```

Similarly, the eigenvectors are returned as a set of column vectors. Therefore, we can access the first eigenvector as follows:

```
>>> eigen(A)$vectors[,1]
[1] 1 0
```

We can draw a few useful properties from eigen-decomposition. Take an $n \times n$ square matrix, $\mathbf{A}$, for example. The number of distinct eigenvalues is, at most, n.

We can also verify the correctness of the resulting eigenvalues and eigenvectors by making use of the condition derived earlier: $\det(\mathbf{A} - \lambda\mathbf{I}) = 0$. For the first eigenvalue and eigenvector, the following code snippet does the verification:

```
>>> det(eigen_rst$values[1] * diag(2) - A)
[1] 0
The second eigenpair can also be verified:
>>> det(eigen_rst$values[2] * diag(2) - A)
[1] 0
```

We can also verify the eigen-decomposition based on the original equation – that is, $\mathbf{Av} = \lambda\mathbf{v}$:

```
>>> A%*%eigen_rst$vector[,1] - eigen_rst$values[1]*eigen_
rst$vector[,1]
     [,1]
[1,]    0
[2,]    0
>>> A%*%eigen_rst$vector[,2] - eigen_rst$values[2]*eigen_
rst$vector[,2]
          [,1]
[1,] -1.110223e-16
[2,]  0.000000e+00
```

Here, the second command returns a very small number, which is due to numerical approximation and can be considered zero. Again, note the use of the scalar-vector multiplication sign, *, and the matrix-vector multiplication sign, %*%.

Now that we have a better understanding of eigenvalues and eigenvectors, let's look at a popular application: PCA.

Introducing principal component analysis

When building an ML model, the dataset that's used to train the model may have redundant information in the predictors. The redundancy in the predictors/columns of the dataset arises from correlated features in the dataset and needs to be taken care of when using a certain class of models. In such cases, PCA is a popular technique to address such challenges as it reduces the feature dimension of the dataset and thus shrinks the redundancy. The problem of **collinearity**, which says that two or more predictors are linearly correlated in a model, could thus be relieved via dimension reduction using PCA.

Collinearity among the predictors is often considered a big problem when building an ML model. Using the Pearson correlation coefficient, it is a number between -1 and 1, where a coefficient near 0 indicates two variables are linearly independent, and a coefficient near -1 or 1 indicates that two variables are linearly related.

When two independent variables are linearly correlated, such as $x_2 = 2x_1$, no extra information is provided by x_2. The perfect correlation between x_1 and x_2 makes x_2 useless in terms of explaining the outcome variable. A natural choice is to remove x_2 from the set of independent variables. However, when the correlation is not perfect, we will lose some amount of information due to the removal.

PCA provides us with another way to combat correlation among the predictors. It allows us to extract meaningful and uncorrelated information from the original dataset. Specifically, it uncovers the hidden and low-dimensional features that underlie the dataset. These low-dimensional hidden features make it convenient for both visualization and interpretation.

To help us understand this technique, we'll start by covering the notion of the variance-covariance matrix.

Understanding the variance-covariance matrix

All ML models sit on a training dataset. In the context of supervised learning, the dataset consists of input-output pairs. The input is also called the design matrix, holding n rows of observations and p columns of features. This $n \times p$ design matrix, **X**, is our focus of study in PCA.

Suppose we would like to know the correlation between each pair of features. The correlation coefficient, ranging from -1 to 1, is calculated based on the covariance of two variables. The covariance is a scalar value that measures the strength of co-movement between two variables. Therefore, the covariance matrix of the design matrix measures the strength of co-movement between each unique pair of features. It is a $p \times p$ square matrix, $cov(\mathbf{X})$, where the entry is located at the ith row, and the jth column represents the covariance value between the x_i and x_j features.

To obtain this covariance matrix, we must de-mean all features – that is, subtract the column-wise mean from each element in a given column. This removes the central tendency and indicates the relative amount of deviation from the mean. This results in the de-meaned $n \times p$ design matrix, ($\mathbf{X} - \bar{\mathbf{X}}$). We must do the following:

1. Transpose the de-meaned design matrix to obtain a $p \times n$ matrix, $(\mathbf{X} - \bar{\mathbf{X}})^T$.

2. Multiply the transposed design matrix with the original design matrix, both de-meaned, to obtain a $p \times p$ square matrix, $(\mathbf{X} - \bar{\mathbf{X}})^T(\mathbf{X} - \bar{\mathbf{X}})$.

3. Divide the result by $n - 1$ to normalize the entries in the matrix (instead of n used in population covariance).

Once we've done this, a $p \times p$ variance-covariance matrix, $\frac{(\mathbf{X} - \overline{\mathbf{X}})^T(\mathbf{X} - \overline{\mathbf{X}})}{n - 1}$, is generated, where the ith diagonal element is the variance of the ith column in the original design matrix.

Let's shed some light on the contents of the variance-covariance matrix, $\frac{(\mathbf{X} - \overline{\mathbf{X}})^T(\mathbf{X} - \overline{\mathbf{X}})}{n - 1}$. We will start by exposing the entries in this matrix, as follows:

$$\frac{(\mathbf{X} - \overline{\mathbf{X}})^T(\mathbf{X} - \overline{\mathbf{X}})}{n - 1} = \begin{bmatrix} \Sigma(x_1 - \overline{x}_1)^2/(n - 1) & \cdots & \Sigma(x_1 - \overline{x}_1)(x_p - \overline{x}_p)/(n - 1) \\ \vdots & \ddots & \vdots \\ \Sigma(x_p - \overline{x}_p)(x_1 - \overline{x}_1)/(n - 1) & \cdots & \Sigma(x_p - \overline{x}_p)^2/(n - 1) \end{bmatrix}$$

The first entry, namely $\Sigma(x_1 - \overline{x}_1)^2/(n - 1)$, is the definition of the sample variance of the x_1 feature. This comes from the dot product between two length-n variables, later divided by $n - 1$. So, the summation works on all n elements in both column vectors.

Moving to the rightmost entry on the first row, we have $\Sigma(x_1 - \overline{x}_1)(x_p - \overline{x}_p)/(n - 1)$ as the covariance between the x_1 and x_p variables. This is exactly how we would calculate the sample covariance between these two variables, where the summation is applied to all n elements in both column vectors. *Figure 8.7* illustrates the calculation process:

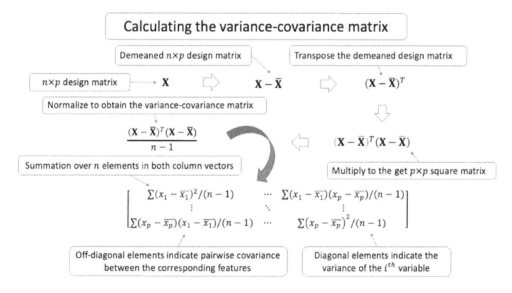

Figure 8.7 – Summarizing the calculation process for the variance-covariance
matrix based on the design matrix in a training dataset

Note that although the $n \times p$ design matrix, $\mathbf{X}$, is not necessarily square, we managed to obtain a $p \times p$ square matrix by multiplying the transposed de-meaned $p \times n$ matrix, $(\mathbf{X} - \overline{\mathbf{X}})^T$, with just the de-meaned $n \times p$ matrix, $(\mathbf{X} - \overline{\mathbf{X}})$.

Let's calculate the variance-covariance of a given matrix:

1. In the following code snippet, we first generate a dummy matrix, where the second column is produced by doubling the first column:

```
X = matrix(c(1:5,2*(1:5)), byrow=FALSE, nrow=5)
>>> X
        [,1] [,2]
[1,]     1    2
[2,]     2    4
[3,]     3    6
[4,]     4    8
[5,]     5   10
```

2. Next, we de-mean both columns:

```
X[,1] = X[,1] - mean(X[,1])
X[,2] = X[,2] - mean(X[,2])
>>> X
        [,1] [,2]
[1,]    -2   -4
[2,]    -1   -2
[3,]     0    0
[4,]     1    2
[5,]     2    4
```

3. Lastly, we can calculate the variance-covariance matrix, like so:

```
>>> t(X)%*%X / (nrow(X)-1)
        [,1] [,2]
[1,]    2.5    5
[2,]    5.0   10
```

The result is a 2x2 matrix, which aligns with our previous discussion.

Note that we can also manually calculate the particular variance or covariance entries. For example, the following command calculates the variance of the second variable:

```
>>> var(X[,2])
[1] 10
```

We can also calculate the covariance between the two variables:

```
>>> cov(X[,1], X[,2])
[1] 5
```

In the following section, we will connect the variance-covariance matrix to PCA.

Connecting to PCA

The variance-covariance matrix, $\frac{(X - \bar{X})^T(X - \bar{X})}{n - 1}$, from the previous section can be used in eigendecomposition, which generates a set of eigenvalues along with the associated eigenvectors. Note that these eigenvalues will be real scalars, and the associated eigenvectors will be orthogonal to each other, each pointing in a distinct direction.

Let's look at a few properties of PCA in connection to these eigenvalues and eigenvectors. First, the total variance of the dataset is the sum of these eigenvalues. Thus, we can rank these eigenvalues in decreasing order, keep only the first few, and take the associated eigenvectors as a reduced set of hidden features for downstream modeling. By doing this, we can achieve dimension reduction.

In addition, these eigenvectors are called **principal components** and point in a particular direction. So, for a specific eigenvector, v_i, the direction it points in can explain λ_i of the total variance of the dataset. Thus, we can explain the cumulative variance (as a percentage) by adding up the corresponding eigenvalues and deciding the trade off between the number of hidden variables to keep and the percentage of the total variance explained.

Now, let's run eigendecomposition on the previous variance-covariance matrix:

```
>>> eigen(t(X)%*%X / (nrow(X)-1))
eigen() decomposition
$values
[1] 1.250000e+01 1.110223e-16

$vectors
          [,1]        [,2]
[1,] 0.4472136 -0.8944272
[2,] 0.8944272  0.4472136
```

The result shows that the first eigenvalue significantly dominates the second one in terms of the values that are output, showing that the first eigenvector is sufficient to explain the total variance of the original design matrix. This makes sense as the second variable is simply double the first variable and thus delivers no additional information. Although the dataset has two columns, it only has one column's worth of information and the second column is completely redundant. The second eigenvector also lies along the same line as the first one, thus having no variability in the eigenspace.

In the next section, we'll introduce a function for performing PCA.

Performing PCA

One function we can use to perform PCA is `prcomp()`. For our exercise, we will use the first four columns of the Iris dataset to perform PCA:

1. First, let's load the dataset:

```
X = iris[,c(1:4)]
>>> head(X)
   Sepal.Length Sepal.Width Petal.Length Petal.Width
1           5.1         3.5          1.4         0.2
2           4.9         3.0          1.4         0.2
3           4.7         3.2          1.3         0.2
4           4.6         3.1          1.5         0.2
5           5.0         3.6          1.4         0.2
6           5.4         3.9          1.7         0.4
```

2. The `prcomp()` function will automatically perform all the aforementioned steps involved in eigendecomposition:

```
X_pca = prcomp(X)
>>> X_pca
Standard deviations (1, .., p=4):
[1] 2.0562689 0.4926162 0.2796596 0.1543862

Rotation (n x k) = (4 x 4):
                       PC1          PC2          PC3         PC4
Sepal.Length    0.36138659  -0.65658877   0.58202985   0.3154872
Sepal.Width    -0.08452251  -0.73016143  -0.59791083  -0.3197231
Petal.Length    0.85667061   0.17337266  -0.07623608  -0.4798390
Petal.Width     0.35828920   0.07548102  -0.54583143   0.7536574
```

Note that the variances that are explained – in other words, eigenvalues – are expressed in terms of standard deviations instead of variance.

Now that we have access to the PCA results, we can visualize these results to make them more intuitive. For this purpose, we will use the `factoextra` package. Remember to install and load this package first.

3. The first function we will use is the `fviz_eig()` function, which shows the percentage of variance explained by each principal component in a scree plot:

```
>>> fviz_eig(X_pca)
```

Running this command generates *Figure 8.8*:

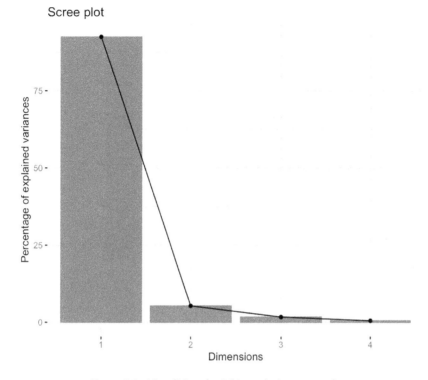

Figure 8.8 – Visualizing the PCA results in a scree plot

4. We can also show a graph for the individual observations using the `fviz_pca_ind()` function, where individuals with similar profiles are grouped:

```
>>> fviz_pca_ind(X_pca,
            col.ind = "cos2", # Color by the quality of
representation
            gradient.cols = c("#00AFBB", "#E7B800", "#FC4E07"),
            repel = TRUE    # Avoid text overlapping
    )
```

Running this command generates *Figure 8.9*. Note that observations with lower quality (less variance explained) are assigned with colors toward the lower half of the palette:

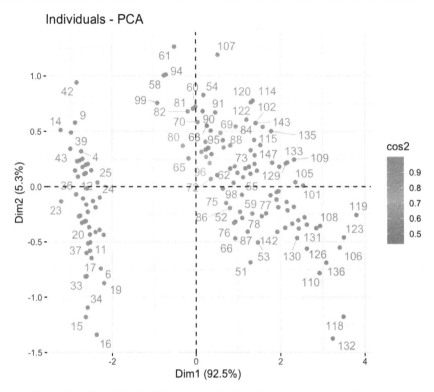

Figure 8.9 – Visualizing individual contributions in explaining the total variance

5. Lastly, we can also visualize the directions of the variables. Positively correlated variables will point to the same side of the plot, while negatively correlated variables will point to opposite sides of the graph:

```
>>> fviz_pca_var(X_pca,
            col.var = "contrib", # Color by contributions to
    the PC
            gradient.cols = c("#00AFBB", "#E7B800", "#FC4E07"),
            repel = TRUE     # Avoid text overlapping
    )
```

Running this command generates *Figure 8.10*:

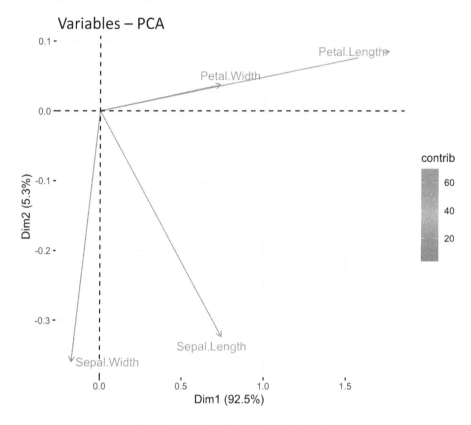

Figure 8.10 – Visualizing variable directions

The graph shows that `Petal.Width` and `Petal.Length` point in almost overlapping directions.

Summary

In this chapter, we covered intermediate linear algebra and its implementations in R. We started by introducing the matrix determinant, a widely used property in numerical analysis. We highlighted the intuition behind the matrix determinant and its connection to matrix rank.

We also covered additional properties, including matrix trace and norm. In particular, we introduced three popular norms: L^1-norm, L^2-norm, and L^∞-norm. We detailed their mathematical constructs and calculation process.

Next, we covered eigendecomposition, which leads to a set of eigenvalues and eigenvectors of a square matrix. We provided a step-by-step derivation and analysis of the core equation, as well as the approach to compute them.

Finally, we covered PCA, a popular technique that's used for dimension reduction. Specifically, we highlighted its role in removing collinearity in the dataset and provided a few ways to compute and visualize PCA results.

In the next chapter, we will switch gears and cover another key branch of mathematics: calculus.

9
Calculus in R

Calculus is a branch of mathematics that studies the relationship between two quantities, connected via a function, from either a micro or a macro perspective. Taking a microscopic lens, such a relationship (often denoted by $y = f(x)$) manifests in the form of a very small change in the output y given an infinitesimal change in the input x. When switching to the macro perspective, the relationship becomes the cumulative change in y as x changes. The micro perspective corresponds to differential calculus, and the macro perspective corresponds to integral calculus, both introduced in this chapter.

By the end of this chapter, you will have grasped essential concepts in differential and integral calculus. Practical implementations in R will also be introduced to leverage R's automatic differentiation and integration capabilities.

In this chapter, we will cover the following topics:

- Introducing calculus
- Working with calculus in R
- Working with integration in R

Technical requirements

To run the code in this chapter, you will need to have the latest version of the `mosaicCalc` package, which is 0.6.0 at the time of writing.

All the code and data for this chapter is available at `https://github.com/PacktPublishing/ The-Statistics-and-Machine-Learning-with-R-Workshop/blob/main/ Chapter_9/working.R`.

Introducing calculus

Calculus is a branch of mathematics that studies the rate of change, such as the slope of a curve at any point. It is a fundamental subject widely used in many areas, including physics, economics, finance, optimization, **artificial intelligence** (**AI**), and more. Calculus was first developed by two gentlemen in the late 17th century: Gottfried Leibniz and Isaac Newton. Newton first developed calculus to analyze physical systems, while Leibniz independently developed the resulting notations we use today. Compared with basic math, which uses operations such as addition or subtraction, calculus applies functions and integrals to study the rate of change. Here, the rate of change can be regarded as the velocity, measuring how fast $f(x)$ changes as x changes. Such changes also have a direction, meaning whether $f(x)$ increases or decreases as x increases.

Similarly, we can measure how fast the rate of change of $f(x)$ changes as x changes. Such a measure is called acceleration, which assesses whether $f(x)$ increases or decreases at an increasing or decreasing rate as x increases.

Differential and integral calculus

The two main branches of calculus are **differential calculus** and **integral calculus**. Differential calculus studies the rate of change of a particular quantity. It examines the rate of change of slopes and curves and studies the sensitivity of an outcome variable y to a very small change in the input variable x. On the other hand, integral calculus studies the volume or **area under the curve** (**AUC**).

As shown in *Figure 9.1*, we plotted a sample function $y = f(x)$ and picked an arbitrary point at $x = x_0$. We then added a very small change Δx to x_0. This change is so small that it will tend toward 0 in the limit—that is, $\Delta x \to 0$. Upon changing x, the dependent variable y will also change, generating a resulting change of Δy. Dividing these two terms gives an indication of the sensitivity of y with respect to x at the point x_0. When $\Delta x \to 0$, we call the sensitivity the derivative of y with respect to x when $x = x_0$. Expressed mathematically, we have $f'(x_0) = \frac{\Delta y}{\Delta x}$ as $\Delta x \to 0$, or simply, $f'(x_0) = \lim_{\Delta x \to 0} \frac{\Delta y}{\Delta x}$.

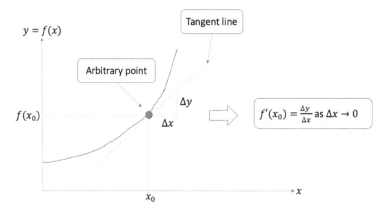

Figure 9.1 – A graph illustrating differential calculus that studies the rate of change at an arbitrary point

The rate of change at point x_0, or the sensitivity of the function $y = f(x)$ at the point $x = x_0$, is expressed as $f'(x_0) = \frac{\Delta y}{\Delta x}$ when Δx is infinitesimally small—that is, $\Delta x \to 0$. On the other hand, integral calculus indicates the AUC $f(x)$ for a specific range $x \in [a, b]$, as shown in *Figure 9.2*:

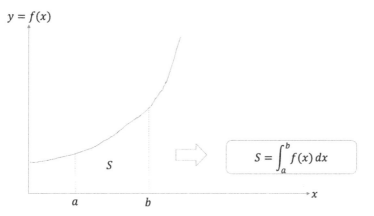

$$S = \int_a^b f(x)\, dx$$

Figure 9.2 – A graph illustrating the integral calculus as the total AUC

We denote the total area as a result of integration $S = \int_a^b f(x)dx$, which reports a positive quantity in this case. This means that the result of integration could also be negative, should the curve be below the x axis.

More on functions

Note that a function is a mapping machine between two sets of elements. For each element in the input set, there is one and only one corresponding element in the output set. The collection of all possible elements in the input set is called the domain, and the collection of all corresponding elements in the output set is called the range. *Figure 9.3* shows four different mapping scenarios. The first three are valid mappings based on the definition of a function, while the last one fails due to the one-to-many mapping relationship:

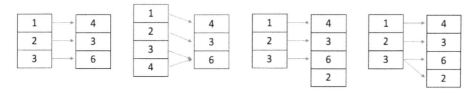

Figure 9.3 – Four different mapping scenarios

The first three mappings are valid functions, while the last one is invalid due to one-to-many mapping in the last element (3->6 and 3->2).

There are multiple names we often use for a function $y = f(x) = x^2$, as follows:

- The variable x can be called an **independent variable**, a **feature**, a **covariate**, or an **input**.

- The variable y can be called a **dependent variable**, an **outcome**, a **target**, a **response**, or an **output**.

- The mapping function $f(x) = x^2$ can be called a **function**, a **mapping**, a **projection**, or a **hypothesis**.

Figure 9.4 summarizes these terms and provides a collection of input-output samples with the corresponding graph. Note that the input can be any number or expressed as a general variable:

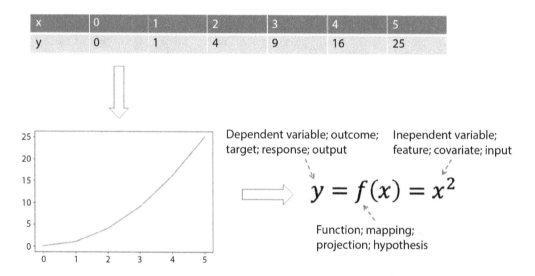

x	0	1	2	3	4	5
y	0	1	4	9	16	25

Dependent variable; outcome; target; response; output

Inependent variable; feature; covariate; input

$$y = f(x) = x^2$$

Function; mapping; projection; hypothesis

The input can be any <u>number</u> or <u>variable</u>

Figure 9.4 – Illustrating a sample mapping function of $y = f(x) = x^2$

Vertical line test

One technique to assess whether a mapping or a curve is a function or not is the vertical line test. In particular, any vertical line in the plane can intersect the graph of a function at most once. That is, for any function $f{:}A \rightarrow B$, each element $x \in A$ is mapped to at most one value $f(x) \in B$. *Figure 9.5* illustrates two curves, where the first curve intersects with the vertical line only once and thus is a valid function, while the second curve intersects with the vertical line twice, thus not satisfying the definition of a function:

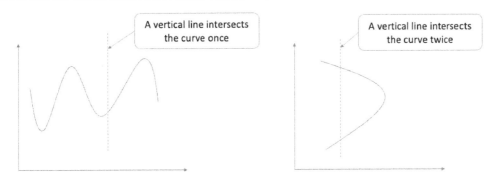

Figure 9.5 – Assessing whether a curve is a function using the vertical line test

Functional symmetry

There are many special properties a function could have, and functional symmetry is one property. For example, a function is called an even function if $f(x) = f(-x)$, such as $f(x) = |x|, f(x) = x^2$, and $f(x) = \cos(x)$. A function is called an odd function if $f(-x) = -f(x)$, such as $f(x) = x^3$.

Increasing and decreasing functions

A function can also be increasing or decreasing. A function is increasing on the interval $[a, b]$ if $f(x_1) > f(x_2)$ for any $x_1 > x_2$, and is decreasing on the interval $[a, b]$ if $f(x_1) < f(x_2)$ for any $x_1 > x_2$.

Slope of a function

The slope m of a line can be calculated based on any two points (x_1, y_1) and (x_2, y_2) using the following definition:

$$m = \frac{y_2 - y_1}{x_2 - x_1}$$

Connecting to the definition of the derivative in *Figure 9.1*, the derivative is essentially the slope of the tangent line at the point $x = x_0$.

Function composition

Function composition refers to the case when a function consists of more than one nested function, denoted as $f(g(x)) = (f \circ g)(x)$. The mapping sequence is this: the input variable x first gets transformed by function $g(x)$, followed by transformation by the function $f(x)$. In $f(g(x))$, $g(x)$ is the interior function and $f(x)$ is the exterior function.

For example, given $f(x) = 2x - 1, g(x) = x^3$, we can obtain $f(g(x))$ as follows:

$$f(g(x)) = 2x^3 - 1$$

Note that the function composition is not commutative—that is, $f(g(x)) \neq g(f(x))$.

Common functions

Let's look at several types of common functions.

Power function

The **power function**, $f(x) = kx^p$, where k and p are constants, and x is the variable. k is also called the coefficient. Examples include the following:

- Cube function: $f(x) = x^3$
- Square root function: $f(x) = \sqrt{x}$
- Cube root function: $f(x) = \sqrt[3]{x}$
- Linear function: $f(x) = x$
- Absolute value function: $f(x) = |x|$
- Square function: $f(x) = x^2$

Polynominal function

The **polynomial function**, $f(x) = a_n x^n + a_{n-1} x^{n-1} + \dots + a_1 x + a_0$, where a_n is the leading coefficient. n is a non-negative integer called the degree of the polynomial, and the coefficients $a_0, \dots, a_n$ are real numbers with $a_n \neq 0$.

Rational function

The **rational function**, $f(x) = \frac{n(x)}{d(x)}$, where both $n(x)$ and $d(x)$ are polynomials, and $d(x) \neq 0$.

Exponential function

The **exponential function**, $f(x) = b^x$, where $b > 0$ and $b \neq 1$.

Logarithmic function

The **logarithmic function**: $f(x) = \log_b x$, where $b > 0$ and $b \neq 1$. Note that when $y = \log_b x$, we have the equivalent form of $x = b^y$.

After getting a quick understanding of the many functions and their properties, let us look at the concept of a limit, which relates to the derivative.

Understanding limits

The limit of a function $f(x)$ says that when the input x is close but not equal to a number c, the value of $f(x = c)$ would be close to a real number L. Mathematically, we can express it as follows:

$$\lim_{x \to c} f(x) = L$$

Or, equivalently, it could be expressed as $f(x) \to L$ as $x \to c$. Note that the value of the limit $\lim_{x \to c} f(x)$ may not necessarily be equal to the value of the function at $x = c$—that is, $f(c)$. Such equality, $\lim_{x \to c} f(x) = f(c)$, happens only when $f(x)$ is a continuous function.

Infinite limit

When the limit $\lim_{x \to c} f(x)$ does not exist as $x \to c$, we say that the function $f(x)$ tends to infinity and thus leads to an infinite limit as $x \to c$. We can also say that $x = c$ is the vertical asymptote of the function $f(x)$.

One example is $f(x) = \frac{1}{x-1}$. We know that the domain is $x \neq 1$. As $x \to 1$, we have $\lim_{x \to 1} \frac{1}{x-1} = \infty$, which breaks down into approaching from the left $\lim_{x \to 1^-} \frac{1}{x-1} = -\infty$ and approaching from the right $\lim_{x \to 1^+} \frac{1}{x-1} = \infty$. Both results jointly lead to $\lim_{x \to 1} \frac{1}{x-1} = \infty$, thus $\lim_{x \to 1} \frac{1}{x-1}$ does not exist. The vertical asymptote is $x = 1$.

Limit at infinity

The limit at infinity is concerned with the value of $f(x)$ when $x \to \infty$ or $x \to -\infty$. This gives the horizontal asymptote of the function—that is, $\lim_{x \to \infty} f(x)$ and $\lim_{x \to -\infty} f(x)$. We call that $y = b$ is a horizontal asymptote if either $\lim_{x \to \infty} f(x) = b$ or $\lim_{x \to -\infty} f(x) = b$.

Let us look at a few examples. For the square function $f(x) = x^2$, we have $\lim_{x \to \infty} x^2 = \infty$ and $\lim_{x \to -\infty} x^2 = \infty$. For the square root function $f(x) = \sqrt{x}$, since the domain is $x \geq 0$, we have $\lim_{x \to \infty} \sqrt{x} = \infty$. For the natural logarithmic function $f(x) = \ln x$, since the domain is $x > 0$, we have $\lim_{x \to \infty} \sqrt{x} = \infty$. For a general result, we have $\lim_{x \to \infty} x^p = \infty$ and $\lim_{x \to \infty} \frac{1}{x^p} = \infty$.

The next section formally introduces derivatives.

Introducing derivatives

For a function $y = f(x)$, the derivative of f at x, denoted as $f'(x)$, is defined as follows:

$$f^{(x)} = \lim_{h \to 0} \frac{f(x+h) - f(x)}{h}, \text{ if the limit exists.}$$

When $f'(x)$ exists for each value of x in the range $[a, b]$, we say that the function f is differentiable over $x \in [a, b]$.

There are multiple applications of a derivative. Besides the previous point on interpreting the derivative as the slope of the tangent line, the most common application is that we can use the derivative to determine if a function is rising/increasing or falling/decreasing at a specific point. It also represents the instantaneous rate of change, or velocity, at a given point x.

A graphical illustration will help our understanding here. *Figure 9.6* provides a sample curve of $f(x)$ with a **secant** line connecting two points $(a, f(a))$ and $(b, f(b))$. Here, a secant line is a straight line that intersects the curve of a function at two or more distinct points. We obtain a second point $(a + h, f$

$(a + h))$ by adding a small amount h to the point $(a, f(a))$. Connecting these two points gives the secant line that crosses the function $f(x)$ by these two points. As $h \to 0$, the secant line will gradually approach the tangent line obtained at $x = a$. Eventually, the secant line will overlap with the tangent line in the limit, and the slope of the secant line, expressed as $\frac{f(a + h) - f(a)}{h}$, will become the slope of the tangent line when they overlap in the limit:

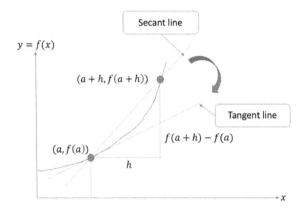

Figure 9.6 – A graph illustrating the derivation of a derivative

In *Figure 9. 6*, the second point $(a + h, f(a + h))$ is produced by adding a small amount h to the point $(a, f(a))$. Connecting these two points gives the secant line. As $h \to 0$, the secant line will gradually approach the tangent line at $x = a$, eventually overlapping with the tangent line in the limit.

The slope of the secant line, also called the average rate of change, is calculated as follows:

$$\frac{f(a + h) - f(a)}{(a + h) - a} = \frac{f(a + h) - f(a)}{h}$$

This is provided that $h \neq 0$. When $h \to 0$, the secant line infinitely approaches the tangent line at $x = a$, giving the instantaneous rate of change at this point (provided that the limit exists):

$$\lim_{h \to 0} \frac{f(a + h) - f(a)}{h}$$

In general, if $y = f(x)$, we can express the derivative as $f'(x)$, y', $\frac{dy}{dx}$, or $\frac{d}{dx}f(x)$.

The following section introduces a few common derivatives.

Common derivatives

Here, we provide a list of derivatives for several common functions:

- Constant function: $\frac{d}{dx}(c) = 0$
- Power function: $\frac{d}{dx}(x^n) = nx^{n-1}$
- Exponential function: $\frac{d}{dx}(e^x) = e^x$, $\frac{d}{dx}(b^x) = (\ln b)b^x, b > 0$

- Logarithmic function: $\frac{d}{dx}(\ln x) = \frac{1}{x}, \frac{d}{dx}(\log_b x) = \frac{1}{(\ln b)x}, b > 0, b \neq 1$

- Sine and cosine functions: $\frac{d}{dx}(\sin x) = \cos x, \frac{d}{dx}\cos x = -\sin x$

Calculating the derivative often involves multiple basic functions. The next section covers a set of properties and rules on derivative calculation.

Common properties and rules of derivatives

We introduce two common properties—the constant multiple property and the sum and difference property—followed by three common rules: the product rule, the quotient rule, and the chain rule. We assume that all limits exist in the following list:

- **Constant multiple property**: For a constant c, if $y = f(x) = ch(x)$, then $f'(x) = ch'(x)$. Expressed differently, we have $y' = ch'$, and $\frac{dy}{dx} = c\frac{dh}{dx}$.

- **Sum and difference property**: If $y = f(x) = h(x) \pm g(x)$, then $f'(x) = h'(x) \pm g'(x)$. Expressed differently, we have $y' = h' + g'$, and $\frac{dy}{dx} = \frac{dh}{dx} + \frac{dg}{dx}$.

- **Product rule**: If $y = f(x) = h(x)g(x)$, then $f'(x) = h'(x)g(x) + h(x)g'(x)$. Expressed differently, we have $y' = h'g + hg'$, and $\frac{dy}{dx} = g\frac{dh}{dx} + h\frac{dg}{dx}$.

- **Quotient rule**: If $y = f(x) = \frac{h(x)}{g(x)}$, then $f'(x) = \frac{g(x)h'(x) - h(x)g'(x)}{[g(x)]^2}$. Expressed differently, we have $y' = \frac{gh' - hg'}{g^2}$, and $\frac{dy}{dx} = \frac{g\frac{dh}{dx} - h\frac{dg}{dx}}{g^2}$.

- **Chain rule**: Given a composite function $y = f(x) = h(g(x))$, we have $f'(x) = h'(g(x))g'(x)$. Equivalently, if we have $y = h(u)$ and $u = g(x)$, then $\frac{dy}{dx} = \frac{dy}{du}\frac{du}{dx}$.

In the next section, we will switch to integral calculus.

Introducing integral calculus

Integration is the inverse of differentiation. For example, differentiating the distance traveling $S(t)$ with respect to time gives the velocity at a specific time point $v(t)$. We can calculate the cumulative distance traveled from time $t = a$ to $t = b$ as $S = \sum_{i=0}^{n-1} v(t_i) \Delta t$, after dividing the time period into n equally spaced intervals. When the number of intervals approaches infinity—that is, $n \to \infty$, we have $S = \lim_{n \to \infty} \sum_{i=0}^{n-1} v(t_i)\Delta t = \int_a^b v(t)dt$.

The result of integration is called an antiderivative, which allows us to reconstruct a function from its derivative. It is the opposite of differentiation. Formally, a function $F(x)$ is an antiderivative of $f(x)$ if $F'(x) = f(x)$.

Figure 9.7 illustrates the equivalence between integration and AUC. We can use integration to find the AUC of a general function $f(x)$. This area can be obtained by adding slices of rectangles that approach zero width. The width of each rectangle is dx, which, by definition, refers to an infinitesimal change in x—that is, differential in x:

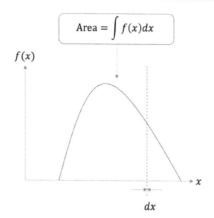

Figure 9.7 – A graph illustrating the equivalence between integration and AUC

Note that an integral can be definite or indefinite. A definite integral has an explicit starting and ending point of integration. For example, if $x \in [a, b]$, the definite integral becomes $\int_a^b f(x)dx$. On the other hand, an indefinite integral has no explicit boundary, giving $\int f(x)dx$, which is used to represent the family of all antiderivatives of $f(x)$.

The following section delves more into indefinite integrals.

Indefinite integrals

Generally, we can represent an indefinite integral as $\int f(x)dx = F(x) + C$ if $F'(x) = f(x)$, where C is a constant and appears here to represent a family of all antiderivative functions whose derivative is $f(x)$. In other words, we have $\frac{d}{dx}\left[\int f(x)dx\right] = f(x)$ and $\int F'(x)dx = F(x) + C$.

Now, let us understand the expression of the indefinite integral $\int f(x)dx$ a bit more. $\int$ is the integral symbol, $f(x)$ is the integrand, and dx is a slice along x, indicating that the antidifferentiation is performed on x. *Figure 9.8* illustrates the naming convention:

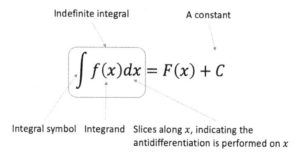

Figure 9.8 – Summarizing the naming convention of indefinite integrals

We will now look at indefinite integrals of common basic functions.

Indefinite integrals of basic functions

The following list provides the indefinite integrals of basic functions. Again, note that the constant C is used to represent a family of all antiderivative functions having the same derivative:

- $\int x^n dx = \frac{1}{n+1} x^{n+1} + C$

- $\int e^x dx = e^x + C$

- $\int b^x dx = \frac{b^x}{\ln b} + C$

- $\int \frac{1}{x} dx = \ln|x| + C$

- $\int \sin x\, dx = -\cos x + C$

- $\int \cos x\, dx = \sin x + C$

Properties of indefinite integrals

This section covers a set of properties that correspond to the derivative function. The following two properties are straightforward linear operations:

- $\int K f(x) dx = K \int f(x) dx$

- $\int [f(x) \pm g(x)] dx = \int f(x) dx \pm \int f(x) dx$

The following list contains a few general formulas to calculate the indefinite integral:

$$\int [f(x)]^n f'(x) dx = \frac{[f(x)]^{n+1}}{n+1} + C, n \neq 1$$

$$\int e^{f(x)} f'(x) dx = e^{f(x)} + C$$

$$\int \frac{1}{f(x)} f'(x) dx = \ln|f(x)| + C$$

Next, we introduce one of the most widely used techniques: integration by parts.

Integration by parts

We know the following based on the product rule introduced earlier:

$$\frac{d}{dx}[f(x)g(x)] = f'(x)g(x) + f(x)g'\left(x\right)$$

We can then integrate both sides and move terms around to get the following:

$$\int f(x)g'(x) dx = f(x)g(x) - \int g(x)f'(x) dx$$

This gives the integration by parts formula. More succinctly, we can define $v = g(x)$ and $u = f(x)$. Thus, $dv = g'(x)dx$ and $du = f'(x)dx$. The preceding equation then becomes this:

$$\int u\, dv = uv - \int v\, du$$

Next, we delve more into definite integrals.

Definite integrals

If $f(x)$ is a continuous function on $x \in [a, b]$ and $F(x)$ is the antiderivative of $f(x)$, that is expressed as $F(x) = \int f(x)dx + C$. The following formula gives the fundamental theorem of calculus:

$$\int_a^b f(x)dx = F(b) - F\left(a\right)$$

This expression says that the AUC $f(x)$ in the range $[a, b]$ can be calculated as the difference between evaluations of the antiderivative function $F(x)$ at the two endpoints.

Definite integrals share mostly the same properties as indefinite integrals, with the following addition:

$$\int_a^c f(x)dx = \int_a^b f(x)dx + \int_b^c f(x)dx$$

The next section will visit the implementation side of things in calculus using R.

Working with calculus in R

In this section, we will use the `mosaicCalc` package to perform calculus-related operations. The functions we will work with are mostly analytical (have explicit expression) and simple in nature. The following code snippet checks if this package is installed and will install the package if the condition evaluates to `true`, followed by importing the package to the current session:

```
if(!require("mosaicCalc")){
   install.packages("mosaicCalc")
}
library(mosaicCalc)
```

To avoid too much imagination, a good way to learn about unknown functions is to plot them out. Let us see how to plot a function using the `mosaicCalc` package.

Plotting basic functions

There are a few input parameters we need to specify upon plotting a function using the `mosaicCalc` package. Overall, we need to specify the expression of the function, the input variable(s), the domain of each input variable used to plot the function, and the values of other parameters set in advance.

There are three common graphing functions for plotting purposes: `slice_plot()`, used to plot functions with only one input variable, `contour_plot()`, used to plot functions with two input variables, and `interactive_plot()`, used to plot interactive graphs. These functions allow us to translate mathematical expressions into code and then plots. Let us explore each of them via the following exercise.

Exercise 9.1 – Plotting basic functions

This exercise will explore the plotting capabilities of the preceding functions, starting with `slice_plot()`. Follow the next steps:

1. Plot the function $y = 2x + 1$ for $x \in [-5,5]$ using the `slice_plot()` function, like so:

    ```
    >>> slice_plot(2*x+1 ~ x, domain(x = range(-5, 5)))
    ```

    Here, we put the function expression $2x + 1$ on the left of the tilde sign (~) and the input variable x on the right. We also specify the domain by passing the boundaries to the `range()` function. Running this command generates the output shown in *Figure 9.9*:

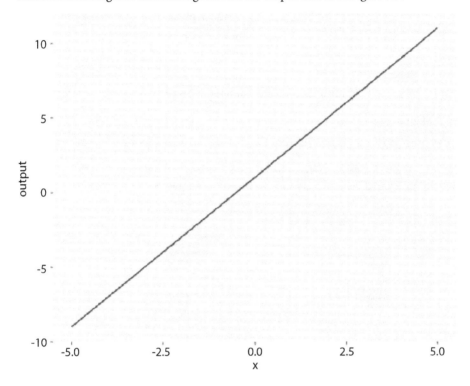

Figure 9.9 – Plotting a straight line using the slice_plot() function

We can also make the function more general by setting the prior parameters for the coefficient m and intercept b of the line $y = mx + b$. This makes the code more general since we only need to specify m and b in the following code snippet to plot any straight line:

```
m = 2
b = 1
slice_plot(m*x+b ~ x, domain(x = range(-5, 5)))
```

Note that an error would occur if no initial value for the input parameter were set, as shown in the following snippet:

```
>>> slice_plot(a*x+b ~ x, domain(x = range(-5, 5)))
Error in slice_plot(a * x + b ~ x, domain(x = range(-5, 5))) :
    Parameter <a> without specified numerical values.
```

In addition, we can use the makeFun() function to give a name to the function to be plotted, as shown in the following code snippet:

```
f = makeFun(2*x+1 ~ x)
slice_plot(f(x) ~ x, domain(x = range(-5, 5)))
```

Once the function is named, we can pass an arbitrary input value to evaluate the function. For example, setting $x = 2$ gives a return of 5, as seen here:

```
>>> f(x=2)
5
```

Next, we look at generating a contour plot for a function with two input variables.

2. Make a contour plot of the equation $z = 2x + 3y$ using the contour_plot() function, like so:

```
>>> contour_plot(2*x + 3*y ~ x & y, domain(x=-5:5, y=-5:5))
```

Here, we use the & sign to indicate more than one input variable and set the corresponding range for both variables. Running this command generates the output shown in *Figure 9.10*:

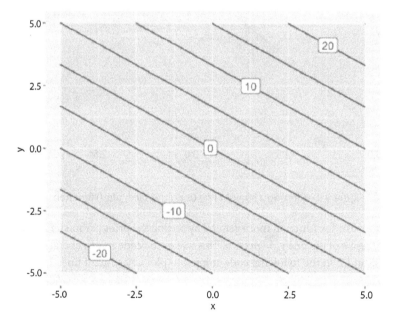

Figure 9.10 – Generating a contour plot of $z = 2x + 3y$ using the contour_plot() function

Next, we look at generating an interactive plot using the `interactive_plot()` function.

3. Generate an interactive 3D plot for the same expression using the `interactive_plot()` function, as follows:

```
>>> interactive_plot(2*x + 3*y ~ x & y, domain(x=-5:5, y=-5:5))
```

Running this command generates the output shown in *Figure 9.11*. Note that the resulting plot is an interactive HTML widget, allowing us to move around and displaying auxiliary information upon mouseover:

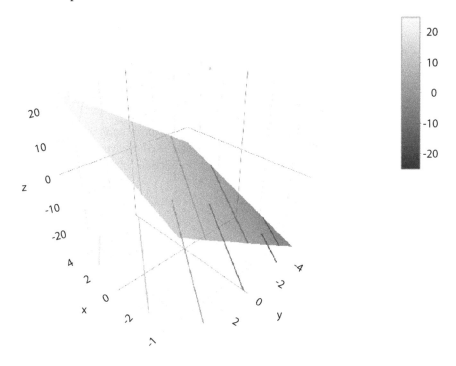

Figure 9.11 – Generating an interactive 3D plot for $z = 2x + 3y$ using the interactive_plot() function

The next section covers working with derivatives via the differentiation operation using the `D()` function.

Working with derivatives

The differentiation operation is completed via the `D()` differentiation operator in R, which inputs a single expression and outputs the derivative function. The input specifies the same expression as required by the plotting function earlier. For example, to specify the function $y = x^2 + 1$, we can pass `x^2+1 ~ x` to the `D()` function, which will then automatically calculate the derivative function $y' = 2x$. We can then assign the result to another variable, which serves as the derivative function and can be used to evaluate the derivative value at any point in the domain.

The following code snippet illustrates the process of obtaining the derivative function. The return of the D () operator is the derivative function, which is $y' = 2x$ after printing it out:

```
f_prime = D(x^2+1 ~ x)
>>> f_prime
function (x)
2 * x
<bytecode: 0x144378300>
```

The following code snippet evaluates two input values. The result shows that the D () function is able to correctly calculate the derivative function and perform an evaluation at an arbitrary input location:

```
>>> f_prime(1)
a
>>> f_prime(2)
4
```

Note that the D () function can carry out the rules mentioned previously upon calculating the derivative function. For example, to obtain the derivative of $y = \sin(x^2 - 5)$, we would invoke the chain rule and calculate the derivative as $y' = 2x\cos(x^2 - 5)$. The D () function completes this for us, as shown in the following code snippet:

```
f_prime = D(sin(x^2-5) ~ x)
>>> f_prime
function (x)
2 * x * cos(x^2 - 5)
```

Let us also verify the quotient rule. In the following code snippet, we pass the function $y = \frac{2x}{x+1}$ to D (), which should ideally return the derivative function $y' = \frac{2(x+1) - 2x}{(x+1)^2}$. The result shows that this is indeed the case:

```
f_prime = D(2*x/(x+1) ~ x)
>>> f_prime
function (x)
{
    .e1 <- 1 + x
    (2 - 2 * (x/.e1))/.e1
}
```

The next section covers the use of symbolic parameters in functions.

Using symbolic parameters

Symbolic parameters offer generality in constructing functions. The same as earlier, we can encode the prior values of parameters before passing them to the expression of the function. The following exercise illustrates this point.

Exercise 9.2 –Using symbolic parameters

In this exercise, we will look at a general function $y = Ax^3 + Bx + 3$, where A and B are constants and x is the only random input variable. The derivative function will be $y' = 3Ax^2 + B$. Follow the next steps:

1. Calculate the derivative of the function $y = Ax^3 + Bx + 3$, as follows:

    ```
    f_prime = D(A*x^3+B*x+3 ~ x)
    >>> f_prime
    function (x, A, B)
    3 * A * x^2 + B
    ```

    We see that the derivative function is correctly calculated.

2. Evaluate the derivative function when $x = 2$, $A = 2$, and $B = 3$, as follows:

    ```
    >>> f_prime(x=2, A=2, B=3)
    27
    ```

    We can also evaluate the function via multiple points, which amounts to plotting the derivative function across a specific range.

3. Plot the derivative function within the range $x \in [-5,5]$ with $A = 2$ and $B = 3$, like so:

    ```
    >>> slice_plot(f_prime(x, A=2, B=3) ~ x, domain(x=range(-5,5)))
    ```

Running this command generates the output shown in *Figure 9.12*:

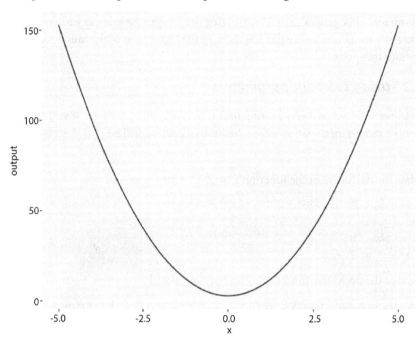

Figure 9.12 – Visualizing the derivative function $y' = 6x^2 + 3$

The next section covers the second derivative.

Working with the second derivative

The second derivative, expressed as $f''(x)$, is just the derivative of the first derivative of the raw function $f(x)$. In the previous example, when $y = Ax^3 + Bx + 3$, we have $y' = 2Ax^2 + B$. Taking another derivative gives $y'' = 4Ax$.

Let us look at how to obtain the second derivative via the following exercise.

Exercise 9.3 – Calculating the second derivative

In this exercise, we will calculate the second derivative of a raw function $f(x)$. The second derivative can be considered as differentiating $f(x)$ with respect to x twice, giving $f''(x) = y'' = \frac{d^2}{dx^2}f(x)$. The double differentiation is achieved by having two x's on the right-hand side of the tilde sign. Follow the next steps:

1. Calculate the second derivative $f''(x)$, as follows:

    ```
    f_pprime = D(A*x^3+B*x+3 ~ x & x)
    >>> f_pprime
    ```

```
function (x, A, B)
6 * A * x
```

The result shows that the second derivative of the draw function is correctly calculated.

2. Evaluate the second derivative function by setting when $x = 2$, $A = 2$, and $B = 3$, as follows:

    ```
    >>> f_pprime(x=2, A=2, B=3)
    24
    ```

In fact, since the second derivative function has nothing to do with the parameter B, any value of B will render the same result. For example, the following code returns the same result even when $B = 1$:

    ```
    >>> f_pprime(x=2, A=2, B=1)
    24
    ```

3. Plot the second derivative function within the range $x \in [-5,5]$ with $A = 2$, like so:

    ```
    >>> slice_plot(f_pprime(x, A=2) ~ x, domain(x=range(-5,5)))
    ```

Running this command generates the output shown in *Figure 9.13*. Here, we note that the second derivative function is a straight line as compared to the bell-shaped curve of the first derivative function. This straight line represents how fast the first derivative function changes:

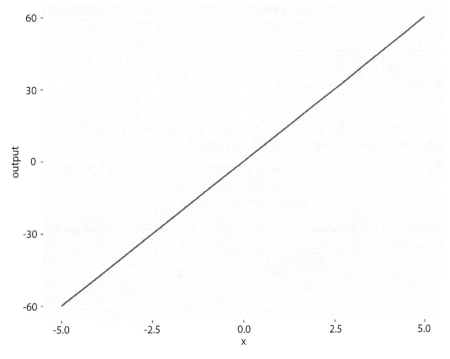

Figure 9.13 – Visualizing the second derivative function $y'' = 12x$

The next section covers more general partial derivative functions.

Working with partial derivatives

The second derivative function can be considered a special case of a partial derivative function. Assume a two-dimensional function $z = f(x, y)$. The second derivative of x is obtained by differentiating x twice. We can also differentiate x first and y later, giving $\frac{d^2}{dxdy} z$ or, equivalently, $\frac{d^2}{dydx} z$. Let us see how this works via the following exercise.

Exercise 9.4 – Calculating the partial derivative

In this exercise, we will calculate three different partial derivatives: $\frac{d^2}{dx^2} z$, $\frac{d^2}{dxdy} z$ (or $\frac{d^2}{dydx} z$), and $\frac{d^2}{dy^2} z$, based on the raw function $z = Ax^2 + Bxy + Cy^2$. Follow the next steps:

1. Calculate the second derivative of z with respect to x. Check whether the result is $\frac{d^2}{dx^2} z = 2A$:

    ```
    f_pprime = D(A*x^2 + B*x*y + C*y^2 ~ x & x)
    >>> f_pprime
    function (x, y, A, B, C)
    2 * A
    ```

    The result shows that the calculation is correct.

2. Calculate the partial derivative of z with regard to x and then y. Check whether the result is $\frac{d^2}{dxdy} z = B$:

    ```
    f_pprime = D(A*x^2 + B*x*y + C*y^2 ~ x & y)
    >>> f_pprime
    function (x, y, A, B, C)
    B
    ```

    The result shows that the calculation is correct. We can also differentiate y first and then x. As shown in the following code snippet, this gives the same result:

    ```
    f_pprime = D(A*x^2 + B*x*y + C*y^2 ~ y & x)
    >>> f_pprime
    B
    ```

3. Calculate the partial derivative of z with respect to x and then y. Check whether the result is $\frac{d^2}{dy^2} z = 2C$:

    ```
    f_pprime = D(A*x^2 + B*x*y + C*y^2 ~ y & y)
    >>> f_pprime
    function (x, y, A, B, C)
    2 * C
    ```

    The result shows that the calculation is correct.

The next section covers calculating integrals, or antiderivatives, using R.

Working with integration in R

Recall that differentiation is performed using the D() function. Assuming $y = Ax^2 + Bx + 3$, we have $y' = f'(x) = 2Ax + B$. We can plot the raw function $f(x)$ and its derivative function $f'(x)$ to facilitate the comparison.

In the following code snippet, we create this expression using the makeFun() function and name the function f:

```
f = makeFun( A*x^2 + B*x + 3 ~ x)
>>> f
function (x, A, B)
A * x^2 + B * x + 3
```

We can do a simple evaluation as follows:

```
>>> f(1, A=1, B=1)
5
```

Now, we obtain the derivative function and store the function in f_prime:

```
f_prime = D(f(x) ~ x)
>>> f_prime
function (x, A, B)
2 * A * x + B
```

This is a new function derived from the original function. We also do a simple evaluation for this function, as follows:

```
>>> f_prime(x=1, A=1, B=1)
3
```

Now, let us plot the raw function $f(x) = x^2 + x + 3$:

```
>>> slice_plot(f(x) ~ x, domain(x = -1:1)) %>%
  gf_labs(title = "Original function f(x)")
```

Running the command generates the output shown in *Figure 9.14*. Here, we have used the gf_labs ()
function to set the title of the graph:

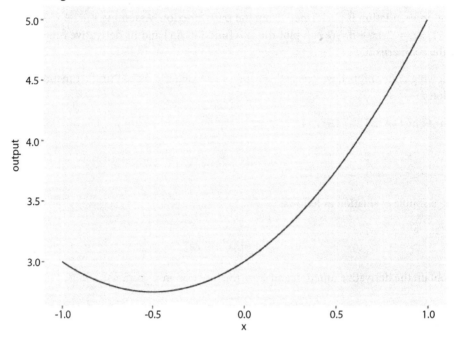

Figure 9.14 – Visualizing the raw function $f(x) = x^2 + x + 3$

We also plot the derivative function $f'(x) = 2x + 1$ as follows:

```
>>> slice_plot(f_prime(x, A=1, B=1) ~ x, domain(x =-1:1), color =
"red") %>%
   gf_labs(title = "Derivative function f'(x)")
```

Running the command generates the output shown in *Figure 9.15*:

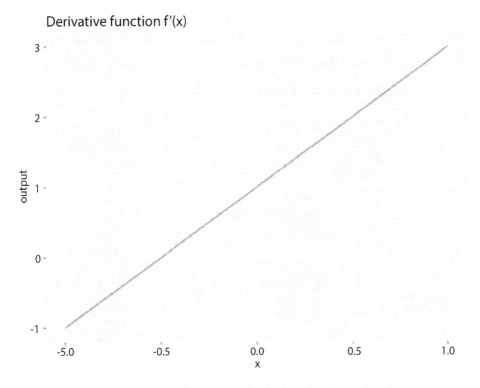

Figure 9.15 – Visualizing the derivative function $f'(x) = 2x + 1$

With the derivative function in place, we can obtain its antiderivative using the antiD() function, as follows:

```
F_integral <- antiD(f_prime(x) ~ x)
>>> F_integral
function (x, A, B, C = 0)
A * x^2 + x * B + C
```

Note that the correct form of the original function is recovered, besides the additional constant C that is added as an additional input argument to the antiderivative function with a default value of 0.

We can evaluate the antiderivative function as well, as follows:

```
>>> F_integral(1, A=1, B=1)
2
```

Taking A=1, B=1, and C=0, the antiderivative function $F(x) = x^2 + x$ is visualized as follows:

```
>>> slice_plot (F_integral (x, A=1, B=1) ~ x, domain (x=-1:1)) %>%
    gf_labs (title = "Antiderivative function F_integral (x) ")
```

Running the command generates the output shown in *Figure 9.16*:

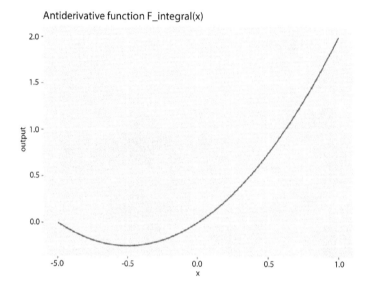

Figure 9.16 – Visualizing the derivative function $f(x) = 2x + 1$

Now, if we differentiate the antiderivative function $F(x) = Ax^2 + Bx + C$ again, we would expect to obtain the same derivative function $f(x) = 2Ax + B$. The following code snippet verifies this result:

```
f_prime2 = D(F_integral (x) ~ x)
>>> f_prime2
function (x, A, B, C = 0)
2 * A * x + B
```

Let us pause a moment and look at the relationship between derivative functions and their antiderivatives.

More on antiderivatives

The anti-differentiation operation that produces $F(x) = \int f'(x)dx + C$ is the reverse of the differentiation that generates $f'(x)$. It represents a family of functions, including the original function $f(x)$. These functions are very much related to each other. Specifically, the derivative function $f'(x)$, which is a derived function, tells the rate of change of the original function $f(x)$ for an arbitrary input point x. It gives a local property of $f(x)$ in a microscopic lens, measuring the sensitivity of $f(x)$ at the current point.

However, we may not always need to derive $f'(x)$ in every scenario. Sometimes, we will work with the derivative function $f'(x)$ as a start, and would be interested in deriving the original function—that is, the antiderivative function $F(x)$. This gives a global property of the raw function $f(x)$, representing the values accumulated within a specific range.

Obtaining the unknown raw function is called anti-differentiation, or integration. The result is called the integral. Depending on whether the integration is evaluated on specific boundaries or not, we have an indefinite integral and a definite integral.

The integration operation generates a family of anti-derivative functions. As in the previous example, the anti-derivative function is $F(x) = Ax^2 + Bx + C$. Setting A=1 and B=1, we will obtain different results at the same input location for different choices of C, as shown in the following code snippet:

```
>>> F_integral(x=1, A=1, B=1, C=0)
2
>>> F_integral(x=1, A=1, B=1, C=1)
3
>>> F_integral(x=1, A=1, B=1, C=2)
4
```

Although these are different raw functions, they all share the same derivative function, as illustrated in the following snippet:

```
>>> D(F_integral(x, A=1, B=1, C=0) ~ x)
function (x)
2 * x + 1
>>> D(F_integral(x, A=1, B=1, C=1) ~ x)
function (x)
2 * x + 1
>>> D(F_integral(x, A=1, B=1, C=2) ~ x)
function (x)
2 * x + 1
```

The family of antiderivative functions, denoted by $F(x) = \int f'(x)dx + C$, thus corresponds to infinitely many functions, including the raw function $f(x)$. These infinitely many antiderivative functions are essentially vertical shifts of $f(x)$, giving $F(x) = f(x) + C$.

Also, note that the derivative function shares the same set of input arguments as the original function, as verified in the following code snippet:

```
>>> f
function (x, A, B)
A * x^2 + B * x + 3
<bytecode: 0x128185eb0>
>>> f_prime
```

```
function (x, A, B)
2 * A * x + B
<bytecode: 0x11fbc6628>
```

However, the antiderivative function requires an additional argument—the constant C, as seen here:

```
>>> F_integral
function (x, A, B, C = 0)
A * x^2 + x * B + C
<bytecode: 0x128303540>
```

The next section looks at how to evaluate the definite integral.

Evaluating the definite integral

Calculating the definite integral requires evaluating the indefinite integral twice: once at the starting point of integration, and once at the end point of integration.

For example, assume the antiderivative function is $F(x) = x^2 + x + C$ for A=1 and B=1. To calculate the indefinite integral $\int_2^3 f(x)dx$, we would evaluate $F(x)$ at both $x = 2$ and $x = 3$ and take their difference, giving $F(3) - F(2)$. The following code snippet shows the result:

```
>>> F_integral(x=3, A=1, B=1) - F_integral(x=2, A=1, B=1)
6
```

Here, one thing to note is that the constant C is assumed to be 0 in both evaluations. In fact, it does not matter which value it assumes, since the constants from both evaluations will always cancel out each other. In other words, we will always have a fixed definite integral, despite the indefinite integral corresponding to infinitely many values at a specific input point.

Summary

In this chapter, we covered the basics of calculus, including differential calculus and integral calculus. In the first section, we introduced an intuitive understanding of these two branches of calculus and covered the fundamentals of common functions and their properties. We started introducing the concept of limit and its connection to the definition of a derivative, followed by covering common derivative rules and properties. We also discussed integral calculus, including indefinite integrals and definite integrals, along with their rules and properties.

The second and third sections touched upon implementations in R. We introduced how to carry out common differentiation and integration using the D() and antiD() functions, with several examples illustrating their usage and conversion between the derivative function and its antiderivative.

In the next chapter, we will enter into the realm of mathematical statistics, starting with the basics of probability.

Part 3: Fundamentals of Mathematical Statistics in R

As we embark on the third part of this book, we transition from the mathematical foundations explored in *Part 2* to the realm of advanced statistical methods and model building. While the preceding parts laid the groundwork in terms of essential statistics and mathematical underpinnings, this part is designed to push your capabilities to a more advanced level, all while utilizing the power of R for practical implementations.

By the end of this part, you'll possess a robust set of advanced statistical and modeling skills. This part provides the advanced knowledge and hands-on experience you need to venture into more advanced topics in data science.

This part has the following chapters:

- *Chapter 10, Probability Basics*
- *Chapter 11, Statistics Estimation*
- *Chapter 12, Linear Regression in R*
- *Chapter 13, Logistic Regression in R*
- *Chapter 14, Bayesian Statistics*

10

Probability Basics

Probability distribution is an essential concept in statistics and machine learning. It describes the underlying distribution that governs the generation of potential outcomes or events in an experiment or random process. There are different types of probability distributions, depending on the specific domain and characteristics of the data. A proper probability distribution is a useful tool in understanding and modeling the behavior of random processes and events, providing convenient tools for decision-making and predictions when developing data-driven predictive and optimization models.

By the end of this chapter, you will understand the common probability distributions and their parameters. You will also be able to use these probability distributions to perform usual tasks such as sampling and probability calculations in R, as well as common sampling distribution and order statistics.

In this chapter, we will cover the following topics:

- Introducing probability distribution
- Exploring common discrete distributions
- Discovering common continuous distributions
- Understanding common sampling distributions
- Understanding order statistics

Technical requirements

To run the code in this chapter, you will need to have the latest versions of the following packages:

- `ggplot2`, 3.4.0
- `dplyr`, 1.0.10

Please note that the versions of the packages mentioned in the preceding list are the latest ones at the time of writing this chapter.

The code and data for this chapter is available at `https://github.com/PacktPublishing/` `The-Statistics-and-Machine-Learning-with-R-Workshop/blob/main/` `Chapter_10/working.R`.

Introducing probability distribution

Probability distribution provides a framework for understanding and predicting the behavior of random variables. Once we know the underlying data-generating probability distribution, we can make more informed decisions about how things are likely to appear, either in a predictive or optimization context. In other words, if the selected probability distribution can model the observed data very well, we have a powerful tool to predict potential future values, as well as the uncertainty of such occurrence.

Here, a random variable is a variable whose value is not fixed and may assume multiple or infinitely many possible values, representing the outcomes (or realizations) of a random event. Probability distributions allow us to represent and analyze the probability of these outcomes, offering a comprehensive view of the underlying uncertainties in various scenarios. A probability distribution takes the random variable, denoted as x, and converts it into a probability, $P(x)$, a floating number valued between 0 and 1. The probability distribution can be a **probability density function (PDF)** or **probability mass function (PMF)** that specifies the probability of observing an outcome for a continuous variable (or discrete variable), or a **cumulative distribution function (CDF)** that provides the total probability that a random variable is less than or equal to a given fixed quantity.

In the following example, we use $f(x)$ to represent the probability density function of x, and $F(x)$ to represent the CDF.

There are two main categories of probability distributions: discrete probability distribution and continuous probability distribution. A discrete probability distribution deals with discrete variables, which are random variables that can assume a limited or countable number of possible values. For example, if we have a probability distribution that specifies the probability of experiencing a rainy day in a week, the underlying random variable is the day of the week and can only take an integer value between 1 and 7. Let's assume there is a total of C possible values for the discrete random variable. For a given possible value, x_i, with $i \in \{1, \dots, C\}$, the corresponding PMF is $f(x_i) \in [0,1]$, and all probabilities should sum to 1, giving $\sum_{i=1}^{C} f(x_i) = 1$. We can think of PMF $f(x)$ as a bar chart that specifies the probability output for each discrete input.

We will cover a few common discrete distributions. For example, the binomial distribution models the number of successes in a fixed number of **Bernoulli** trials with the same probability of success, the **Poisson** distribution models the number of events within a fixed interval of time or space, and the geometric distribution models the number of trials required for the first success in a sequence of Bernoulli trials. These will be covered later in this chapter.

Continuous probability distributions, on the other hand, involve continuous variables, which can take an infinite number of values within a specified range, denoted as $\mathcal{X}$. The random variable, $x \in \mathcal{X}$, is now continuous, and the corresponding PDF $f(x) \in [0,1]$ satisfies $\int f(x)dx = 1$ for $x \in \mathcal{X}$, where we have switched the summation sign to an integral to account for an infinite amount of possible values of the continuous variable, x. We can think of PDF $f(x)$ as a line plot that specifies the probability output for each continuous input.

We will cover a few widely used continuous distributions, starting with the normal (or Gaussian) distribution, which describes the distribution of many natural quantities. Other examples of continuous distributions include the exponential distribution, which models the time between independent events in a Poisson process, and the uniform distribution, which assigns equal probability to all outcomes within a specified range.

Figure 10.1 summarizes these two types of probability distributions:

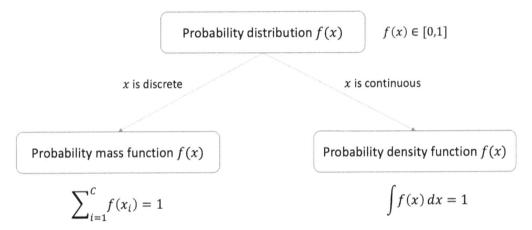

Figure 10.1 – Summarizing the two categories of probability distributions. Both distributions sum to 1

Note that each probability distribution has an associated closed-form expression with a corresponding set of parameters. We will highlight the expression and parameters for each distribution next, starting with the discrete distributions.

Exploring common discrete probability distributions

Discrete probability distributions are characterized by their corresponding PMFs, which assign a probability to each possible outcome of the input random variable. The sum of the probabilities for all possible outcomes in a discrete distribution equals 1, leading to $\sum_{i=1}^{C} f(x_i) = 1$. This also means that one of the outcomes *must* occur, giving $f(x_i) > 0, \forall i = 1, ..., C$.

Discrete probability distributions are vital in various fields, such as finance. They are commonly used for statistical analyses, including hypothesis testing, parameter estimation, and predictive modeling. We can use discrete probability distributions to quantify uncertainties, make predictions, and gain insights into the underlying data-generating process of the observed outcomes.

Let's start with the most fundamental discrete distribution: the Bernoulli distribution.

The Bernoulli distribution

The **Bernoulli distribution** is a fundamental discrete probability distribution that specifies the behavior of a binary random variable in a single Bernoulli trial. Here, a **Bernoulli trial** is a single experiment with only two possible outcomes, which can also be labeled as "success" and "failure." It is the simplest discrete probability distribution and serves as the basis for more complex distributions, such as binomial and geometric distributions.

The Bernoulli distribution is widely used in modeling scenarios with binary outcomes, such as coin tosses, yes/no survey questions, or the presence/absence of a specific feature in a dataset. For example, in statistical hypothesis testing, the Bernoulli distribution is often used in scenarios such as comparing the success rates of two treatments for a medical condition. In finance, the Bernoulli distribution can be used to model binary outcomes, such as a stock's price going up or down.

As a convention, the two outcomes in a Bernoulli distribution are often encoded as 1 for success and 0 for failure. The Bernoulli distribution is characterized by a single parameter, p, which represents the probability of success. In other words, we have $f(x = 1) = p$. Similarly, since total probabilities sum to 1, the probability of failure is given by $1 - p$, giving $f(x = 0) = 1 - p$.

We can express the PMF of a Bernoulli distribution as follows:

$$f(x) = \begin{cases} p, & \text{if } x = 1 \\ 1 - p, & \text{if } x = 0 \end{cases}$$

Alternatively, we can express $f(x)$ in a more compact form, as follows. It is easy to verify that these two representations are equivalent:

$$f(x = i) = p^i(1 - p)^{1-i} \text{ for } i \in \{0,1\}$$

Note that $p \in [0,1]$.

As for the mean, μ (first moment), and variance, σ^2 (second moment), which characterize the Bernoulli distribution, we have the following:

$$\mu = p$$

$$\sigma^2 = p(1 - p)$$

We'll simulate and analyze Bernoulli-distributed random variables using R in the following exercise.

Exercise 10.1 – simulating and analyzing Bernoulli-distributed random variables

In this exercise, we will simulate and analyze Bernoulli-distributed random variables using the `rbinom()` function:

1. Simulate a single Bernoulli trial with a success probability of 0.6:

    ```
    # The probability of success
    p = 0.6
    ```

```
# Produce a random Bernoulli outcome
outcome = rbinom(1, size = 1, prob = p)
>>> print(outcome)
0
```

Here, the outcome will be displayed as either 0 or 1. We can control the random seed to ensure the reproducibility of the results:

```
set.seed(8)
>>> rbinom(1, size = 1, prob = p)
1
```

2. Generate five random Bernoulli outcomes with the same probability of success:

```
# Number of experiments
n = 5
# Generate corresponding outcomes
outcomes = rbinom(n, size = 1, prob = p)
>>> print(outcomes)
1 0 0 1 0
```

3. Calculate the mean and variance of the Bernoulli distribution:

```
# Get mean and variance
mean_bernoulli = p
var_bernoulli = p * (1 - p)
>>> cat("Mean:", mean_bernoulli, "\nVariance:", var_bernoulli)
Mean: 0.6
Variance: 0.24
```

Here, we use the cat() function to concatenate and print out the results.

4. Analyze the results of multiple Bernoulli trials in terms of the observed/empirical probability of success:

```
# Number of successes
num_successes = sum(outcomes)
# Empirical probability of success
empirical_p = num_successes / n
>>> cat("Number of successes:", num_successes, "\nEmpirical
probability of success:", empirical_p)
Number of successes: 2
Empirical probability of success: 0.4
```

Since we only sampled 5 times, the resulting empirical probability of success (0.4) is very far from the true probability (0.6). We can enlarge the size of random trials to get a more reliable estimate:

```
n = 1000
num_successes = sum(rbinom(n, size = 1, prob = p))
```

```
empirical_p = num_successes / n
>>> cat("Number of successes:", num_successes, "\nEmpirical
probability of success:", empirical_p)
Number of successes: 600
Empirical probability of success: 0.6
```

Using a total of 1000 trials, we can now reproduce the exact true probability of success.

The next section reviews the binomial distribution.

The binomial distribution

The **binomial distribution** extends the Bernoulli distribution to multiple repeated trials, where each trial is a single Bernoulli experiment that assumes a value of 1 or 0. Specifically, it is a discrete probability distribution that specifies the count of successes in a given number of Bernoulli trials. These trials are independent and share the same probability of success.

There are two parameters for the binomial distribution PMF:

- The number of trials, n

- The probability of success, p, in each trial

> **Note**
> We still assume only two possible outcomes: success (1) or failure (0). The probability of failure can also be represented as q, where $q = 1 - p$.

The PMF of the binomial distribution with a total of k successes is given in the following formula:

$$P(x = k) = C(n,k)p^k(1 - p)^{n-1}$$

Here, $C(n,k)$ represents the number of combinations of choosing k successes from n trials, which can be calculated using the binomial coefficient formula:

$$C(n,k) = \frac{n!}{k!(n-k)!}$$

The first two moments are as follows:

$$\mu = np$$

$$\sigma^2 = np(1 - p)$$

Using the PMF of the binomial distribution, we can compute the probability of observing a specific number of successes in a specific number of trials.

The binomial distribution has some important relationships with other probability distributions. For instance, as *n* approaches infinity and *p* remains constant, the binomial distribution converges to the normal distribution, a continuous probability distribution to be introduced later.

Let's go through an exercise to get familiar with the functions related to binomial distribution.

Exercise 10.2 – simulating and analyzing binomial random variables

In this exercise, we will simulate and analyze binomial-distributed random variables using the dbinom() and pbinom() functions:

1. Use the dbinom() function to calculate the probability of observing 0 to 10 successes based on a binomial distribution with a success probability of 0.5 and a total of 10 trials:

    ```
    n = 10 # Number of trials
    p = 0.5 # Probability of success
    # Get binomial probabilities for different occurrences of
    successes
    binom_probs = dbinom(0:n, n, p)
    >>> binom_probs
     [1] 0.0009765625 0.0097656250 0.0439453125 0.1171875000
     [5] 0.2050781250 0.2460937500 0.2050781250 0.1171875000
     [9] 0.0439453125 0.0097656250 0.0009765625
    ```

    Here, we use 0:n to create a list of integers from 0 to 10, each of which will then get passed to the dbinom() function to evaluate the corresponding probability.

2. Create a bar plot of the binomial probabilities using the barplot() function:

    ```
    >>> barplot(binom_probs, names.arg = 0:n, xlab = "Number of
    Successes", ylab = "Probability", main = "Binomial Distribution
    (n = 10, p = 0.5)")
    ```

    Running this code generates *Figure 10.2*, which shows that the middle occurrence of 5 has the highest probability:

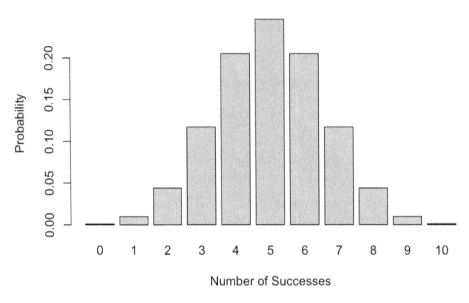

Figure 10.2 – Visualizing the binomial distribution with n=10 and p=0.5

3. Calculate the cumulative binomial probabilities using the pbinom() function:

```
cum_binom_probs <- pbinom(0:n, n, p)
>>> cum_binom_probs
[1] 0.0009765625 0.0107421875 0.0546875000 0.1718750000
[5] 0.3769531250 0.6230468750 0.8281250000 0.9453125000
[9] 0.9892578125 0.9990234375 1.0000000000
```

The result shows that the cumulative binomial probabilities from the CDF are calculated as the cumulative sum of the previous element-wise probabilities in the PMF.

We can also use the CDF to compute the probability of observing a specific value or higher.

4. Calculate the probability of obtaining at least seven successes:

```
prob_at_least_7_successes = 1 - pbinom(6, n, p)
>>> prob_at_least_7_successes
0.171875
```

Here, the probability of obtaining at least 7 successes is calculated by taking the complement of observing 6 or fewer successes – that is, we have $P(X \geq 7) = 1 - P(X \leq 6)$.

Let's go through another application-related exercise to put these calculations into perspective.

Exercise 10.3 – calculating winning probabilities

In this exercise, we will calculate the winning probabilities of a sports team:

1. Suppose the sports team has a probability of 80% of winning a match. If there are a total of five matches, what is the probability of winning at least four matches?

```
n = 5
p = 0.8
prob_at_least_4_wins = 1 - pbinom(3, n, p)
>>> prob_at_least_4_wins
0.73728
```

2. Calculate the probability of winning, at most, three matches:

```
prob_at_most_3_wins = pbinom(3, n, p)
>>> prob_at_most_3_wins
0.26272
```

Note that this probability is a complement to the previous probability of winning at least four matches. We can verify this relationship as follows:

```
>>> prob_at_most_3_wins == 1 - prob_at_least_4_wins
TRUE
```

In the following section, we'll pause to discuss the normal approximation to the binomial distribution. This relationship is widely used in statistical analysis and has its roots in the central limit theorem.

The normal approximation to the binomial distribution

The normal (or **Gaussian**) approximation to the binomial distribution says that the binomial distribution can be approximated by a normal distribution that shares the same mean value ($\mu = np$) and variance value ($\sigma^2 = np(1 - p)$). Such normal approximation to the binomial distribution becomes more accurate as the number of experiments (n) increases, and the success probability (p) does not approach 0 or 1. As a rule of thumb, we often use the normal approximation when both $np \geq 10$ and $nq \geq 10$.

To use the normal approximation, we need to standardize the binomial random variable, x, by converting it into the form of a standard normal variable, z, based on the following formula:

$$z = \frac{x - \mu}{\sigma}$$

Here, $\mu = np$ and $\sigma = \sqrt{np(1 - p)}$. This is also called the **z-score**. Going through such standardization is a common practice when trying to compare different quantities on the same scale. We can make use of the standard normal distribution (to be introduced later) to work on the corresponding PDF or CDF, depending on the specific task. In this case, we can use the standard normal distribution (that is, $N(0,1)$) to approximate the probabilities associated with the binomial distribution.

Let's look at a concrete example. Suppose we toss a coin 100 times (assuming a fair coin with an equal probability of landing with a head or a tail) and would like to compute the probability of obtaining between 40 and 60 heads (both inclusive). Let x be the random variable that denotes the number of heads. We know that x assumes a binomial distribution with $n = 100$ and $p = 0.5$.

To check whether the normal approximation is appropriate, we calculate $np = 100*0.5 = 50 > 10$ and $np(1 - p) = 100*0.5*0.5 = 25 > 10$. We also verify the same using R, as shown in the following code snippet:

```
n = 100
p = 0.5
# check conditions for normal approximation
>>> n*p > 10
TRUE
>>> n*p*(1-p) > 10
TRUE
```

Both conditions evaluate TRUE. Now, we can standardize the upper and lower limits (60 and 40, respectively) to convert them into a standardized score. To do this, we need to obtain the parameters of the binomial distribution, followed by applying the standardization formula, $z = \frac{x-\mu}{\sigma}$, to get the z-score. The following code snippet completes the standardization:

```
# compute mean and std
mu = n*p
>>> mu
50
std = sqrt(n*p*(1-p))
>>> std
5
# compute P(lower_limit <= X <= upper_limit)
lower_limit = 40
upper_limit = 60
# Using z score
standard_lower_limit = (lower_limit - mu) / std
standard_upper_limit = (upper_limit - mu) / std
>>> standard_lower_limit
-2
>>> standard_upper_limit
2
```

With the standardized z-score, we can now calculate the original probability based on the standard normal distribution alone. In other words, we have the following:

$$P(40 \leq x \leq 60) = P\left(\frac{40-50}{5} \leq \frac{x-50}{5} \leq \frac{60-50}{5}\right) = P\left(-2 \leq z \leq 2\right)$$

As shown in the following code snippet, we can now call the pnorm() function to calculate the CDF at points $z = -2$ and $z = 2$, whose difference gives the final probability:

```
# approximate using standard normal cdf
>>> pnorm(standard_upper_limit) - pnorm(standard_lower_limit)
0.9544997
```

Let's also calculate the corresponding probability using the binomial distribution to see how close the normal approximation is. The following code snippet uses the pbinom() function to obtain the CDF of the binomial at both limits and then takes the difference to give the total probability of observing an outcome in this range:

```
# use binomial distribution
>>> pbinom(upper_limit, n, p) - pbinom(lower_limit, n, p)
0.9539559
```

The result shows that the approximation is accurate up to the second decimal place. Therefore, the normalization approximation provides an alternative approach to calculating the probabilities if directly using the binomial distribution is inconvenient. However, although the normal approximation is a powerful tool, we still need to check the required conditions to ensure a good approximation.

Figure 10.3 summarizes the two approaches to calculating the total probability of observing a specific range of values. We can calculate it by taking the difference in the CDF between the two boundaries of the range. Alternatively, we can rely on the normal distribution to approximate the binomial distribution and calculate the total probability after obtaining the standardized z-score based on these parameters, assuming the conditions are satisfied:

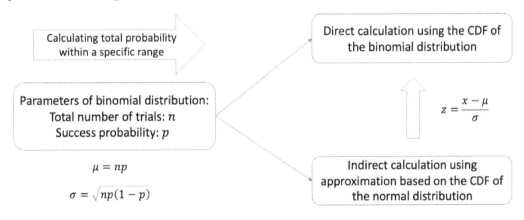

Figure 10.3 – Summarizing the normal approximation to the binomial distribution
when calculating the total probability of observing a specific range of values

We will review the Poisson distribution in the next section.

The Poisson distribution

Another popular discrete probability distribution is the Poisson distribution, which describes the number of events within a fixed interval of time or space. It has a single constant parameter that specifies the average rate of occurrence. Specifically, we must denote λ as the average rate of occurrence of events in the given interval. We can express the PMF of the Poisson distribution as follows:

$$P(x = k) = \frac{\lambda^k e^{-\lambda}}{k!}$$

Here, $P(x = k)$ denotes the probability of experiencing a total of k occurrences in the fixed interval, e is Euler's number (approximately 2.71), and $k!$ is the factorial of k (the product of all positive integers up to k).

With this equation, we can calculate the probability of observing any (integer) number of events within the given fixed interval. We can further calculate the mean and variance of the Poisson distribution as follows:

$$\mu = \lambda$$

$$\sigma^2 = \lambda$$

The Poisson distribution is often used to model rare events that occur independently and at a constant average rate. Some real-world applications of Poisson processes include the number of hotel bookings received at the front desk per hour and the number of emails arriving in an inbox within a day.

Let's go through an exercise to get familiar with common probability calculations related to the Poisson distribution.

Exercise 10.4 – simulating and analyzing Poisson-distributed random variables

In this exercise, we will use R to work with the Poisson distribution, including calculating the probabilities (using the PMF), plotting the distribution, and generating random samples. Specifically, we will calculate the Poisson probabilities for an average rate of occurrence of 5 events per interval ($\lambda = 5$), plot the PMF, and calculate the cumulative probabilities. We will also generate a random sample of 100 observations from this Poisson distribution:

1. Calculate the probabilities of observing 0 to 15 occurrences/events per interval based on a Poisson distribution parameterized by $\lambda = 5$ using the dpois() function:

    ```
    lambda = 5 # distribution parameter
    # Calculate probabilities for each scenario
    pois_probs = dpois(0:15, lambda)
    >>> pois_probs
    [1] 0.0067379470 0.0336897350 0.0842243375 0.1403738958
    [5] 0.1754673698 0.1754673698 0.1462228081 0.1044448630
    ```

```
 [9]  0.0652780393 0.0362655774 0.0181327887 0.0082421767
[13]  0.0034342403 0.0013208616 0.0004717363 0.0001572454
```

2. Create a bar plot of the Poisson probabilities:

```
>>> barplot(pois_probs, names.arg = 0:15, xlab = "Number of
Events", ylab = "Probability", main = "Poisson Distribution
(lambda = 5)")
```

Running this code generates *Figure 10.4*. As expected, the peak probabilities appear around five times:

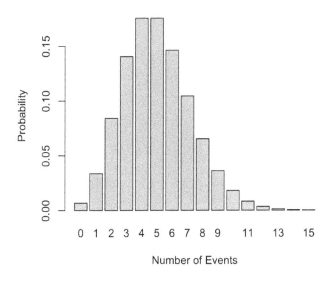

Figure 10.4 – Visualizing the PMF of the Poisson distribution as a bar plot

3. Calculate cumulative Poisson probabilities for each number of events (0 to 15) using the ppois() function:

```
cum_pois_probs = ppois(0:15, lambda)
>>> cum_pois_probs
 [1]  0.006737947 0.040427682 0.124652019 0.265025915
 [5]  0.440493285 0.615960655 0.762183463 0.866628326
 [9]  0.931906365 0.968171943 0.986304731 0.994546908
[13]  0.997981148 0.999302010 0.999773746 0.999930992
```

Let's also plot the CDF in a bar chart:

```
>>> barplot(cum_pois_probs, names.arg = 0:15, xlab = "Number of
Events", ylab = "Cumulative Probability", main = "CDF of Poisson
Distribution (lambda = 5)")
```

Running this code generates *Figure 10.5*. Note that the CDF curve increases sharply around the mean occurrence of five times and gradually saturates as we move toward the right end of the graph:

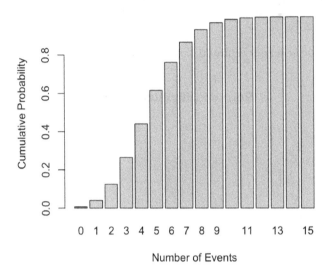

Figure 10.5 – Visualizing the CDF of the Poisson distribution as a bar plot

4. Generate 100 random samples from this Poisson distribution using the rpois() function:

```
pois_samples = rpois(100, lambda)
>>> pois_samples
  [1]  8  5  8  4  3  4  6  2  5  6  3  3  7  8  8  7
 [17]  5  9  6  1  4  2  7  7  5  5  5  2  7  4  6  5
 [33]  4  4  3  0  8  5  4  4  7  5 11  6  5  4  8  8
 [49]  2  5  6  2  3  4  6  4  6  2  5  3  6  0  5  8
 [65]  7  1  8  4  4  4  4  5  4  4  4  5  5  6  3  4
 [81]  3  0  8  9  2  3  4 13  2  6  8  9  6  4  7  7
 [97]  8  6  3  5
```

As expected, the majority of the occurrences are around five times.

Poisson approximation to binomial distribution

As it turns out, we can also use the Poisson distribution to approximate the binomial distribution under specific conditions. For instance, when the number of trials (n) in a binomial distribution is large and the success probability (p) is small, the binomial distribution can be approximated by a Poisson distribution with $\lambda = np$.

Let's use an example to demonstrate how to apply the Poisson approximation to the binomial distribution. Suppose we have a binomial distribution with $n = 1000$ and $p = 0.01$. We want to find the probability of observing exactly 15 successes:

1. We can start with the binomial probability and calculate the corresponding probability using the dbinom() function after specifying the parameters:

```
# Binomial parameters
n = 1000
p = 0.01
# Probability of observing 15 successes
binom_prob = dbinom(15, n, p)
>>> binom_prob
0.03454173
```

Next, we must calculate the approximate Poisson parameter, $\lambda = np$:

```
lambda_approx = n * p
>>> lambda_approx
10
```

Now, we can calculate the Poisson probability of observing 15 successes:

```
pois_approx_prob <- dpois(15, lambda_approx)
>>> pois_approx_prob
0.03471807
```

The result suggests that the approximation is quite accurate to the third decimal point.

One more interesting property is that adding up several independent Poisson-distributed random variables also produces a Poisson distribution, which is parameterized by the sum of the corresponding individual λ values. For example, if x_1 and x_2 are independent Poisson random variables with λ_1 and λ_2, respectively, their sum ($y = x_1 + x_2$) also follows a Poisson distribution with $\lambda = \lambda_1 + \lambda_2$. This gives a convenient property when working with the sum of multiple Poisson-distributed random variables.

In the next section, we'll cover another widely used discrete distribution: the geometric distribution.

The geometric distribution

The geometric distribution is a discrete probability distribution that describes the number of trials required for the first success in a sequence of independent Bernoulli trials, each with the same probability of success. Similar to the binomial distribution, the geometric distribution is a collection of multiple independent Bernoulli trials, although the subject of interest is the first occurrence of success in the sequence of trials. Here, the first occurrence of success means that all previous trials need to be non-success, and the current trial is the first success among multiple trials performed so far.

It is commonly used to model the waiting time until an event occurs or the number of attempts needed to achieve a desired outcome. Examples include the number of attempts to pass a driving test until success, the number of times we observe continuous sunny days, and the number of coin flips needed to obtain the first head. The geometric distribution is quite useful when modeling the waiting time or the number of attempts needed to achieve the first success in a sequence of independent Bernoulli trials.

The geometric distribution is defined by a single parameter, p, which represents the probability of success on each Bernoulli trial. The PMF of the geometric distribution is given by the following formula:

$$P(x = k) = (1 - p)^{k-1} p$$

This formula specifies the probability of observing the first success on the kth trial. Therefore, the probability is calculated as a joint probability of observing k individual events, where the first $k - 1$ events are non-success with a joint probability of $(1 - p)^{k-1}$, and the last event is a success with a probability of p.

The mean and variance parameters of a geometric distribution can be expressed as follows:

$$\mu = \frac{1}{p}$$
$$\sigma^2 = \frac{1 - p}{p^2}$$

Note that the geometric distribution is **memoryless**, meaning that the probability of success in the next trial does not depend on the past trials. In other words, the waiting time until the first success does not change, regardless of the number of past trials that have already been conducted.

Let's go through an exercise to simulate and analyze a random variable while following the geometric distribution.

Exercise 10.5 – simulating and analyzing geometrically-distributed random variables

In this exercise, we will use R to work with the geometric distribution, including calculating the PMF and CDF probabilities, plotting the distribution, and generating random samples:

1. For a geometric distribution with a success probability of $p = 0.25$, calculate geometric probabilities for each number of trials (from 1 to 10) using the dgeom() function:

    ```
    # Parameters
    p = 0.25 # Probability of success
    # Get geometric probabilities
    geom_probs = dgeom(0:9, p)
    >>> geom_probs
    [1] 0.25000000 0.18750000 0.14062500 0.10546875 0.07910156
    0.05932617
     [7] 0.04449463 0.03337097 0.02502823 0.01877117
    ```

Note that the 0 : 9 argument represents the number of failures in the dgeom() function.

2.   Create a bar plot for these geometric probabilities:

```
>>> barplot(geom_probs, names.arg = 1:10, xlab = "Number of
Trials", ylab = "Probability", main = "Geometric Distribution (p
= 0.25)")
```

Running this code generates *Figure 10.6*. As expected, the probability of obtaining a longer sequence of continuous failures decreases as the number of trials increases:

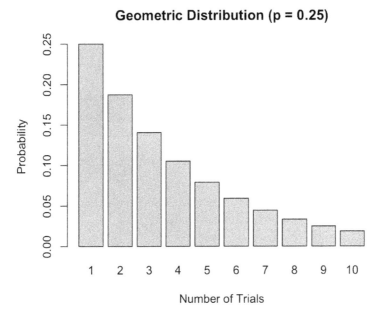

Figure 10.6 – Visualizing the PMF of the geometric distribution as a bar plot

3.   Calculate cumulative geometric probabilities for the previous trials using the pgeom() function:

```
cum_geom_probs = pgeom(0:9, p)
>>> cum_geom_probs
 [1]  0.2500000 0.4375000 0.5781250 0.6835938 0.7626953
 [6]  0.8220215 0.8665161 0.8998871 0.9249153 0.9436865
```

4.   Generate 100 random samples from this geometric distribution using the rgeom() function:

```
geom_samples = rgeom(100, p)
>>> geom_samples
  [1]  0  0  0  2 10  1  1 10  0  0  1  1  5  3  1  0  2  0  0
 [20]  4  0  1  4  2  3  2  2  2  4  1  6 12  4  1  7  3  1  1
 [39]  0  2  1  2  3  0  8  0  0  2 10  3  2  8  0  3  1  2  3
```

```
[58]   0   0   1   7   0   0   3   4  11   8   8   2   0   5   1   1   1   3   1
[77]   3   1   3   3   6   0   0   7   1   0   0   1   0   1   0   0   0   3   0
[96]   0   4  25   0   3
```

The result shows a decreasing frequency as the numbers grow large, which matches the PMF of the geometric distribution.

Let's go through another application-related exercise on the probability of finding bugs in a computer program.

Exercise 10.6 – simulating and analyzing geometrically-distributed random variables

In this exercise, we will visit a real-world example that involves a software tester trying to find bugs in a computer program. In this example, the software tester finds bugs in a program with a probability of 0.1 on each attempt, and the attempts are independent. The company wants to know the probability of finding the first bug within the first five attempts, as well as the expected number of attempts needed to find the first bug:

1. Calculate the probability of finding the first bug within the first five attempts using the pgeom() function:

    ```
    p = 0.1 # Probability of finding a bug on each attempt
    # Calculate the CDF for up to 5 attempts
    prob_within_5_attempts = pgeom(4, p)
    >>> prob_within_5_attempts
    0.40951
    ```

    Here, note that we use 4 since the dgeom() function uses zero-based indexing.

    As the pgeom() function returns the CDF of a specific input, we can equivalently calculate this probability by summing up all previous probabilities up to the current input, as shown here:

    ```
    >>> sum(dgeom(0:4, p))
    0.40951
    ```

2. Calculate the expected number of attempts needed to find the first bug using the mean (expected value) of the geometric distribution:

    ```
    mean_attempts <- 1 / p
    >>> mean_attempts
    10
    ```

3. Visualize the probabilities of finding the first bug within different numbers of attempts (from 1 to 20) in a bar chart:

    ```
    geom_probs <- dgeom(0:19, p)
    # Create a bar plot of probabilities
    ```

```
barplot(geom_probs, names.arg = 1:20, xlab = "Number of
Attempts", ylab = "Probability", main = "Geometric Distribution
(p = 0.1)")
```

Running this code generates *Figure 10.7*. This figure suggests that rare events (those requiring a continuous stream of failures) assume lower probabilities as we move toward the right:

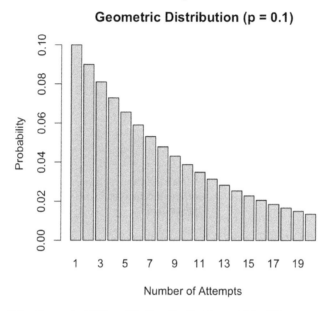

Figure 10.7 – Visualizing the probabilities of finding the first bug within different numbers of attempts

As this figure suggests, the probability of observing the first bug decreases as the number of attempts increases.

Comparing different discrete probability distributions

The discrete probability distributions introduced so far in this chapter are essential tools to model scenarios where the outcome variable takes on discrete values. Each discrete distribution has specific assumptions, properties, and applications. *Figure 10.8* provides a summary takeaway and analysis by comparing the main characteristics of the binomial, Poisson, and geometric distributions:

	Binomial distribution	Poisson distribution	Geometric distribution
Usage	Models the number of successes in a fixed number of independent Bernoulli trials with a constant probability of success	Models the number of events occurring in a fixed interval of time or space, given a constant average rate of occurrence	Models the number of trials required for the first success in a sequence of independent Bernoulli trials, each with the same probability of success
Parameters	n (number of trials) and p (probability of success)	λ (average rate of occurrence)	p (probability of success)
PMF	$P(x = k)$ $= C(n, k)p^k(1-p)^{n-1}$	$P(x = k) = \dfrac{\lambda^k e^{-\lambda}}{k!}$	$P(x = k) = (1-p)^{k-1}p$
Mean	$\mu = np$	$\mu = \lambda$	$\mu = \dfrac{1}{p}$
Variance	$\sigma^2 = np(1-p)$	$\sigma^2 = \lambda$	$\sigma^2 = \dfrac{1-p}{p^2}$
Applications	Coin tosses, quality control, opinion polling	Phone call arrivals, email arrivals, accidents at an intersection	Waiting time until an event occurs, number of attempts to achieve a desired outcome

Figure 10.8 – Summarizing and comparing different discrete distributions

With different discrete distributions at hand, it is important to understand the specific assumptions and requirements of each distribution to select the appropriate one for a given problem. For example, the binomial distribution is suitable for modeling the number of successes in a given number of experiments, the Poisson distribution is suitable for modeling the number of events occurring in a fixed period, and the geometric distribution is often used to model the number of trials required to achieve the first success.

In practice, these discrete probability distributions can be used to analyze various real-world scenarios, make predictions, and optimize processes. Getting a good understanding of the characteristics and applications of each distribution allows you to choose the right distribution for the specific problem and perform relevant analyses using R.

The next section introduces continuous distributions, including normal distribution, exponential distribution, and uniform distribution.

Discovering common continuous probability distributions

Continuous probability distributions model the probability of random variables that assume any value within a specific continuous range. In other words, the underlying random variable is continuous instead of discrete. These distributions describe the probabilities of observing values that fall within a continuous interval, rather than equal to individual discrete outcomes in a discrete probability distribution. Specifically, in a continuous probability distribution, the probability of the random variable equal to any specific value is typically zero, since the possible outcomes are uncountable. Instead, probabilities for continuous distributions are calculated for intervals or ranges of values.

We can use a PDF to describe a continuous distribution. This corresponds to the PMF of a discrete probability distribution. The PDF defines the probability of observing a value within an infinitesimally small interval around a given point. The area under the PDF curve over a specific range represents the probability of the random variable falling within that range – that is, the probabilities are calculated for intervals or ranges of values by integrating the PDF over the desired range. In contrast, probabilities are assigned to individual points in a PMF, and the probabilities are calculated for a set of discrete values by summing up their individual probabilities.

In addition, the visualization for continuous probability distributions is also different. Compared to the bar chart used for the PMF of discrete probabilities distributions, the PDF of continuous distributions is plotted as smooth curves called density plots. The area under the curve over a specific range represents the probability of the random variable falling within that range.

Figure 10.9 summarizes the main differences between discrete and continuous probability distributions:

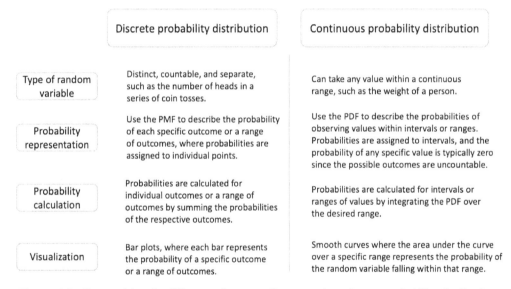

	Discrete probability distribution	Continuous probability distribution
Type of random variable	Distinct, countable, and separate, such as the number of heads in a series of coin tosses.	Can take any value within a continuous range, such as the weight of a person.
Probability representation	Use the PMF to describe the probability of each specific outcome or a range of outcomes, where probabilities are assigned to individual points.	Use the PDF to describe the probabilities of observing values within intervals or ranges. Probabilities are assigned to intervals, and the probability of any specific value is typically zero since the possible outcomes are uncountable.
Probability calculation	Probabilities are calculated for individual outcomes or a range of outcomes by summing the probabilities of the respective outcomes.	Probabilities are calculated for intervals or ranges of values by integrating the PDF over the desired range.
Visualization	Bar plots, where each bar represents the probability of a specific outcome or a range of outcomes.	Smooth curves where the area under the curve over a specific range represents the probability of the random variable falling within that range.

Figure 10.9 – Summarizing the differences between discrete and continuous probability distributions

In summary, discrete and continuous probability distributions model different types of random variables. Discrete probability distributions represent countable outcomes, while continuous distributions represent uncountable possibilities within a continuous range. These differences between these two types of distributions determine how we select the appropriate distribution for a given problem and perform relevant analyses.

In the next section, we'll introduce the most widely used continuous probability distribution: the normal probability distribution.

The normal distribution

The **normal probability distribution**, also called the **Gaussian distribution**, is a continuous probability distribution that models scenarios where the continuous outcomes are symmetrically distributed around the mean. It is the most widely used probability distribution in practice as many natural and social phenomena tend to follow a normal distribution as a result of the central limit theorem.

Two parameters are used to characterize the normal distribution: the mean (μ) and the standard deviation (σ). The mean represents the central tendency or the average value of the distribution. The outcomes around the center of the distribution get the highest probability. The standard deviation describes the dispersion or spread of the data from the mean, serving as a measure of variability in the distribution.

The PDF of the normal distribution is specified as follows:

$$f(x) = \frac{1}{\sqrt{2\pi}\,\sigma} e^{-\frac{(x-\mu)^2}{2\sigma^2}}$$

We can also write $x \sim N(\mu, \sigma^2)$, which reads as the random variable, x, follows a normal distribution parameterized by μ and σ^2.

Graphically, the normal distribution looks like a bell-shaped curve, with most values concentrated near the mean and fewer values toward the extremes at both ends. An empirical rule, called the 68-95-99.7 rule, says that approximately 68% of the values lie within 1 standard deviation of the mean, 95% within 2 standard deviations, and 99.7% within 3 standard deviations.

A particular normal distribution that is commonly used is the standard normal distribution, which is written as $z \sim N(0, 1)$ – that is, a standard normal distribution has $\mu = 0$ and $\sigma = 0$. A special property is that we can transform any normally distributed random variable into a standard normal variable using the following formula:

$$z = \frac{x - \mu}{\sigma}$$

We can then use the standard normal table (called a Z-table) to find probabilities and percentiles of interest.

Let's go through an exercise to practice the calculations related to the normal distribution.

Exercise 10.7 – simulating and analyzing normal random variables

In this exercise, we will simulate and analyze normal-distributed random variables:

1. Calculate the probability density of the standard normal distribution using the `dnorm()` function from `-4` to `4` with a step size of `0.1`:

    ```
    # Parameters
    mu = 0       # Mean
    ```

```
sigma = 1    # Standard deviation
# Get the probability density for different x
x = seq(-4, 4, by = 0.1)
normal_density = dnorm(x, mu, sigma)
```

Here, we use the `seq()` function to create a vector of equally spaced values and extract the corresponding probability for each of the input values using the `dnorm()` function.

2. Plot the normal distribution as a continuous curve using the `plot()` function:

```
# Plot the normal distribution
>>> plot(x, normal_density, type = "l", xlab = "x", ylab =
"Probability Density", main = "Normal Distribution (μ = 0, σ =
1)")
```

Running this code generates *Figure 10.10*, which shows that the PDF is centered around the mean, 0, and has a standard deviation of 1 as the spread:

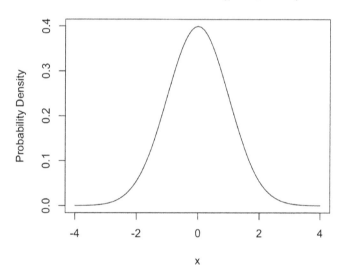

Figure 10.10 – Visualizing the density plot of the standard normal distribution

3. Calculate the cumulative probabilities of the normal distribution using the `pnorm()` function:

```
# Get cumulative probabilities for different x
normal_cum_prob <- pnorm(x, mu, sigma)
```

Similarly, we can plot the CDF as follows:

```
>>> plot(x, normal_cum_prob, type = "l", xlab = "x", ylab =
"Cumulative Probability Density", main = "Cumulative Normal
Distribution (μ = 0, σ = 1)")
```

Running this code generates *Figure 10.11*:

Cumulative Normal Distribution (μ = 0, σ = 1)

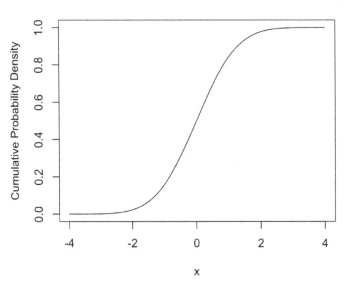

Figure 10.11 – Visualizing the cumulative density function of the standard normal distribution

4. Generate random samples from a normal distribution using the rnorm() function:

```
# Generate 100 random samples from a normal distribution with μ
= 0 and σ = 1
normal_samples <- rnorm(100, mu, sigma)
```

5. Find the 90th quantile (inverse cumulative probability) for a given probability using the qnorm() function:

```
# Find the quantile corresponding to the 90th percentile
quantile_90 <- qnorm(0.9, mu, sigma)
>>> quantile_90
1.281552
```

Let's look at another exercise for solving practical problems using the normal distribution.

Exercise 10.8 – calculating probabilities with the normal distribution

Let's assume a company manufactures batteries with an average life span of 100 hours and a standard deviation of 10 hours. Let's also assume that the lifespan of the batteries follows a normal distribution:

1. Simulate a dataset of 1,000 batteries:

```
set.seed(8)
mean_lifespan = 100
sd_lifespan = 10
n = 1000
lifespans = rnorm(n, mean_lifespan, sd_lifespan)
```

Here, we use the `rnomr()` function to sample from the given normal distribution randomly. We also specify the random seed for reproducibility.

2. Calculate the probability that a randomly chosen battery will last more than 120 hours:

```
threshold = 120
probability = 1 - pnorm(threshold, mean_lifespan, sd_lifespan)
>>> probability
0.02275013
```

Here, we use the `pnorm()` function to calculate the total probability of being smaller than `120`, then take the complement to get the probability of being larger than `120`.

As expected, the probability of deviating from the mean by two standard deviations is quite small.

3. Plot the PDF of the lifespans with the area under the curve above `120` hours shaded:

```
df <- data.frame(lifespan = lifespans)
df_density <- density(lifespans)
df_shaded <- data.frame(x = df_density$x, y = df_density$y)
df_shaded <- df_shaded[df_shaded$x > threshold,]

ggplot(df, aes(x=lifespan)) +
  geom_density(fill="lightblue") +
  geom_vline(xintercept = threshold, linetype="dashed",
color="red") +
  geom_area(data = df_shaded, aes(x=x, y=y), fill="orange",
alpha=0.5) +
  theme_minimal() +
  labs(title="Lifespan of batteries", x="Lifespan (hours)",
y="Probability Density")
```

Here, we build a DataFrame, `df`, to store the sample value and the corresponding density that was obtained using the `density()` function. We then subset to get the corresponding DataFrame, `df_shaded`, for the area to be shaded. In `ggplot`, we use the `geom_density()` function to plot the density curve, `geom_vline()` to add a vertical line indicating the threshold, and `geom_area()` to shade the area toward the right.

Running this code generates *Figure 10.12*:

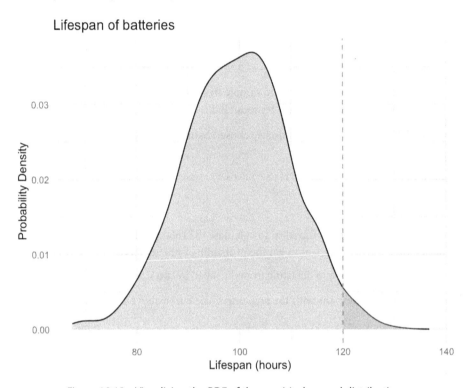

Figure 10.12 – Visualizing the PDF of the empirical normal distribution
with the area above the threshold shaded

The next section covers another continuous probability distribution: the exponential distribution.

The exponential distribution

The **exponential distribution** is a continuous probability distribution that's commonly used to model the time or space between events in a Poisson process. As covered in the previous section, a Poisson process models events that occur independently and at a constant average rate. The exponential distribution is often employed to describe the waiting time between rare events, such as the time between phone calls at a call center.

The PDF of the exponential distribution is given by the following formula:

$$f(x) = \lambda e^{-\lambda x}, x \geq 0$$

Here, λ is the rate parameter, which represents the average number of events that occur per unit of time or space, and e is the base of the natural logarithm.

A defining characteristic of an exponential distribution is the memoryless property, which says that the probability of an event occurring in the future is independent of the time that has already elapsed since the last event. This makes the exponential distribution a suitable choice for modeling waiting times between independent and rare events.

The mean and standard deviation of the exponential distribution are as follows:

$$\mu = \frac{1}{\lambda}$$
$$\sigma^2 = \frac{1}{\lambda^2}$$

The exponential distribution contains a single parameter, λ, and models the average number of events occurring per unit of time or space for the waiting time between events. The same parameter, λ, in the context of modeling a Poisson process, refers to the average number of events occurring in a fixed interval, in terms of either time or space. Both distributions are used to model different aspects of the same Poisson process.

Now, let's go through an exercise to review the probability calculations related to the exponential distribution.

Exercise 10.9 – calculating probabilities with the exponential distribution

In this exercise, we will look at generating random samples while following an exponential distribution and calculate and visualize the total probability above a certain threshold:

1. Generate a random sample of 1,000 data points from an exponential distribution with a rate parameter of `0.01` using the `rexp()` function:

    ```
    set.seed(8) # Set seed for reproducibility
    lambda = 0.01
    sample_size = 1000
    exponential_sample = rexp(sample_size, rate = lambda)
    ```

2. Calculate the probability that the waiting time between events is more than 150 units using the `pexp()` function:

    ```
    threshold = 150
    probability_above_threshold = 1 - pexp(threshold, rate = lambda)
    >>> probability_above_threshold
    0.2231302
    ```

3. Plot the PDF and shade the area under the curve for waiting times greater than the threshold:

```
# Create a data frame for the waiting times
waiting_times = seq(0, max(exponential_sample), length.out =
1000)
density_values = dexp(waiting_times, rate = lambda)
df = data.frame(waiting_times, density_values)
# Filter data for the shaded region
df_shaded = df[df$waiting_times > threshold,]
# Plot the PDF of the exponential distribution
ggplot(df, aes(x = waiting_times, y = density_values)) +
  geom_line() +
  geom_area(data = df_shaded, aes(x = waiting_times, y =
density_values), fill = "orange", alpha = 0.5) +
  geom_vline(xintercept = threshold, linetype = "dashed", color
= "red") +
  theme_minimal() +
  labs(title = "Exponential Distribution (λ = 0.01)", x =
"Waiting Time", y = "Probability Density")
```

Here, we create a DataFrame for different waiting times generated in the random samples, obtain the corresponding densities using the dexp() function, and build the df and df_shaded DataFrames to be used in ggplot.

Running this code generates *Figure 10.13*:

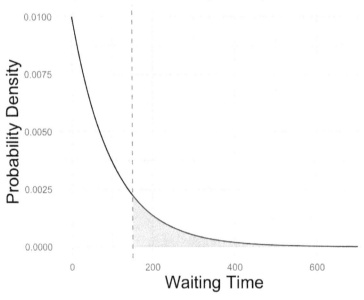

Figure 10.13 – Visualizing the empirical PDF of the exponential
distribution with the area above the threshold shaded

The next section introduces uniform distribution.

Uniform distribution

As the name suggests, uniform distribution is a continuous probability distribution where all outcomes within a given range are equally likely. The PDF, which appears as a straight line (illustrated next), is characterized by two parameters, the lower bound (a) and the upper bound (b), which define the range of possible values. All values within this range have the same probability of occurrence, and values outside the range have a probability of zero.

The PDF of the uniform distribution is given by the following formula:

$$f(x) = \begin{cases} \frac{1}{b-a}, & \text{if } x \in [a, b] \\ 0, & \text{otherwise} \end{cases}$$

The parameters of a uniform distribution are as follows:

$$\mu = \frac{a+b}{2}$$
$$\sigma^2 = \frac{(b-a)^2}{12}$$

Let's go through a similar exercise to analyze uniformly distributed random variables.

Exercise 10.10 – calculating probabilities with uniform distribution

In this exercise, we will look at generating random samples, calculating probabilities, and plotting the PDF of a uniform distribution:

1. Generate a random sample of 10,000 data points from a uniform distribution with a lower bound (a) of 2 and an upper bound (b) of 10 using the `runif()` function:

    ```
    set.seed(8) # Set seed for reproducibility
    a = 2
    b = 10
    sample_size = 10000
    uniform_sample = runif(sample_size, min = a, max = b)
    ```

2. Calculate the probability that a value selected from the uniform distribution is greater than 7 using the `punif()` function:

    ```
    threshold = 7
    probability = 1 - punif(threshold, min = a, max = b)
    >>> probability
    0.375
    ```

Here, we use the `punif()` function to calculate the CDF between a and `threshold`. We can also approximate this probability using the empirical samples:

```
probability2 = sum(uniform_sample > t
hreshold) / length(uniform_sample)
>>> probability2
0.3771
```

We can see that the approximation is quite close, and it will be even closer when the sample size gets larger.

3. Plot its PDF using `ggplot()`:

```
library(ggplot2)
# Create a data frame for the distribution
x_values = seq(a, b, length.out = 1000)
density_values = dunif(x_values, min = a, max = b)
df = data.frame(x_values, density_values)
# Plot the PDF of the uniform distribution
ggplot(df, aes(x = x_values, y = density_values)) +
  geom_line() +
  theme_minimal() +
  labs(title = "Uniform Distribution (a = 2, b = 10)", x =
"Value", y = "Probability Density")
```

Here, we build a sequence of placeholders on the *x* axis. Then, we obtain the corresponding densities and combine them into a DataFrame for plotting. Running this code generates *Figure 10.14*:

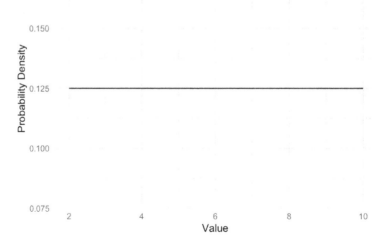

Figure 10.14 – Visualizing the PDF of a uniform distribution

Uniform distribution can be used to generate normally distributed random samples. Let's look at how this can be done.

Generating normally distributed random samples

So far, we have learned how to sample from a Gaussian distribution using the `rnorm()` function. It helps to look at how these random samples are generated under the hood. This specific technique is known as the inverse transform method, which relies on the inverse CDF or quantile function of the target distribution (in this case, the normal distribution).

The process involves three steps. First, we will generate random samples from a uniform distribution, usually $U(0,1)$, where all values between 0 and 1 are equally likely. Next, for each uniform random sample, we will locate the corresponding value in the target distribution using the inverse CDF (quantile function) of the standard normal distribution. Lastly, we will apply the scale-location transformation to convert the standard normal random sample into a random sample from the target normal distribution.

This method relies on the property that the CDF of a continuous random variable is a function that maps the sample space (the domain of the PDF) to the range of $[0, 1]$. The inverse CDF is the inverse of the CDF and maps the interval, $[0, 1]$, back to the sample space. By passing the uniform random samples into the inverse CDF, we reverse the mapping and obtain random samples that follow the target distribution, subject to additional transformation.

Let's look at this process in greater detail. Suppose we want to sample from a normal distribution, N (μ, σ^2). How do we generate random samples from this particular distribution? One approach is to generate a random sample, x, from a standard normal distribution $N(0,1)$, and then apply the scale-location transformation to obtain the final sample, $\sigma x + \mu$. By scaling the sample based on σ, followed by adding μ, the resulting sample will follow a normal distribution with mean, μ, and variance, σ^2.

The first step of sampling from a standard normal distribution is the key, whereas the second step is a simple and deterministic transformation. The approach we introduced earlier is to transform a uniformly distributed variable using the inverse CDF of a standard normal distribution. For example, if U is uniformly distributed on $[0,1]$, then $\Phi^{-1}(U)$ follows $N(0,1)$, where Φ^{-1} is the inverse of the cumulative function of a standard normal distribution.

An illustrative example is shown in *Figure 10.15*. First, we sample a random point from the uniform distribution on $[0,1]$. Next, we use the inverse CDF of the standard normal distribution to obtain the corresponding sample in the CDF space, given the fact that the CDF monotonically maps an arbitrary input value to an output on $[0,1]$. Mathematically, the random sample of the standard normal is given by $x = \Phi^{-1}(U)$. Considering the one-to-one mapping relationship between the PDF and the CDF of the standard normal, we could obtain the same input, x, in the PDF space as well. We would also expect most of the samples to be centered around the mean. Finally, we apply the scale-location transformation to convert to the random sample of the normal distribution with the desired mean and variance:

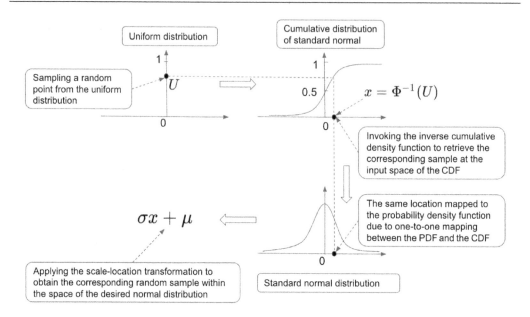

Figure 10.15 – Obtaining a random sample from the desired univariate Gaussian distribution

Let's look at a concrete example of generating normally distributed random samples using the inverse transform method. In the following code snippet, we first set the seed for reproducibility and define the parameters of the target normal distribution. We then generate 5 samples while following the uniform distribution between 0 and 1. Finally, we calculate the corresponding quantiles for the uniform sample using the inverse CDF (the quantile function, qnorm()) of the normal distribution:

```
set.seed(8) # Set seed for reproducibility
# Define the target normal distribution parameters
mu = 5
sigma = 2
# Generate uniform random variables
n = 5
uniform_sample = runif(n)
# Calculate the corresponding quantiles for the uniform sample using
the inverse CDF (quantile function) of the normal distribution
normal_sample = qnorm(uniform_sample, mean = mu, sd = sigma)
>>> normal_sample
[1] 4.830828 3.372006 6.680800 5.780755 4.073034
```

We can also apply the scale-location transformation to obtain the same random samples, as shown in the following code snippet:

```
normal_sample2 = qnorm(uniform_sample, mean = 0, sd = 1)
>>> normal_sample2 * sigma + mu
[1] 4.830828 3.372006 6.680800 5.780755 4.073034
```

The next section covers sampling distributions.

Understanding common sampling distributions

A sampling distribution is a probability distribution of a sample statistic based on many samples drawn from a population. In other words, it is the distribution of a particular statistic (such as the mean, median, or proportion) calculated from many sets of samples from the same population, where each set has the same size. There are two things to take note of here. First, the sampling distribution is not about the random samples drawn from the PDF. Instead, it is a distribution that's made from an aggregate statistic, which comes from another distribution drawn from the PDF. Second, we would need to sample from the PDF in multiple rounds to create the sampling distribution, where each round consists of multiple samples from the PDF.

Let's look at an exercise in R to illustrate the concept of the sampling distribution using the sample mean as the statistic of interest. We will generate samples from a population whose distribution is given and calculate the sample means. Then, we will create a histogram of the sample means to visualize the sampling distribution of the sample mean.

Exercise 10.11 – generating a sampling distribution

In this exercise, we will first generate a population of samples from a normal distribution. We will then sample from this population in multiple rounds, with each round consisting of multiple samples. Finally, we will extract the mean of each round of samples and plot them together in a histogram:

1. Generate 100,000 samples from $N(50,10)$. Check the summary of the samples using `summary()`:

    ```
    set.seed(8) # Set seed for reproducibility
    # Define the population parameters
    population_mean = 50
    population_sd = 10
    population_size = 100000
    # Generate the population using a normal distribution
    population <- rnorm(population_size, mean = population_mean, sd
    = population_sd)
    >>> summary(population)
       Min. 1st Qu.  Median    Mean 3rd Qu.    Max.
      7.597  43.261  50.051  50.027  56.781  89.365
    ```

2. Define a function to sample 50 numbers from the previous population and return the mean of the samples:

    ```
    # Define the sample size in each round
    sample_size_per_round = 50
    # Function to draw a sample and calculate its mean
    get_sample_mean <- function(population, sample_size_per_round) {
    ```

```
    sample <- sample(population, size = sample_size_per_round,
    replace = FALSE)
    return(mean(sample))
}
```

Here, we use the `sample()` function to sample 50 numbers from the population of samples created earlier without replacement. We can test out this function and observe a different return in each run, which is also close to the population mean of 50:

```
>>> get_sample_mean(population, sample_size_per_round)
50.30953
>>> get_sample_mean(population, sample_size_per_round)
48.9098
```

3. Repeat this function for a total of 1,000 times to obtain a corresponding set of 1,000 sample means:

```
# Generate multiple rounds of sample means
num_rounds = 1000 # the number of rounds to sample
sample_means = replicate(num_rounds, get_sample_mean(population,
sample_size))
>>> summary(sample_means)
   Min. 1st Qu.  Median    Mean 3rd Qu.    Max.
  49.76   49.96   50.02   50.03   50.09   50.34
```

Here, we use the `replicate()` function to apply the `get_sample_mean()` function repeatedly for a specified number of times. It is often used in simulations, resampling methods, or any situation where you need to perform the same operation multiple times and collect the results.

The result shows a sample mean from the sampling distribution that is very close to the population mean.

4. Visualize the sampling distribution of the sample mean using a histogram:

```
library(ggplot2)
sampling_distribution_df = data.frame(sample_means)
ggplot(sampling_distribution_df, aes(x = sample_means)) +
    geom_histogram(aes(y = after_stat(density)), bins = 30, color
= "black", fill = "lightblue") +
    geom_density(color = "red", lwd = 1.2) +
    theme_minimal() +
    labs(title = "Sampling Distribution of the Sample Mean",
        x = "Sample Mean",
        y = "Density")
```

Running this code generates *Figure 10.16*:

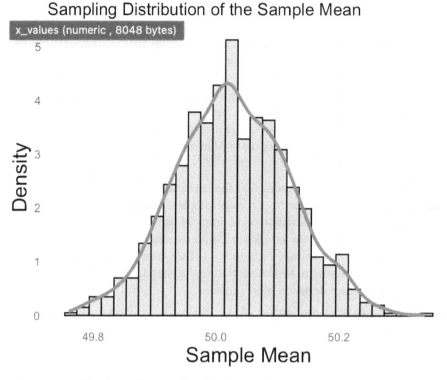

Figure 10.16 – Visualizing the sampling distribution of the sample mean in a histogram

In the next section, we'll introduce common sampling distributions that are used in statistical estimation and hypothesis testing.

Common sampling distributions

Several common sampling distributions are widely used in statistical inference. These distributions arise from the properties of different statistics calculated from random samples drawn from a population. Some of the most important sampling distributions include the following:

- **Sampling distribution of the sample mean**: When the population follows a normal distribution, the sampling distribution of the sample also assumes a normal distribution, where the mean is the same and the standard deviation is a scaled version of the population standard deviation. According to the **central limit theorem** (to be introduced in the next chapter), the sampling distribution of the sample mean approaches a normal distribution as more samples are collected, regardless of what the raw population distribution is. In other words, when the sample size is large (usually $n > 30$), the sampling distribution of the sample mean approaches a normal distribution, even if the original population distribution is not necessarily normal. This property allows us to use the normal distribution to make inferences for the sample mean, such as hypothesis testing and confidence interval construction.

- **Sampling distribution of the sample proportion**: The distribution of the sample proportion (the proportion of successes in a group of samples) in a large number of independent Bernoulli trials (binary outcomes) follows a normal distribution, given a sufficiently large sample size and a success probability not too close to 0 or 1. In this case, the mean stays the same and the standard deviation is a transformed population proportion, adjusted by the sample size. In particular, under certain conditions ($np > 10$ and $n(1 - p) > 10$ based on our previous introduction on approximating the binomial distribution via normal distribution), the sampling distribution of the sample proportion can be approximated by a normal distribution with the same mean and a standard deviation given by $\sigma = \sqrt{p(1 - p)/n}$.

- **t-distribution**: The **t-distribution** is also called the **Student's t-distribution**. It is used when estimating the population mean from a small sample (typically $n<30$) drawn from a normal population, whose standard deviation is not provided. The t-distribution is similar in shape to the normal distribution but has a thicker tail, with the exact shape determined by the degree of freedom (df), which is typically equal to $n - 1$. The t-distribution can be used to calculate the t-scores, which are used in hypothesis testing and confidence interval construction when the population standard deviation is not provided.

- **Chi-square distribution**: The **chi-square distribution** is used in hypothesis testing and the construction of confidence intervals for the population variance. It can also be used to test independence in contingency tables. It is a family of right-skewed distributions defined by a single parameter, the degree of freedom. In the context of hypothesis testing for the population variance, the test statistic assumes a chi-square distribution with $n - 1$ degrees of freedom. In the context of contingency tables, the test statistic follows a chi-square distribution with $(r - 1)(c - 1)$ degrees of freedom, where r denotes the number of rows and c denotes the number of columns in the table.

- **F-distribution**: The **F-distribution** is used in the **analysis of variance** (**ANOVA**) to test the null hypothesis that the means of multiple groups are equal. It is a family of right-skewed distributions that are defined by two parameters, the df for the numerator ($df1$) and the denominator ($df2$). For example, in the context of one-way ANOVA, the test statistic follows an F-distribution with $df1=k - 1$ and $df2=N - k$.

Overall, these sampling distributions are important tools in various inferential procedures, such as making statistical inferences about the parameters of the population distribution based on sample data via hypothesis testing or constructing confidence intervals. The four types of sampling distributions are also a subset of a bigger list in the playbook of statistical inferences, each looking at different assumptions, sample statistics, and inference procedures.

Let's look at an exercise on constructing the confidence interval for the population mean using the t-distribution.

Exercise 10.12 – estimating the population mean using the t-distribution

In this exercise, we will use a small sample to estimate the population mean and construct the confidence interval of the estimate using the t-distribution in R:

1. Generate 10 samples from a normal distribution, $N(50,10)$, using the `rnorm()` function:

    ```
    sample_size = 10
    mu = 50
    sigma = 10
    samples = rnorm(sample_size, mean = mu, sd = sigma)
    >>> samples
    [1] 41.57424 39.61629 59.86689 58.94655 43.43934 28.41854
    67.05759 50.36661 51.61680 37.71842
    ```

2. Calculate the sample mean and standard deviation:

    ```
    sample_mean = mean(samples)
    sample_sd = sd(samples)
    >>> sample_mean
    47.86213
    >>> sample_sd
    11.85024
    ```

    We will now use the sample mean and standard deviation to estimate the population mean.

3. Calculate the 95% confidence interval using the t-distribution:

    ```
    alpha = 0.05
    t_critical = qt(1 - alpha/2, df = sample_size - 1)   # t-value
    for a two-tailed test with alpha = 0.05 and df = n - 1
    margin_of_error_t = t_critical * (sample_sd / sqrt(sample_size))
    ci_t = c(sample_mean - margin_of_error_t, sample_mean + margin_
    of_error_t)
    >>> ci_t
    39.38497 56.33928
    ```

    Here, we use the `qt()` function to find the critical t-value corresponding to a two-tailed test with a significance level of 0.05 and degrees of freedom, $n - 1$. Then, we calculate the margin of error by multiplying the critical t-value by the standard error of the sample mean (note that we need to divide by the square root of sample size) and construct the confidence interval by adding and subtracting the margin of error from the sample mean.

We will introduce constructing the confidence interval around a sample estimate in more detail in the next chapter. For now, it suffices to understand how a sampling distribution could be derived and used to produce estimates about the profile of the population.

The next section covers another interesting and important topic: order statistics.

Understanding order statistics

Order statistics are the values of a collection of samples when arranged in ascending or descending order. These ordered samples provide useful information about the distribution and characteristics of the sampled data. Usually, the kth order statistic is the kth smallest value in the sorted sample.

For example, for a collection of samples of size n, the order statistics are denoted as $X_1, X_2, ..., X_n$, where X_1 is the smallest value (the minimum), X_n is the largest value (the maximum), and X_k represents the kth smallest value in the sorted sample.

Let's look at how to extract order statistics in R.

Extracting order statistics

Extracting the order statistics of a collection of samples could involve two types of tasks. We may be interested in collecting samples in an ordered fashion, which can be achieved using the sort () function. Alternatively, we may be interested in extracting a specific order statistic from the ordered collection of samples, such as finding the third-largest sample in the collection or calculating a particular quantile of the collection.

Let's look at an example of such extraction in R.

Exercise 10.13 – extracting order statistics

In this exercise, we will generate a collection of normally distributed random samples and look at sorting the samples and extracting particular order statistics from them:

1. Generate 10 random samples from a normal distribution with a mean of 50 and a standard deviation of 10:

    ```
    set.seed(8)
    samples = rnorm(10, mean = 50, sd = 10)
    >>> samples
    [1] 49.15414 58.40400 45.36517 44.49165 57.36040 48.92119
    48.29711 39.11668 19.88948 44.06826
    ```

2. Sort them in ascending order using the sort () function:

    ```
    sorted_samples = sort(samples)
    >>> sorted_samples
    [1] 19.88948 39.11668 44.06826 44.49165 45.36517 48.29711
    48.92119 49.15414 57.36040 58.40400
    ```

We can see that these samples are now ordered in ascending order. We can switch to descending order by setting `decreasing = T` in the `sort ()` function:

```
>>> sort(samples, decreasing = T)
[1] 58.40400 57.36040 49.15414 48.92119 48.29711 45.36517
44.49165 44.06826 39.11668 19.88948
```

3. Find the minimum value (first order statistic):

```
min_value = sorted_samples[1]
>>> min_value
19.88948
```

4. Find the maximum value (last order statistic):

```
max_value = sorted_samples[length(sorted_samples)]
>>> max_value
58.404
```

Here, we obtain the index of the last entry using the length of the collection of samples.

5. Find the third order statistic (that is, `k = 3`):

```
k = 3
kth_order_stat = sorted_samples[k]
>>> kth_order_stat
44.06826
```

6. Calculate the median (50th percentile):

```
median_value = median(samples)
>>> median_value
46.83114
```

Note that the median (or any other order statistics) stays the same in the raw samples or the sorted samples:

```
>>> median(sorted_samples)
46.83114
```

7. Calculate the 25th and 75th percentiles (first and third quartiles) using the `quantile ()` function:

```
quartiles = quantile(samples, probs = c(0.25, 0.75))
>>> quartiles
    25%      75%
44.1741 49.0959
```

Again, applying the same function to the ordered samples gives the same results:

```
>>> quantile(sorted_samples, probs = c(0.25, 0.75))
    25%      75%
44.1741 49.0959
```

In the next section, we'll cover a very important use of order statistics in finance: the value at risk.

Calculating the value at risk

The **value at risk (VaR)** is a widely used risk management metric in finance that estimates the potential loss in a portfolio or investment over a specified period for a given confidence level, such as 95% or 99%. It is used to quantify the downside risk and allocate capital accordingly. Here, the confidence level represents the probability that the potential loss will not exceed the calculated VaR. A higher confidence level indicates a more conservative estimate of the risk.

There are different approaches to calculating the VaR. We are going to focus on the simplest approach – that is, using **historical simulation**. This method uses historical data to simulate potential losses. First, we must sort the historical returns in ascending order. The VaR is then calculated as the percentile of return corresponding to the specified confidence level. This is a simple and intuitive approach, although it assumes that the historical behavior of the returns is representative of the future.

Let's go through an exercise to illustrate how to calculate the VaR.

Exercise 10.14 – calculating the VaR

In this exercise, we will discuss how to calculate the VaR, a measure of the potential loss in the value of a portfolio over a specified period for a given confidence level:

1. Generate daily returns in a year (252 trading days) from a normal distribution with a mean of 0.08 and a standard deviation of 0.05:

    ```
    # Set a seed for reproducibility
    set.seed(8)
    # Generate a random sample of daily returns from a normal
    distribution
    sample_size = 252  # Number of trading days in a year
    mu = 0.08          # Mean daily return
    sigma = 0.05       # Standard deviation of daily returns
    daily_returns = rnorm(sample_size, mean = mu, sd = sigma)
    ```

We can check the summary of the daily returns as follows, which shows that the daily return could be as high as 20% and as low as -7%. This means that although the underlying asset is profitable on average (with an expected return of 8%), there is still a significant risk in terms of daily fluctuations. The VaR gives us a measure of how large the risk is in extreme situations:

```
>>> summary(daily_returns)
    Min.   1st Qu.    Median     Mean   3rd Qu.      Max.
 -0.07073   0.04480   0.07926   0.07799   0.11424   0.20195
```

2. Calculate the VaR at the 95% confidence level for a total portfolio value of 1 million USD:

```
confidence_level = 0.95
portfolio_value = 1000000  # Portfolio value in USD
sorted_returns = sort(daily_returns)
VaR_index = ceiling(sample_size * (1 - confidence_level))
VaR = sorted_returns[VaR_index]
VaR_amount = portfolio_value * (1 - (1 + VaR))
>>> VaR
-0.006301223
>>> VaR_amount
6301.223
```

Here, we sort the daily returns in ascending order and store them in `sorted_returns`. Then, we obtain the index of the bottom 5% quantile in `VaR_index`, which is then used to retrieve the corresponding VaR of daily returns. Finally, we convert the percentage return into the loss in portfolio value in `VaR_amount`. Note that although the result is a negative number, we often report it as a positive number to indicate the potential loss (or even more) that could occur in an extreme situation.

3. Visualize the daily returns as a density plot and shade the area of the VaR:

```
library(dplyr)
daily_returns_df <- data.frame(DailyReturns = daily_returns)
# Create the density plot
density_plot <- ggplot(daily_returns_df, aes(x = DailyReturns))
+
  geom_density(fill = "blue", alpha = 0.5) +
  geom_vline(aes(xintercept = VaR), linetype = "dashed", color =
"red") +
  labs(x = "Daily Returns", y = "Density", title = "Density Plot
of Daily Returns with VaR") +
  theme_minimal()
# Add shaded area below the VaR to the density plot
density_data <- ggplot_build(density_plot)$data[[1]] %>%
  as.data.frame() %>%
```

```
  filter(x < VaR)
density_plot +
  geom_ribbon(data = density_data, aes(x = x, ymin = 0, ymax =
y), fill = "red", alpha = 0.5)
```

Here, we first convert the daily returns into a DataFrame, which is used by ggplot() to plot a density curve using geom_density(). We also add a vertical line representing the VaR using geom_vline(). To shade the area below the VaR, we use ggplot_build() to filter the data and use geom_ribbon() to color the region that satisfies the filtering condition.

Running this code generates *Figure 10.17*:

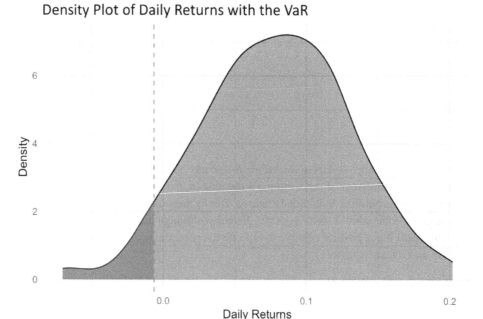

Figure 10.17 – Visualizing the daily returns and the VaR area

Thus, we can quantify the VaR based on the empirical distribution of the observed daily returns.

Summary

In this chapter, we covered common probability distributions. We started by introducing discrete probability distributions, including the Bernoulli distribution, the binomial distribution, the Poisson distribution, and the geometric distribution. We followed by covering common continuous probability distributions, including the normal distribution, the exponential distribution, and the uniform distribution. Next, we introduced common sampling distributions and their use in statistical inferences for population statistics. Finally, we covered order statistics and their use in calculating the VaR in the context of daily stock returns.

In the next chapter, we will cover statistical estimation procedures, including point estimation, the central limit theorem, and the confidence interval.

11
Statistical Estimation

In this chapter, we will introduce you to a range of statistical techniques that enable you to make inferences and estimations using both numerical and categorical data. We will explore key concepts and methods, such as hypothesis testing, confidence intervals, and estimation techniques, that empower us to make generalizations about populations from a given sample.

By the end of this chapter, you will grasp the core concepts of statistical inference and be able to perform hypothesis testing in different scenarios.

In this chapter, we will cover the following main topics:

- Statistical inference for categorical data
- Statistical inference for numerical data
- Constructing the bootstrapped confidence interval
- Introducing the central limit theorem used in t-distribution
- Constructing the confidence interval for the population mean using the t-distribution
- Performing hypothesis testing for two means
- Introducing ANOVA

To run the code in this chapter, you will need to have the latest versions of the following packages:

- `dplyr`, 1.0.10
- `ggplot2`, 3.4.0
- `socviz`, 1.2
- `infer`, 1.0.4

Please note that the versions mentioned in the preceding list are the latest ones at the time I am writing this book. All the code and data for this chapter is available at `https://github.com/PacktPublishing/The-Statistics-and-Machine-Learning-with-R-Workshop/blob/main/Chapter_11/working.R`.

Statistical inference for categorical data

A categorical variable has distinct categories or levels, rather than numerical values. Categorical data is common in our daily lives, such as gender (male or female, although a modern view may differ), type of property sales (new property or resale), and industry. The ability to make sound inferences about these variables is thus essential for drawing meaningful conclusions and making well-informed decisions in diverse contexts.

Being a categorical variable often means we cannot pass it to a **machine learning** (**ML**) model without additional preprocessing. Take the industry variable, for example. Instead of passing the categorical values (`string` values such as `"finance"` or `"technology"`) to the model, a common approach is to one-hot encode the variable into multiple columns, with each column corresponding to a specific industry, indicating a binary value of 0 or 1.

In this section, we will explore various statistical techniques designed specifically to handle categorical data, enabling us to derive valuable insights and make inferences about populations based on available samples. We will also discuss important concepts, such as proportions, independence, and goodness of fit, which form the foundation for understanding and working with categorical variables, covering both cases with a single parameter and multiple parameters.

Let us start by discussing the inference for a single parameter.

Statistical inference for a single parameter

A population parameter, the subject of interest and to be inferred, is a fixed quantity that describes a particular statistical attribute of a population, including the mean, proportion, or standard deviation. This quantity often stays hidden from us. For example, in order to get the most popular major in a university, we need to count the number of enrolled students in each major across the whole university and then return the major with the biggest count.

In the context of statistical inference for a single parameter, we aim to estimate this unknown parameter or test hypotheses about its value based on the information gathered from a sample. In other words, we would use statistical inference tools to infer unknown population parameters based on the known sample at hand. In the previous example, we would infer the most popular major of the whole university by a limited sample of students enrolled in a specific academic year.

Let us first explore the **General Social Survey** (**GSS**) dataset.

Introducing the General Social Survey dataset

The GSS is a comprehensive dataset widely used by researchers and policymakers to understand social, cultural, and political trends in the United States. The GSS has been continued by the **National Opinion Research Center** (**NORC**) at the University of Chicago since 1972, with the objective of collecting data on a broad range of topics, including attitudes, behaviors, and opinions on various issues.

Let us load the GSS dataset from the `socviz` package (remember to install this package via `install.packages("socviz")`):

```
library(socviz)
data(gss_lon)
```

The GSS dataset is now stored in the `gss_lon` variable, which contains a total of 62,466 rows and 25 columns, as shown here:

```
>>> dim(gss_lon)
62466    25
```

The GSS dataset contains numerous variables that cover diverse topics, such as education, income, family structure, political beliefs, and religious affiliation. Let us examine the structure of the dataset using the `glimpse()` function from the `dplyr` package, designed to help you quickly explore and understand the structure of the data:

```
>>> glimpse(gss_lon)
```

Figure 11.1 shows a screenshot of the first few variables returned.

```
> glimpse(gss_lon)
Rows: 62,466
Columns: 25
$ year      <dbl> 1972, 1972, 1972, 1972, 1972, 1972, 1972, 1972, 1972, 19…
$ id        <dbl> 1, 2, 3, 4, 5, 6, 7, 8, 9, 10, 11, 12, 13, 14, 15, 16, 1…
$ ballot    <labelled> NA, NA, NA, NA, NA, NA, NA, NA, NA, NA, NA, NA, NA,…
$ age       <labelled> 23, 70, 48, 27, 61, 26, 28, 27, 21, 30, 30, 56, 54,…
$ degree    <fct> Bachelor, Lt High School, High School, Bachelor, High Sc…
$ race      <fct> White, White, White, White, White, White, White, White,…
$ sex       <fct> Female, Male, Female, Female, Female, Male, Male, Male, …
$ siblings  <fct> 3, 4, 5, 5, 2, 1, 6+, 1, 2, 6+, 6+, 6+, 2, 2, 0, 6+, 0, …
```

Figure 11.1 – Showing the first few rows of the result from running the glimpse() function

Next, we will look at calculating a specific statistic based on a categorical variable.

Calculating the sample proportion

The `siblings` column in the dataset is a categorical variable that tracks the number of siblings in the family. In the following exercise, we would like to calculate the proportion of survey respondents whose family has two siblings in the latest year, 2016.

Exercise 11.1 – calculating the sample proportion of siblings

In this exercise, we first obtain a summary of the `siblings` column and subset the dataset to focus on the year 2016, which will then be used to calculate the proportion of surveys with a specific number of siblings in the family:

1. Obtain a summary of the `siblings` column using the `summary()` function:

    ```
    >>> summary(gss_lon$siblings)
        0     1     2     3     4     5    6+   NA's
     3047 10152 11313  9561  7024  5066 14612  1691
    ```

    The result suggests that most surveys are conducted for families with six siblings or more!

2. Subset the dataset for the year 2016:

    ```
    gss2016 = gss_lon %>% filter(year == 2016)
    Plot the count of siblings in a bar chart using ggplot().
    ggplot(gss2016, aes(x = siblings)) +
      geom_bar() +
      labs(title = "Frequency count of siblings", x = "Number of
    siblings", y = "Count") +
      theme(text = element_text(size = 16))
    ```

    Running the code generates the chart in *Figure 11.2*.

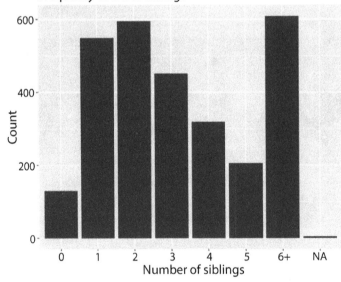

Figure 11.2 – Visualizing the frequency count of the number of siblings in a bar chart

3. Calculate the proportion of surveys with two siblings:

```
p_hat = gss2016 %>%
  summarize(prop_2sib = mean(siblings=="2", na.rm=TRUE)) %>%
  pull()
>>> p_hat
0.208246
```

Here, we use the `summarize()` function to calculate the mean of a series of binary values, which corresponds to the proportion of surveys with two siblings. We then use the `pull()` function to obtain the proportion from the resulting DataFrame.

We use the sample proportion to estimate the population statistic. In other words, we calculate the proportion of families with two siblings based on the available samples to approximate the corresponding proportion if we were to calculate the same based on all the data in the population. Such an estimate comes with a confidence interval that quantifies the list of possible values for the population proportion.

The next section shows how to calculate the confidence interval for the sample proportion.

Calculating the confidence interval

The confidence interval is an important tool in making inferences about the population parameters based on sample data. A confidence interval provides an estimated range within which a population parameter, such as proportion, is likely to be found with a specified confidence level, such as 95%. When working with sample proportions, calculating confidence intervals allows us to understand the true proportion in the population better and gauge the uncertainty associated with the estimation of the population proportion.

We can use the following steps to calculate the confidence interval:

1. Calculate the sample proportion, $\hat{p}$ (pronounced as p-hat). This is the value we calculated based on the sample data in 2016. In other contexts, $\hat{p}$ is calculated by dividing the number of successes (for the attribute of interest) by the total sample size.

2. Determine the desired level of confidence, commonly denoted as $(1 - \alpha)$ x 100%, where α represents the level of significance. In other words, it is the probability of rejecting the null hypothesis when it is true. The most frequently used confidence levels are 90%, 95%, and 99%.

3. Calculate the standard error of the sample proportion, which is given by the following formula:

$$SE = \sqrt{\frac{\hat{p}(1 - \hat{p})}{n}}$$

Here, the standard error also corresponds to the standard deviation of the sample proportion, which is assumed to follow a Bernoulli distribution with a success probability of $\hat{p}$ (recall the introduction of Bernoulli distribution in the previous chapter). Such calculation relies on two assumptions: the observations in the samples are independent and there are sufficient observations in the sample. A common rule of thumb for checking the second assumption is to ensure both $n\hat{p} > 10$ and $n(1 - \hat{p}) > 10$.

Alternatively, instead of assuming a Bernoulli distribution, we can use the bootstrap procedure to estimate the standard error without any distributional assumption. Bootstrap is a non-parametric method that involves resampling the data with replacement to create new samples, calculating the statistic of interest (in this case, the proportion) for each resampled dataset, and estimating the standard error from the variability of the calculated statistics across the resampled datasets.

4. Find the critical value (z-score) corresponding to the preset confidence level. This can be done using the qnorm() function, which gives us the quantiles of the standard normal distribution.

5. Compute the **margin of error (ME)** as the product of the standard error and the critical value:

$$ME = SE * z\_score$$

6. Calculate the confidence interval by adding and subtracting the ME from the sample proportion, giving the following:

$$Lower\ limit = \hat{p} - ME$$

$$Upper\ limit = \hat{p} + ME$$

The confidence interval provides a list of possible values for the population proportion according to the specific confidence level. See *Figure 11.3* for a summary of the calculation process.

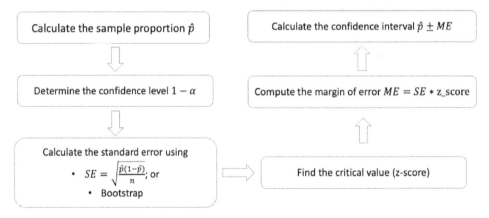

Figure 11.3 – Summarizing the process of calculating the confidence interval based on sample proportion

Let us stay with the bootstrap procedure a little longer. Without assuming any specific distribution, the bootstrap procedure is a flexible approach that can provide more accurate estimates of the standard error, especially for small sample sizes or when the data is not well behaved. However, It can be computationally intensive, especially for large datasets or when many bootstrap replications are generated.

Figure 11.4 provides a schematic overview of the bootstrap procedure. Let's review:

1. First, we start with the whole dataset and specify the variable of interest, which is the siblings variable in this case. This is achieved via the specify() function.

2.   Next, we draw samples from the variable with replacement, where the new sample will be the same size as the original dataset. Such resampling introduces randomness to the resulting dataset.

3.   We repeat the process many times, leading to a collection of bootstrapped artificial datasets using the `generate()` function.

4.   For each replicated dataset, we will calculate the sample statistic of interest, which is the proportion of observations with two siblings in this case. This is done via the `calculate()` function.

5.   These sample statistics derived using repeated sampling of the original dataset will then form a distribution, called the bootstrapped distribution (plotted via `ggplot()`), whose standard deviation (extracted via the `summarize()` function) will be a good approximation of the standard error.

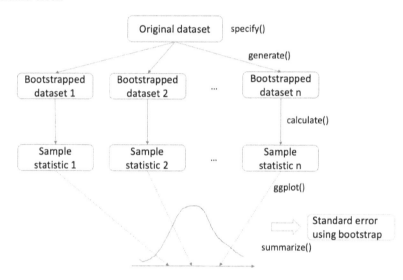

Figure 11.4 – The schematic overview of obtaining the standard error using the bootstrap procedure

The bootstrapped samples convey different levels of uncertainty in the sample statistic and jointly form a density distribution of multiple artificial sample statistics. The standard deviation of the bootstrapped distribution then gives the standard error of the sample statistic. Note that functions such as `specify()`, `generate()`, and `calculate()` all come from the `infer` package in R. Remember to install this package before continuing with the following code.

Let us go through the following exercise to understand the bootstrap procedure for calculating the confidence interval.

Exercise 11.2 – calculating the confidence interval via bootstrap

In this exercise, we will explore calculating the confidence interval of the sample proportion. The confidence interval includes the list of estimates within which the true population proportion may assume, given the observed samples. It is a way to quantify the uncertainty in estimating the population proportion based on the actual observations. Besides a step-by-step walk-through of the calculation process using bootstrap, we will also compare the result with the alternative approach using the assumed Bernoulli distribution:

1. Build a set of bootstrapped sample statistics using the specify-generate-calculate procedure from the `infer` package described earlier. Remember to build a binary variable to indicate the binary condition of having an observation with two siblings:

    ```
    library(infer)

    gss2016 = gss2016 %>%
      mutate(siblings_two_ind = if_else(siblings=="2","Y","N")) %>%
      filter(!is.na(siblings_two_ind))

    bs = gss2016 %>%
      specify(response = siblings_two_ind,
              success = "Y") %>%
      generate(reps = 500,
               type = "bootstrap") %>%
      calculate(stat = "prop")
    ```

Here, we first create a binary indicator variable using the `if_else()` function to denote whether the family in the current survey has two siblings. We also remove rows with NA values in this column. Next, we use the `specify()` function to indicate the `siblings_two_ind` variable of interest and the level that corresponds to a success. We then use the `generate()` function to generate 500 bootstrapped samples, and use the `calculate()` function to obtain the corresponding sample statistic (proportion of success) in each bootstrapped sample by setting `stat = "prop"`.

Let us observe the contents in the bootstrapped sample statistics:

```
>>> bs
Response: siblings_two_ind (factor)
# A tibble: 500 × 2
   replicate  stat
       <int> <dbl>
 1         1 0.205
 2         2 0.209
 3         3 0.218
 4         4 0.189
 5         5 0.207
 6         6 0.205
 7         7 0.221
 8         8 0.214
 9         9 0.212
10        10 0.212
```

```
# … with 490 more rows
# i Use `print(n = ...)` to see more rows
```

The result shows that the bs object is a tibble DataFrame with 500 rows (corresponding to the total number of the bootstrapped sample) and 2 columns. The first column (replicate) denotes the number of bootstrapped samples, and the second column (stat) indicates the proportion of success (that is, the number of rows with siblings_two_ind==2 divided by the total number of rows) in the bootstrapped sample.

2. Plot the bootstrapped sample statistics in a density plot:

```
>>> ggplot(bs, aes(x = stat)) +
  geom_density() +
  labs(title = "Density plot of the sample proportions", x =
"Sample proportion", y = "Density") +
  theme(text = element_text(size = 16))
```

Running the code generates the plot in *Figure 11.5*. The spread of this distribution, which relates to the standard deviation, directly determines the magnitude of the standard error. Also, if we were to increase the number of bootstrapped samples, we would expect a smoother density curve.

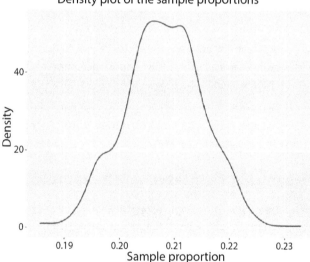

Figure 11.5 – Visualizing the density plot of all bootstrapped sample proportions

1. Calculate the standard error as the standard deviation of the empirical distribution based on the bootstrapped sample proportions:

```
SE = bs %>%
  summarise(sd(stat)) %>%
```

```
      pull()
>>> SE
0.007181953
```

Here, we use the `sd()` function to calculate the standard deviation of the `stat` column in `bs`, and then return the value via the `pull()` function. The standard error will then be scaled by the predetermined z-score and subtracted from and added to the original sample proportion to obtain the confidence interval.

2. Calculate the confidence interval of the original sample proportion with a 95% confidence interval:

```
>>> c(p_hat - 2*SE, p_hat + 2*SE)
0.1938821 0.2226099
```

Here, since a 95% confidence level corresponds to a z-score of 2, we multiply it with the standard error before subtracting from and adding to the original sample proportion (`p_hat`) to obtain the confidence interval.

3. Calculate the confidence interval using the structure information by assuming a Bernoulli distribution for the probability of success:

```
SE2 = sqrt(p_hat*(1-p_hat)/nrow(gss2016))
>>> c(p_hat - 2*SE2, p_hat + 2*SE2)
0.193079 0.223413
```

Here, we use the explicit form of the variance of the Bernoulli distribution to calculate the standard error. The result shows a fairly similar confidence interval compared with the one obtained using the bootstrap approach.

The confidence interval provides a measure of uncertainty for our estimate of the unknown population proportion using the observed sample proportion. Let us look at how to interpret the confidence interval in the next section.

Interpreting the confidence interval of the sample proportion

Interpreting the confidence interval of the sample proportion involves understanding the meaning of the interval and the associated confidence level. In our previous example, the bootstrap approach reports a confidence interval of `[0.1938821, 0.2226099]`. There are two levels of interpretation for this confidence interval.

First, the range of the confidence interval suggests that the true population proportion of families with two siblings is likely to fall between 19.39% and 22.26%. This range is based on the sample data and estimates the uncertainty in the true proportion.

Second, the 95% confidence interval means that if we were to conduct the survey many times (either in 2016 or other years), we would generate different random samples of the same size, based on which we can calculate the 95% confidence interval for each sample. Among these artificial samples, we will obtain a collection of intervals, and approximately 95% of them would include the true population proportion within the interval.

Note that the confidence interval is still an estimate, and the true population proportion may fall outside the calculated interval. However, the confidence interval provides a useful way to quantify the uncertainty in the estimate and gives a list of plausible values for the true population proportion based on the observed samples.

The next section introduces hypothesis testing for the sample proportion.

Hypothesis testing for the sample proportion

Hypothesis testing for the sample proportion is very much related to the confidence interval introduced in a previous section, which captures the level of uncertainty in the estimate for the unknown proportion based on the population data. Naturally, a sample with fewer observations leads to a wide confidence interval. Hypothesis testing for the sample proportion aims to determine whether there is enough evidence in a sample to support or reject a claim about the population proportion. The process starts with a null hypothesis (H0), which represents the baseline assumption about the population proportion. Correspondingly, there is an alternative hypothesis (H1) that represents the claim or statement we are testing against the null hypothesis. Hypothesis testing then compares the observed sample proportion to a specified null hypothesis in order to assess whether we have enough evidence to reject the null hypothesis in favor of the alternative hypothesis.

Let us go through an overview of the procedure involved in carrying out hypothesis testing:

1. **Formulate the hypothesis**. In this step, we set up the null hypothesis (H0) and alternative hypothesis (H1). The null hypothesis often says there is no effect, and the situation remains the status quo, as indicated by an equality sign in H0. On the other hand, the alternative hypothesis states that there is an effect or difference, as indicated by an inequality sign in H1. For example, we can set the following hypotheses for H0 and H1:

    *   H0: $p = p_0$ (the population proportion is equal to a specified value, p_0)

    *   H1: $p \neq p_0$ (the population proportion is not equal to p_0)

2. **Choose a significance level (α)**. The significance level is a probability threshold we use to reject the null hypothesis when it is true. Widely used significance levels include 0.05 (5%) and 0.01 (1%).

3. **Calculate the test statistic**. Now that we observe a sample proportion based on the actual data, we can calculate the probability of observing such a sample proportion *if* the null hypothesis were true. This starts with calculating the test statistic (z-score) for the sample proportion using the following formula:

$$z = \frac{\hat{p} - p_0}{\sqrt{p_0(1 - p_0)/n}}$$

where $\hat{p}$ is the sample proportion, p_0 is the population proportion assuming the null hypothesis, and n is the sample size. There are two things to note there. First, the denominator resembles the standard deviation based on the sample proportion covered earlier. Indeed, we are assuming a Bernoulli distribution with a success probability of p_0. With a total of n observations, the standard deviation for the sample proportion variable is $\sqrt{p_0(1 - p_0)/n}$. Second, the whole term corresponds to the process of converting a number into a z-score of a specific distribution, a topic covered in the previous chapter. Here, we assume a normal distribution with mean p_0 and standard deviation $\sqrt{p_0(1 - p_0)/n}$. We can then convert the observed sample proportion $\hat{p}$ to the corresponding z-score for ease of calculation later on.

Note that we can also use the bootstrap approach to calculate the empirical p-value under the null hypothesis.

4. **Determine the p-value**. The z-score is a measure that falls on a standard Gaussian distribution. It is a test statistic, and we are often interested in the probability of observing the test statistic at this or an even more extreme value. This is called the p-value, denoted as $\hat{p}$, when we assume the null hypothesis is true. In other words, we try to assess how likely it is to observe some phenomenon, assuming the null hypothesis is true. If the probability of observing $\hat{p}$ or an even more extreme number is very small, we have confidence that the null hypothesis is false, and we can reject H0 in favor of H1.

Note that for a two-tailed test, we can also calculate the p-value using the standard normal distribution and doubling the single-side probability:

$$\text{p-value} = 2P(Z > |z|)$$

5. **Make a decision**. Finally, we compare the p-value to the preset significance level (α) and use the following rule to make a decision.

If the p-value $\leq \alpha$, reject the null hypothesis in favor of the alternative hypothesis. Doing so suggests that there is enough evidence to suggest that the population proportion differs from the hypothesized proportion p_0.

If the p-value $> \alpha$, fail to reject the null hypothesis. This means that there is not enough evidence to suggest that the population proportion is different from p_0.

Conducting the hypothesis testing follows a similar process. The only difference is the use of the `hypothesise()` function (placed after `specify()`), which serves as a null hypothesis. We then perform the same bootstrap procedure to obtain a density plot of the bootstrapped sample proportions, followed by calculating the total probability of obtaining a proportion at least as extreme as the one indicated in the null hypothesis.

Let us go through an exercise to review the process of performing hypothesis testing for the sample proportion.

Exercise 11.3 – performing hypothesis testing for the sample proportion

In this exercise, we will set up a hypothetical population proportion in a null hypothesis and test the validity of this hypothesis based on the observed sample proportion:

1. Plot the frequency count of families with and without two siblings in 2016 in a bar plot:

    ```
    gss2016 %>%
      ggplot(aes(x = siblings_two_ind)) +
      geom_bar() +
      labs(title = "Frequency count of families with two siblings",
     x = "Have two siblings", y = "Count") +
      theme(text = element_text(size = 16))
    ```

    Running the code generates the plot in *Figure 11.6*, which shows that families with two siblings account for around ¼ of all families.

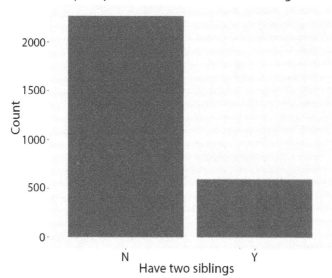

Figure 11.6 – Visualizing the frequency count of families with two siblings

2. Calculate the sample proportion of families with two siblings:

    ```
    p_hat = gss2016 %>%
      summarize(mean(siblings_two_ind=="Y")) %>%
      pull()
    ```

```
>>> p_hat
0.208246
```

Here, we first build a series of binary outcomes using `siblings_two_ind=="Y"`. Taking the average of this column gives the ratio of TRUE values, which gets executed in a `summarize()` context. We then extract the value of the sample proportion using `pull()`.

3. Use the specify-hypothesise-generate-calculate procedure to generate a collection of bootstrapped sample proportions under the null hypothesis, which specifies a population proportion of 0.19:

```
null = gss2016 %>%
   specify(response = siblings_two_ind,
             success = "Y") %>%
   hypothesise(null = "point",
                p = 0.19) %>%
   generate(reps = 500,
             type = "draw") %>%
   calculate(stat = "prop")
>>> null
Response: siblings_two_ind (factor)
Null Hypothesis: point
# A tibble: 500 × 2
    replicate  stat
    <fct>      <dbl>
  1 1          0.179
  2 2          0.193
  3 3          0.176
  4 4          0.181
  5 5          0.181
  6 6          0.198
  7 7          0.191
  8 8          0.189
  9 9          0.194
 10 10         0.189
# ... with 490 more rows
# i Use `print(n = ...)` to see more rows
```

4. Generate the density plot of the bootstrapped sample proportions along with the proportion suggested by the null hypothesis via a vertical line:

```
ggplot(null, aes(x = stat)) +
   geom_density() +
   geom_vline(xintercept = p_hat,
                color = "red") +
```

```
    labs(title = "Density plot using bootstrap", x = "Sample
proportion", y = "Density") +
    theme(text = element_text(size = 16))
```

Running the code generates the plot in *Figure 11.7*. The probability of observing a value at least as extreme as the one indicated by the red line (according to the null hypothesis) is thus the total area under the density curve toward the right of the red line. We then double the result to account for the opposite direction.

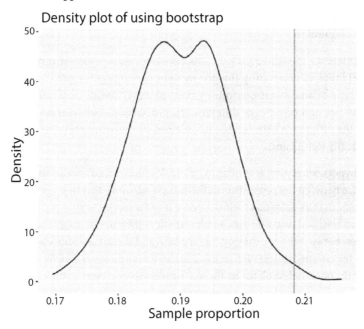

Figure 11.7 – Visualizing the density plot of the bootstrapped sample proportions for hypothesis testing

5. Calculate the p-value:

```
>>> null %>%
    summarise(mean(stat > p_hat)) %>%
    pull() * 2
0.02
```

Since this result is less than the preset significance level of 5%, we have sufficient evidence to favor the alternative hypothesis and reject the null hypothesis. In other words, the assumed 19% is statistically different from the true population proportion with a confidence level of up to 95%. We can therefore draw the conclusion that the true population proportion is not 19%.

The next section looks at the inference for the difference in sample proportions between two categorical variables.

Inference for the difference in sample proportions

The setting now is that we have two categorical variables. Take gender and degree, for example. The data will report a proportion of degree holders for both females and males. A natural question to ask is whether males are more likely to get a degree than females. A particular dataset will report a snapshot of these proportions, which may or may not suggest a higher percentage of degree holders are males. The tools from hypothesis testing could then come in to answer the following question: if males are a higher proportion of degree holders in the dataset, is such difference statistically significant? In other words, are males more likely to get a degree than females, or vice versa? This section attempts to answer this type of question.

Inference for the difference in sample proportions between two categorical variables (for example, gender and degree) involves comparing the proportions of samples for each level in two different populations. This type of analysis is commonly used in experiments or observational studies to determine the existence of a significant difference in proportions between two groups. The main goal is to estimate the difference between the population proportions and determine whether this difference is statistically significant.

The procedure for hypothesis testing is similar to before. We first formulate the null hypothesis, which assumes no difference between the proportion of the two populations, that is, $p_1 = p_2$, or $p_1 - p_2 = 0$. The alternative hypothesis then states that their difference is not zero; that is, $p_1 - p_2 \neq 0$. Next, we choose a specific significance level and calculate the sample statistic (difference in sample proportion, including the pooled proportion between the two categorical variables) and the test statistic (via either a closed-form expression based on the assumed distribution or using the bootstrap method). Finally, we obtain the p-value and decide whether the observed result under the null hypothesis possesses statistical significance or not.

Let us go through a concrete exercise following our previous example.

Exercise 11.4 – performing hypothesis testing for the difference in sample proportions

In this exercise, we focus on how to conduct hypothesis testing for the difference in the sample proportion between gender and status of higher degree. Here, we define a higher degree as a bachelor's and above. The proportion of higher-degree holders will likely differ between the male and female groups, and we will test whether such a difference is significant given the observed data:

1. Add a binary column called `higher_degree` to the previous DataFrame, `gss2016`, to indicate the status of higher degree, including bachelor's and above:

    ```
    gss2016 = gss2016 %>%
      mutate(higher_degree = if_else(degree %in%
    c("Bachelor","Graduate"), "Y", "N"))
    ```

2. Print the ratio between the two levels for `gender` and `higher_degree`:

```
>>> table(gss2016$higher_degree)
   N    Y
2008  854
>>> table(gss2016$sex)
  Male Female
  1274   1588
```

3. Plot these counts in a bar chart:

```
ggplot(gss2016, aes(x = sex, fill=higher_degree)) +
  geom_bar() +
  labs(title = "Frequency count for gender and degree", x =
"Gender", y = "Count") +
  theme(text = element_text(size = 16))
```

Running the code generates the chart in *Figure 11.8*.

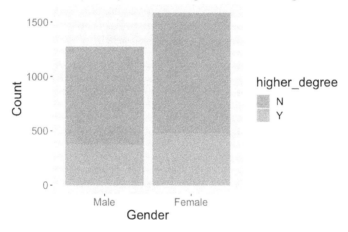

Figure 11.8 – Visualizing the frequency count of gender and higher-degree status

We can also plot them in percentages by specifying `position = "fill"` in the `geom_bar()` function:

```
ggplot(gss2016, aes(x = sex, fill=higher_degree)) +
  geom_bar(position = "fill") +
  labs(title = "Sample proportions for gender and degree", x =
"Gender", y = "Ratio") +
  theme(text = element_text(size = 16))
```

Running the code generates the chart in *Figure 11.9*, which suggests no obvious difference in the proportion of higher-degree holders between the male and female groups.

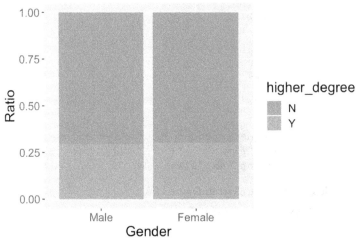

Figure 11.9 – Visualizing the frequency count of gender and higher-degree status

4. Calculate the difference in sample proportions of higher-degree holders between males and females:

```
p_hats = gss2016 %>%
  group_by(sex) %>%
  summarise(mean(higher_degree=="Y", na.rm=TRUE)) %>%
  pull()
d_hat = diff(p_hats)
>>> d_hat
0.007288771
```

The result also shows that the difference is quite small, with the female group being 0.7% higher than the male group (refer to the slightly higher blue bar of the female group in the previous figure). Let us see whether such a difference is statistically significant.

5. Generate one bootstrap sample set under the null hypothesis, which states that there is no difference in the ratio of higher-degree holders between the male and female groups:

```
gss2016 %>%
  specify(
    response = higher_degree,
    explanatory = sex,
    success = "Y"
  ) %>%
```

```
  hypothesise(null = "independence") %>%
  generate(reps = 1, type = "permute")
Response: higher_degree (factor)
Explanatory: sex (factor)
Null Hypothesis: independence
# A tibble: 2,862 × 3
# Groups:   replicate [1]
   higher_degree sex     replicate
   <fct>         <fct>       <int>
 1 N             Male            1
 2 N             Male            1
 3 Y             Male            1
 4 N             Female          1
 5 N             Female          1
 6 N             Female          1
 7 Y             Male            1
 8 N             Female          1
 9 N             Male            1
10 N             Male            1
# … with 2,852 more rows
# i Use `print(n = ...)` to see more rows
```

Here, we use `higher_degree` as the response variable and `sex` as the explanatory variable in a logistic regression setting (to be introduced in *Chapter 13*). Under the null hypothesis, we randomly sample from the original dataset and create a new artificial dataset of the same shape.

6. Repeat the same bootstrap sampling procedures 500 times and calculate the difference in sample proportions of higher-degree holders between female and male groups (note the sequence here) for each set of bootstrapped samples:

```
null = gss2016 %>%
  specify(
    higher_degree ~ sex,
    success = "Y"
  ) %>%
  hypothesise(null = "independence") %>%
  generate(reps = 500, type = "permute") %>%
  calculate(stat = "diff in props", order = c("Female", "Male"))
>>> null
Response: higher_degree (factor)
Explanatory: sex (factor)
Null Hypothesis: independence
# A tibble: 500 × 2
   replicate      stat
```

```
           <int>       <dbl>
   1           1    0.00870
   2           2    0.00587
   3           3   -0.00120
   4           4    0.0228
   5           5    0.00446
   6           6   -0.00827
   7           7   -0.0366
   8           8    0.0129
   9           9    0.0172
  10          10   -0.00261
  # … with 490 more rows
  # i Use `print(n = ...)` to see more rows
```

7. Plot these bootstrapped sample statistics in a density curve and plot the observed difference as a vertical red line:

```
ggplot(null, aes(x = stat)) +
  geom_density() +
  geom_vline(xintercept = d_hat, color = "red") +
  labs(x = "Difference in sample proportion (female - male)", y
= "Count") +
  theme(text = element_text(size = 16))
```

Running the code generates the plot in *Figure 11.10*, which shows that the red line is not located toward the extreme side of the empirical distribution. This suggests that the p-value, which will be calculated next, may be high.

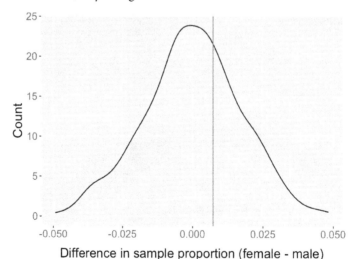

Figure 11.10 – Showing the density plot for the bootstrapped sample statistics and observed differences

8. Compute the two-tailed p-value:

```
null %>%
  summarize(pval = 2 * mean(stat > d_hat)) %>%
  pull()
0.608
```

The result shows a pretty high p-value, which suggests that we lack sufficient evidence to reject the null hypothesis. In other words, there is not enough information to suggest that the proportion of higher-degree holders between males and females is different.

The hypothesis testing relies on a predefined significance level. That significance level, denoted as α, has something to do with the statistical error of the procedure. The next section introduces two common types of statistical error when performing hypothesis testing.

Type I and Type II errors

There are two types of errors when conducting hypothesis testing and making a decision about the null hypothesis (H0) and the alternative hypothesis (H1). They are called Type I and Type II errors.

The Type I error refers to false positives. It happens when the null hypothesis is true but mistakenly rejected. In other words, we find evidence in our sample data that suggests a significant effect or difference exists and we favor the alternative hypothesis, even though it does not actually exist in the population. We denote the probability of experiencing a Type I error as α. It is also called the significance level, which was set to 0.05 in the previous example. A 5% significance level means that there is a 5% chance of rejecting the null hypothesis when it is true. The significance level thus represents the probability of committing a false positive error.

The Type II error focuses on the false negative case. It occurs when we fail to reject a false null hypothesis. In other words, we do not find evidence in our sample data to reject the null hypothesis, even though it does exist in the population. The probability of making a Type II error is denoted by β, which is also referred to as the power of the test. The complement of the power, denoted as $1 - \beta$, represents the probability of rejecting the null hypothesis when it is false.

Type I errors involve falsely rejecting the null hypothesis, while Type II errors involve failing to reject the null hypothesis when false. Both types of errors are important considerations in hypothesis testing because they can lead to incorrect conclusions. To minimize the risk of these errors, we can make a careful choice regarding the significance level (α) and also ensure that their study has sufficient power ($1 - \beta$). The power of a test depends on the sample size, the effect size (which is a quantitative measure of the magnitude of an empirical relationship between variables), and the chosen significance level. Larger sample sizes and larger effect sizes both increase the power of a test, reducing the likelihood of Type II errors.

Figure 11.11 provides an overview of the different types of outcomes in a hypothesis test. Note that the false positive and false negative are related to the quality of the decision. Depending on the type of a false decision, we would classify the errors as either Type I or Type II errors.

		Decision	
		Do not reject H0	Reject H0
Truth	H0 true	Good	False positive Type I error
	H0 false	False negative Type II error	Good

Figure 11.11 – Overview of different types of outcomes in a hypothesis test

The next section introduces the chi-square test, which tests the independence of two categorical variables.

Testing the independence of two categorical variables

To check the independence of two categorical variables, the process involves checking the existence of a statistically significant relationship between them. One common procedure is the chi-square test for independence. It works by comparing the observed frequencies in a contingency table with the expected frequencies under the assumption of independence.

Let us first review the contingency table for two categorical variables.

Introducing the contingency table

A contingency table, also known as a cross-tabulation or crosstab, is a table used to display the frequency distribution of two or more categorical variables. It summarizes the relationships between the variables by showing how their categories intersect or co-occur in the data. It provides a good summary of the relationships between categorical variables.

Let us stick with the example of the relationship between gender and degree. This time, we will look at all types of degrees, as shown in the following code:

```
>>> table(gss2016$degree)
Lt High School    High School Junior College          Bachelor
      Graduate
           328           1459            215               536
            318
```

To indicate its relationship with gender, we can plot the degree together with gender in a stacked bar plot as before:

```
ggplot(gss2016, aes(x = sex, fill=degree)) +
  geom_bar() +
  labs(title = "Frequency count for gender and degree", x = "Gender",
y = "Count") +
  theme(text = element_text(size = 16))
```

Running the code generates the plot in *Figure 11.12*.

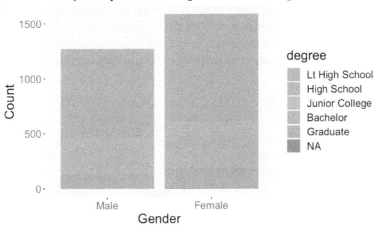

Figure 11.12 – Visualizing the relationship between gender and degree in a bar plot

However, the figure provides no information on the exact count for each category. To obtain the exact frequency for each category of the two variables, we can use the contingency table:

```
tab = gss2016 %>%
  select(sex, degree) %>%
  table()
>>> tab
        degree
sex     Lt High School High School Junior College Bachelor Graduate
  Male             147         661             89      243      132
  Female           181         798            126      293      186
```

Here, we used the `table()` function to generate the contingency table after selecting both `sex` and `degree`.

The next section introduces the chi-square test to test for the independence between these two categorical variables.

Applying the chi-square test for independence between two categorical variables

The chi-square test is a statistical test used to decide a possibly significant relationship (dependence) between two categorical variables in a collection of observed samples. It can be used to test for independence or goodness of fit. In this chapter, we focus mainly on the test for independence between two categorical variables. The test compares the observed frequencies with the expected ones in a contingency table, assuming that the variables are independent. If the observed and expected frequencies are significantly different, the test suggests that the variables are not independent; in other words, they are dependent on each other.

Following the same approach as before, we can generate an artificial bootstrapped dataset to obtain a sample statistic, called the chi-square statistic. This dataset is generated by permuting the original dataset under the assumption of independence in the null hypothesis. In the following code, we generate one permutated dataset of the same shape as the original dataset, assuming independence under the null hypothesis:

```
perm_1 = gss2016 %>%
  # Specify the variables of interest
  specify(degree ~ sex) %>%
  # Set up the null hypothesis
  hypothesize(null = "independence") %>%
  # Generate a single permuted dataset
  generate(reps = 1, type = "permute")
>>> perm_1
Response: degree (factor)
Explanatory: sex (factor)
Null Hypothesis: independence
# A tibble: 2,856 × 3
# Groups:    replicate [1]
   degree          sex      replicate
   <fct>           <fct>        <int>
 1 Junior College Male             1
 2 Bachelor        Male             1
 3 High School     Male             1
 4 High School     Female           1
 5 High School     Female           1
 6 High School     Female           1
 7 Graduate        Male             1
 8 Bachelor        Female           1
 9 High School     Male             1
10 High School     Male             1
# … with 2,846 more rows
# i Use `print(n = ...)` to see more rows
```

Next, we create 500 permutated datasets and extract the corresponding chi-square statistic:

```
null_spac = gss2016 %>%
  specify(degree ~ sex) %>%
  hypothesize(null = "independence") %>%
  generate(reps = 500, type = "permute") %>%
  calculate(stat = "Chisq")
>>> null_spac
Response: degree (factor)
Explanatory: sex (factor)
Null Hypothesis: independence
# A tibble: 500 × 2
   replicate  stat
       <int> <dbl>
 1         1  3.50
 2         2  1.11
 3         3 14.0
 4         4  4.62
 5         5  1.41
 6         6  1.41
 7         7  9.69
 8         8  4.17
 9         9  5.97
10        10  2.86
# … with 490 more rows
# i Use `print(n = ...)` to see more rows
```

To run the test, we obtain the expected frequency for each cell in the contingency table under the assumption of independence between the categorical variables. The expected frequency for a cell is computed as (*row sum * column sum*) / *overall sum*:

```
# calculate expected frequency table
row_totals = rowSums(tab)
col_totals = colSums(tab)
overall_total = sum(tab)
expected = outer(row_totals, col_totals) / overall_total
>>> expected
        Lt High School High School Junior College Bachelor Graduate
Male           146.084    649.8067        95.7563 238.7227 141.6303
Female         181.916    809.1933       119.2437 297.2773 176.3697
```

Here, we first obtain the row-wise and column-wise sum, as well as the total sum. We then use the outer() function to obtain the outer product between these two vectors, which is then scaled by the total sum to obtain the expected frequency count in each cell.

Now, we compute the observed chi-square statistic based on the available samples:

```
# Compute chi-square statistic
observed_chi_square = sum((tab - expected)^2 / expected)
>>> observed_chi_square
2.536349
```

We can then plot the observed chi-square statistic within the density curve of previous bootstrapped sample statistics to get a sense of where the observed statistic is located, based on which we will be able to calculate the corresponding p-value:

```
ggplot(null_spac, aes(x = stat)) +
  geom_density() +
  geom_vline(xintercept = observed_chi_square, color = "red") +
  labs(title = "Density curve of bootstrapped chi-square statistic", x
= "Chi-square statistic", y = "Density") +
  theme(text = element_text(size = 16))
```

Running the code generates the plot in *Figure 11.13*, which shows a high p-value.

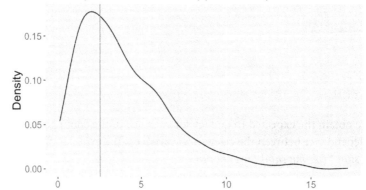

Figure 11.13 – Visualizing the density curve of bootstrapped chi-square statistics and the observed statistic

Now we can calculate the p-value. As shown in the following code, the p-value of 0.72 is indeed quite high, and thus there is no sufficient evidence to reject the null hypothesis:

```
>>> null_spac %>%
  summarize(pval = 2 * mean(stat < observed_chi_square)) %>%
  pull()
0.72
```

In the next section, we will shift to look at statistical inference for numerical data.

Statistical inference for numerical data

In this section, we will switch to look at statistical inference using numerical data. We will cover two approaches. The first approach relies on the bootstrapping procedure and permutes the original dataset to create additional artificial datasets, which can then be used to derive the confidence intervals. The second approach uses a theoretical assumption on the distribution of the bootstrapped samples and relies on the t-distribution to achieve the same result. We will learn how to perform a t-test, derive a confidence interval, and conduct an **analysis of variance** (**ANOVA**).

As discussed earlier, bootstrapping is a non-parametric resampling method that allows us to estimate the sampling distribution of a particular statistic, such as the mean, median, or proportion, as in the previous section. This is achieved by repeatedly drawing random samples with replacement from the original data. By doing so, we can calculate confidence intervals and perform hypothesis tests without relying on specific distributional assumptions.

Additionally, the t-distribution is a probability distribution used for hypothesis testing if the sample size is small and the standard deviation of the population data remains unknown. It is a more general approach that assumes the bootstrapped samples follow a specific distribution. We will then use this distribution to estimate confidence intervals and perform the hypothesis test.

The t-test is a widely used statistical test that allows us to compare the mean values of two groups or test whether the mean of a single group is equal to a specific value. This time, our interest is the mean of a group since the variable is numeric. The test relies on the t-distribution and takes into account the sample sizes, sample means, and sample variances.

Confidence intervals offer a list of possible values, where the true population statistic, such as the mean or proportion, is likely to lie, with a specified level of confidence (specified by the significance level α).

Finally, ANOVA extends the t-test used when there are more than two groups to compare. ANOVA helps us determine possible significant differences among the group means by dividing the total variability of the observed data into two parts: between-group variability and within-group variability. It tests the null hypothesis that the mean values of all groups are equal. If the null hypothesis is rejected, we can continue to identify which specific group means differ from each other.

Let us start with generating a bootstrap distribution for the median.

Generating a bootstrap distribution for the median

As discussed earlier, when building a bootstrap distribution for a single statistic, we first generate a collection of bootstrap samples via sampling with replacement, and then record the relevant statistic (in this case, the median) of each distribution.

Let us go through an exercise to build the collection of bootstrap samples.

Exercise 11.5 – generating a bootstrap distribution for the sample median

In this exercise, we will apply the same specify-generate-calculate workflow using the `infer` package to generate a bootstrap distribution for the sample median using the `mtcars` dataset.

Load the `mtcars` dataset and view its structure:

```
data(mtcars)
>>> str(mtcars)
'data.frame':  32 obs. of  11 variables:
 $ mpg : num  21 21 22.8 21.4 18.7 18.1 14.3 24.4 22.8 19.2 ...
 $ cyl : num  6 6 4 6 8 6 8 4 4 6 ...
 $ disp: num  160 160 108 258 360 ...
 $ hp  : num  110 110 93 110 175 105 245 62 95 123 ...
 $ drat: num  3.9 3.9 3.85 3.08 3.15 2.76 3.21 3.69 3.92 3.92 ...
 $ wt  : num  2.62 2.88 2.32 3.21 3.44 ...
 $ qsec: num  16.5 17 18.6 19.4 17 ...
 $ vs  : num  0 0 1 1 0 1 0 1 1 1 ...
 $ am  : num  1 1 1 0 0 0 0 0 0 0 ...
 $ gear: num  4 4 4 3 3 3 3 4 4 4 ...
 $ carb: num  4 4 1 1 2 1 4 2 2 4 ...
```

The result shows that we have a dataset with 32 rows and 11 columns. In the following steps, we will use the mpg variable and generate a bootstrap distribution of its median:

1. Generate 10,000 bootstrap samples according to the mpg variable and obtain the median of all samples:

```
bs <- mtcars %>%
  specify(response = mpg) %>%
  generate(reps = 10000, type = "bootstrap") %>%
  calculate(stat = "median")
>>> bs
Response: mpg (numeric)
# A tibble: 10,000 × 2
   replicate  stat
       <int> <dbl>
1          1  21.4
2          2  22.2
3          3  20.4
4          4  17.8
5          5  19.2
6          6  19.2
7          7  18.4
8          8  20.4
```

```
   9             9  19.0
  10            10  21.4
# … with 9,990 more rows
# i Use `print(n = ...)` to see more rows
```

Here, we specify stat = "median" in the calculate() function to extract the median in each bootstrap sample.

2. Plot the bootstrap distribution as a density curve of the bootstrapped sample statistics:

```
ggplot(bs, aes(x = stat)) +
  geom_density() +
  labs(title = "Density plot for bootstrapped median", x =
"Median", y = "Probability") +
  theme(text = element_text(size = 16))
```

Running the code generates the plot in *Figure 11.14*.

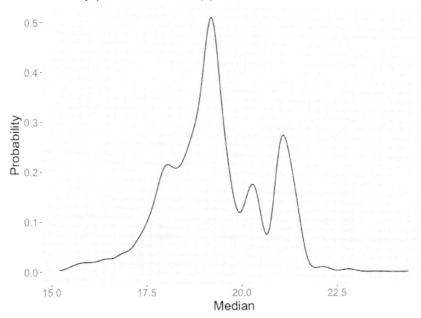

Figure 11.14 – Visualizing the density curve of the bootstrapped sample median

The next section looks at constructing the bootstrapped confidence interval.

Constructing the bootstrapped confidence interval

We have looked at how to construct the bootstrapped confidence interval using the standard error method. This involves adding and subtracting the scaled standard error from the observed sample statistic. It turns out that there is another, simpler method, which just uses the percentile of the bootstrap distribution to obtain the confidence interval.

Let us continue with the previous example. Say we would like to calculate the 95% confidence interval of the previous bootstrap distribution. We can achieve this by calculating the upper and lower quantiles (97.5% and 2.5%, respectively) of the bootstrap distribution. The following code achieves this:

```
>>> bs %>%
  summarize(
    l = quantile(stat, 0.025),
    u = quantile(stat, 0.975)
  )
# A tibble: 1 × 2
      l      u
  <dbl>  <dbl>
1  16.6   21.4
```

Let us also calculate the bootstrap confidence interval using the standard error method, as shown in the following code:

```
SE = bs %>%
  summarise(sd(stat)) %>%
  pull()
observed_median = median(mtcars$mpg)
>>> c(observed_median - 2*SE, observed_median + 2*SE)
16.64783 21.75217
```

As expected, the result is close to the one obtained using the percentile method. However, the standard error method is a more accurate method than the percentile method.

The next section covers re-centering a bootstrap distribution upon testing a null hypothesis.

Re-centering a bootstrap distribution

The bootstrap distribution from the previous section is generated by randomly sampling the original dataset with replacement. Each set of bootstrap samples maintains the same size as the original sample sets. However, we cannot directly use this bootstrap distribution for hypothesis testing.

Upon introducing a null hypothesis, what we did in the previous hypothesis test section for two categorical variables is re-generated a new bootstrap distribution under the null hypothesis. We then place the observed sample statistic as a vertical red line along the bootstrap distribution to calculate the p-value, representing the probability of experiencing a phenomenon at least as extreme as the observed sample statistic. The only additional step is to generate the bootstrap distribution under the null hypothesis.

When generating bootstrap samples under the null hypothesis, the main idea is to remove the effect we are testing for and create samples, assuming the null hypothesis is true. In other words, we create samples that would be expected if there were no difference between the groups. For example, when comparing means between two groups, we would subtract the overall mean from each observation to center the data around 0 before performing the random sampling with replacement.

There is another way to achieve this. Recall that the original bootstrap distribution, by design, is centered around the observed sample statistic. Upon introducing the null hypothesis, we could simply move the original bootstrap distribution to be centered around the statistic in the null hypothesis, which is the null value. This shifted bootstrap distribution represents the same distribution if we were to remove the effect in the original dataset and then perform bootstrap sampling again. We can then place the observed sample statistic along the shifted bootstrap distribution to calculate the corresponding p-value, which represents the ratio of simulations that generate a sample statistic at least as favorable to the alternative hypothesis as the actual sample statistic. *Figure 11.15* demonstrates this process.

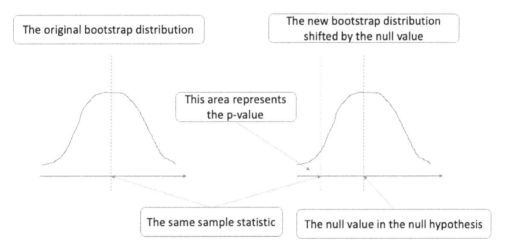

Figure 11.15 – Shifting the bootstrap distribution to be centered around the null value

Let us generate the bootstrap distribution for hypothesis testing for the previous example. We want to test the null hypothesis with a population median of 16 for the mpg variable. The following code generates the bootstrapped sample statistics, where we specify the null value via med = 16 and the point estimate with null = "point" in the hypothesize() function:

```
bs = mtcars %>%
  specify(response = mpg) %>%
  hypothesize(null = "point", med = 16) %>%
```

```
   generate(reps = 10000, type = "bootstrap") %>%
   calculate(stat = "median")
>>> bs
Response: mpg (numeric)
Null Hypothesis: point
# A tibble: 10,000 × 2
   replicate  stat
       <int> <dbl>
 1         1  16
 2         2  16.2
 3         3  18
 4         4  16.2
 5         5  16
 6         6  16.2
 7         7  16
 8         8  16
 9         9  17.8
10        10  16
# … with 9,990 more rows
# i Use `print(n = ...)` to see more rows
```

Now, we plot these bootstrapped sample statistics in a density plot, along with the observed sample statistic as a vertical red line:

```
ggplot(bs, aes(x = stat)) +
  geom_density() +
  geom_vline(xintercept = median(mtcars$mpg), color = "red") +
  labs(title = "Density curve of bootstrapped median", x = "Sample
median", y = "Density") +
  theme(text = element_text(size = 16))
```

Running the code generates the plot in *Figure 11.16*, which shows a small p-value.

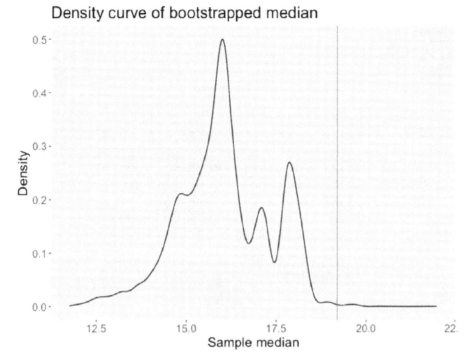

Figure 11.16 – Density plot of bootstrapped sample medians
and observed sample median (vertical red line)

In the next section, we will cover another distribution-based inference approach based on the **central limit theorem (CLT)**.

Introducing the central limit theorem used in t-distribution

The CLT says that the distribution from the sum (or average) of many independent and identically distributed random variables would jointly form a normal distribution, regardless of the underlying distribution of these individual variables. Due to the CLT, normal distribution is often used to approximate the sampling distribution of various statistics, such as the sample mean and the sample proportion.

The t-distribution is related to the CLT in the context of statistical inference. When we're estimating a population mean from a sample, we often have no access to the true standard deviation of the population. Instead, we resort to the sample standard deviation as an estimate. In this case, the sampling distribution of the sample mean doesn't follow a normal distribution, but rather a t-distribution. In other words, when we extract the sample mean from a set of observed samples, and we are unsure of the population standard deviation (as is often the case when working with actual data), the sample mean can be modeled as a realization from the t-distribution.

The t-distribution is a family of continuous probability distributions that are symmetric and bell-shaped, which shows similarity to the normal distribution. However, the t-distribution shows heavier tails, which accounts for the greater uncertainty due to estimating the population standard deviation from the observed data. That is, observations of a t-distribution are more likely to fall into distant tails (such as beyond two standard deviations away from the mean) than the normal distribution. The shape of the t-distribution relies on the **degrees of freedom (df)**, which depends on the sample size and determines the thickness of the tails. As more samples are collected, the df moves up, and the t-distribution gradually approximates the normal distribution.

We briefly covered the qt () function used to find the cutoffs under the t-distribution in the previous chapter. Now let us go through an exercise to get more familiar with calculations related to the t-distribution.

Exercise 11.6 – understanding the t-distribution

In this exercise, we will use the pt () function to find probabilities under the t-distribution. For a given cutoff quantile value, q, and a given df, the pt (q, df) function gives us the probability under the t-distribution with df for values of t less than q. In other words, we have $P(t_{df} < T) =$ pt (q = T, df). We can also use the qt () function to find the quantiles for a specific probability under the t-distribution. That is, if $P(t_{df} < T) = p$, then $T =$ qt (p, df):

1. Find the probability under the t-distribution with 10 df below T=3:

    ```
    x = pt(3, df = 10)
    >>> x
    0.9933282
    ```

2. Find the probability under the t-distribution with 10 df above T=3:

    ```
    y = 1 - x
    >>> y
    0.006671828
    ```

    Note that we first calculate the probability of being below a specific cutoff value under the t-distribution, and then take the complement to find the probability above the threshold.

3. Find the probability under the t-distribution with `100` df above `T=3`:

```
z = 1 - pt(3, df = 100)
>>> z
0.001703958
```

Since `df=100` has a better approximation to the normal distribution than `df=10`, the resulting probability, z, is thus smaller than y.

4. Find the 95th percentile of the t-distribution with 10 df:

```
d = qt(0.95, df = 10)
>>> d
1.812461
```

5. Find the cutoff value that bounds the upper end of the middle 95th percentile of the t-distribution with `10` df:

```
e = qt(0.975, df = 10)
>>> e
2.228139
```

Here, the upper end of the middle 95th percentile refers to the 97.5th percentile.

6. Find the cutoff value that bounds the upper end of the middle 95th percentile of the t-distribution with `100` df:

```
f = qt(0.975, df = 100)
>>> f
1.983972
```

The next section discusses how to construct the confidence interval for the population mean using the t-distribution.

Constructing the confidence interval for the population mean using the t-distribution

Let us review the process of statistical inference for the population mean. We start with a limited sample, from which we can derive the sample mean. Since we want to estimate the population mean, we would like to perform statistical inference based on the observed sample mean and quantify the range where the population statistic may exist.

For example, the average miles per gallon, shown in the following code, is around 20 in the `mtcars` dataset:

```
>>> mean(mtcars$mpg)
20.09062
```

Given this result, we won't be surprised to encounter another similar dataset with an average mpg of 19 or 21. However, we would be surprised if the value is 5, 50, or even 100. When assessing a new collection of samples, we need a way to quantify the variability of the sample mean across multiple samples. We have learned two ways to do this: use the bootstrap approach to simulate artificial samples or use the CLT to approximate such variability. We will focus on the CLT approach in this section.

According to the CLT, the sample mean of any sampling distribution would be approximately normally distributed, regardless of the original distribution. In other words, we have the following:

$$\bar{x} \sim N\left(mean = \mu, SE = \frac{\sigma}{\sqrt{n}}\right)$$

Note that this is a theoretical distribution we are unable to obtain. For example, the population standard deviation, σ, stays unknown, and we only have access to the observed samples. Instead, we would estimate the standard error using the sample standard deviation, s, giving the following:

$$\bar{x} \sim N\left(mean = \mu, SE = \frac{s}{\sqrt{n}}\right)$$

We would then employ the t-distribution of $n - 1$ degree of freedom to make an inference for the population mean as it gives thicker tails due to the additional uncertainty introduced by s.

In addition, note that the approximation using the CLT relies on a few assumptions. For example, the samples need to be independent of each other. This is often satisfied when the samples are randomly selected, or if the samples account for less than 10% of the total population if they are selected without replacement. The sample size also needs to be larger to account for potential skewness in the samples.

We can construct the 95% confidence interval using the t.test() function, as shown in the following code:

```
# Construct 95% CI for avg mpg
>>> t.test(mtcars$mpg)
  One Sample t-test

data:  mtcars$mpg
t = 18.857, df = 31, p-value < 2.2e-16
alternative hypothesis: true mean is not equal to 0
95 percent confidence interval:
 17.91768 22.26357
sample estimates:
mean of x
 20.09062
```

Here, we are performing a one-sample t-test, where the default null hypothesis states that the population mean is 0. The result shows a very small p-value, suggesting that we could reject the null hypothesis in favor of the alternative hypothesis; that is, the population mean is not 0. The 95% confidence interval (between 17.91768 and 22.26357) is also constructed based on the t-distribution with a df of 31 and a t-statistic of 18.857.

The next section reviews the hypothesis testing for two means using both bootstrap simulation and t-test approximation.

Performing hypothesis testing for two means

In this section, we will explore the process of comparing two sample means using hypothesis testing. When comparing two sample means, we want to determine whether a significant difference exists between the means of two distinct populations or groups.

Suppose now we have two groups of samples. These two groups could represent a specific value before and after treatment for each sample. Our objective is thus to compare the sample statistics of these two groups, such as the sample mean, and determine whether the treatment has an effect. To do this, we can perform a hypothesis test to compare mean values from the two independent distributions using either bootstrap simulation or t-test approximation.

When using the t-test in the hypothesis test to compare the mean values of two independent samples, the two-sample t-test assumes normal distribution for the data, and that the variances of the two populations are equal. However, in cases where these assumptions may not hold, alternative non-parametric tests or resampling methods, such as bootstrap, can be employed to make inferences about the population means.

Let us go through an exercise to see these two methods of hypothesis testing in play.

Exercise 11.7 – comparing two means

In this exercise, we will explore two approaches (t-test and bootstrap) to compare two sample means and calculate the confidence interval of the difference in sample means:

1. Generate a dummy dataset that consists of two groups of samples:

```
# Define two samples
sample1 = c(10, 12, 14, 16, 18)
sample2 = c(15, 17, 19, 21, 23)

# Combine samples into a data frame
data = tibble(
  value = c(sample1, sample2),
  group = factor(rep(c("Group 1", "Group 2"), each =
length(sample1)))
)
>>> data
# A tibble: 10 × 2
   value group
   <dbl> <fct>
```

```
 1     10 Group 1
 2     12 Group 1
 3     14 Group 1
 4     16 Group 1
 5     18 Group 1
 6     15 Group 2
 7     17 Group 2
 8     19 Group 2
 9     21 Group 2
10     23 Group 2
```

Here, we created a `tibble` DataFrame with the `value` column indicating the sample observation and the `group` column indicating the group number. We would like to assess the difference in the sample mean between these two groups.

2. Perform bootstrap sampling `1000` times and calculate the bootstrap statistics under the null hypothesis that these two groups are independent of each other, and there is no difference in their means:

```
bootstrap_results = data %>%
  specify(response = value, explanatory = group) %>%
  hypothesize(null = "independence") %>%
  generate(reps = 1000, type = "bootstrap") %>%
  calculate(stat = "diff in means", order = c("Group 1", "Group
2"))
>>> bootstrap_results
Response: value (numeric)
Explanatory: group (factor)
Null Hypothesis: independence
# A tibble: 1,000 × 2
   replicate   stat
       <int> <dbl>
 1         1 -7.5
 2         2 -6.17
 3         3 -5
 4         4 -2.20
 5         5 -8.05
 6         6 -4.2
 7         7 -3.5
 8         8 -6.67
 9         9 -4.2
10        10 -6.90
# … with 990 more rows
# i Use `print(n = ...)` to see more rows
```

Here, the pipeline for the hypothesis test starts by specifying the response (`value`) and explanatory (`group`) variables, setting up the null hypothesis, generating bootstrap samples under the null hypothesis, and then calculating the test statistic (in this case, the difference in means) for each bootstrap sample. The null hypothesis states that we assume the sample mean values for both groups come from the same population, and that any observed difference is merely due to chance.

3. Calculate the confidence interval based on the bootstrap statistics:

```
ci = bootstrap_results %>%
   filter(!is.na(stat)) %>%
   get_confidence_interval(level = 0.95, type = "percentile")
>>> ci
# A tibble: 1 × 2
   lower_ci upper_ci
      <dbl>    <dbl>
1        -9    -1.17
```

4. Perform a two-sample t-test using the `t.test()` function:

```
t_test_result = t.test(sample1, sample2)
>>> t_test_result
    Welch Two Sample t-test

data:  sample1 and sample2
t = -2.5, df = 8, p-value = 0.03694
alternative hypothesis: true difference in means is not equal to
0
95 percent confidence interval:
 -9.6120083 -0.3879917
sample estimates:
mean of x mean of y
       14        19
```

The result shows that the 95% confidence interval based on the t-distribution is close but still different from the one obtained via bootstrap sampling. We can also perform the t-test by passing in the model form:

```
t_test_result2 = t.test(value ~ group, data = data)
>>> t_test_result2
    Welch Two Sample t-test

data:  value by group
t = -2.5, df = 8, p-value = 0.03694
alternative hypothesis: true difference in means between group
Group 1 and group Group 2 is not equal to 0
```

```
95 percent confidence interval:
 -9.6120083 -0.3879917
sample estimates:
mean in group Group 1 mean in group Group 2
                    14                     19
```

The next section introduces ANOVA, or the analysis of variance.

Introducing ANOVA

ANOVA is a statistical hypothesis testing method used to compare the means of more than two groups, which extends the two-sample t-test discussed in the previous section. The goal of ANOVA is to test potential significant differences among the group means (the between-group variability) while accounting for the variability within each group (the within-group variability).

ANOVA relies on the F-statistic in hypothesis testing. The F-statistic is a ratio of two estimates of variance: the between-group variance and the within-group variance. The between-group variance measures the differences among the group means, while the within-group variance represents the variability within each group. The F-statistic can be calculated based on these two group variances.

In hypothesis testing, the null hypothesis for ANOVA states that all group means are equal, and any observed differences are due to chance. The alternative hypothesis, on the other hand, suggests that at least one group's mean differs from the others. If the F-statistic is sufficiently large, the between-group variance is significantly greater than the within-group variance, which provides evidence against the null hypothesis.

Let us look at a concrete example. We first load the `PlantGrowth` dataset, which contains the weights of plants after they have been subjected to three different treatments:

```
data(PlantGrowth)
>>> str(PlantGrowth)
'data.frame':   30 obs. of  2 variables:
 $ weight: num  4.17 5.58 5.18 6.11 4.5 4.61 5.17 4.53 5.33 5.14 ...
 $ group : Factor w/ 3 levels "ctrl","trt1",..: 1 1 1 1 1 1 1 1 1 1
 ...
```

Next, we perform the one-way ANOVA test using the same specify-hypothesize-generate-calculate procedure. Specifically, we first specify the response variable (`weight`) and the explanatory variable (`group`). We then set up the null hypothesis, stating no difference in the means of the groups, using `hypothesize(null = "independence")`. Next, we generate 1,000 permuted datasets using `generate(reps = 1000, type = "permute")`. Finally, we calculate the F-statistic for each permuted dataset using `calculate(stat = "F")`:

```
anova_results = PlantGrowth %>%
  specify(response = weight, explanatory = group) %>%
```

```
   hypothesize(null = "independence") %>%
   generate(reps = 1000, type = "permute") %>%
   calculate(stat = "F")
>>> anova_results
Response: weight (numeric)
Explanatory: group (factor)
Null Hypothesis: independence
# A tibble: 1,000 × 2
   replicate  stat
       <int> <dbl>
 1         1 0.162
 2         2 0.198
 3         3 1.18
 4         4 0.328
 5         5 1.21
 6         6 3.00
 7         7 1.93
 8         8 0.605
 9         9 0.446
10        10 1.10
# … with 990 more rows
# i Use `print(n = ...)` to see more rows
```

Last, we can calculate the p-value using the observed F-statistic and the distribution of the F-statistics obtained from the permuted datasets. When the p-value is smaller than the preset significance level (for example, 0.05), we could reject the null hypothesis and say that there is a significant difference among the means of the groups:

```
p_value = anova_results %>%
   get_p_value(obs_stat = anova_results, direction = "right") %>%
   pull()
>>> p_value
0.376
```

The result suggests that we do not have enough confidence to reject the null hypothesis.

Summary

In this chapter, we covered different types of statistical inferences for hypothesis testing, targeting both numerical and categorical data. We introduced inference methods for a single variable, two variables, and multiple variables, using either proportion (for categorical variable) or mean (for numerical variable) as the sample statistic. The hypothesis testing procedure, including both the parametric approach using model-based approximation and the non-parametric approach using bootstrap-based simulations, offers valuable tools such as the confidence interval and p-value. These tools allow us to make a decision about whether we can reject the null hypothesis in favor of the alternative hypothesis. Such a decision also relates to the Type I and Type II errors.

In the next chapter, we will cover one of the most widely used statistical and ML models: linear regression.

12

Linear Regression in R

In this chapter, we will introduce linear regression, a fundamental statistical approach that's used to model the relationship between a target variable and multiple explanatory (also called independent) variables. We will cover the basics of linear regression, starting with simple linear regression and then extending the concepts to multiple linear regression. We will learn how to estimate the model coefficients, evaluate the goodness of fit, and test the significance of the coefficients using hypothesis testing. Additionally, we will discuss the assumptions underlying linear regression and explore techniques to address potential issues, such as nonlinearity, interaction effect, multicollinearity, and heteroskedasticity. We will also introduce two widely used regularization techniques: the ridge and **Least Absolute Shrinkage and Selection Operator (lasso)** penalties.

By the end of this chapter, you will learn the core principles of linear regression, its extensions to regularized linear regression, and the implementation details involved.

In this chapter, we will cover the following topics:

- Introducing linear regression
- Introducing penalized linear regression
- Working with ridge regression
- Working with lasso regression

To run the code in this chapter, you will need to have the latest versions of the following packages:

- `ggplot2`, 3.4.0
- `tidyr`, 1.2.1
- `dplyr`, 1.0.10
- `car`, 3.1.1
- `lmtest`, 0.9.40
- `glmnet`, 4.1.7

Please note that the versions of the packages mentioned in the preceding list are the latest ones at the time of writing this chapter. All the code and data for this chapter are available at `https://github.com/PacktPublishing/The-Statistics-and-Machine-Learning-with-R-Workshop/blob/main/Chapter_12/working.R`.

Introducing linear regression

At the core of linear regression is the concept of fitting a straight line – or more generally, a hyperplane – to the data points. Such fitting aims to minimize the deviation between the observed and predicted values. When it comes to simple linear regression, one target variable is regressed by one predictor, and the goal is to fit a straight line that best mimics the relationship between the two variables. For multiple linear regression, there is more than one predictor, and the goal is to fit a hyperplane that best describes the relationship among the variables. Both tasks can be achieved by minimizing a measure of deviation between the predictions and the corresponding targets.

In linear regression, obtaining an optimal model means identifying the best coefficients that define the relationship between the target variable and the input predictors. These coefficients represent the change in the target associated with a single unit change in the associated predictor, assuming all other variables are constant. This allows us to quantify the magnitude (size of the coefficient) and direction (sign of the coefficient) of the relationship between the variables, which can be used for inference (highlighting explainability) and prediction.

When it comes to inference, we often look at the relative impact on the target variable given a unit change to the input variable. Examples of such explanatory modeling include how marketing spend affects quarterly sales, how smoker status affects insurance premiums, and how education affects income. On the other hand, predictive modeling focuses on predicting a target quantity. Examples include predicting quarterly sales given the marketing spend, predicting the insurance premium given a policyholder's profile information, such as age and gender, and predicting income given someone's education, age, work experience, and industry.

In linear regression, the expected outcome is modeled as a weighted sum of all the input variables. It also assumes that the change in the output is linearly proportional to the change in any input variable. This is the simplest form of the regression method.

Let's start with **simple linear regression (SLR)**.

Understanding simple linear regression

SLR is a powerful and widely used statistical model that specifies the relationship between two continuous variables, including one input and one output. It allows us to understand how a response variable (also referred to as the dependent or target variable) changes as the explanatory variable (also called the independent variable or the input variable) varies. By fitting a straight line to the observed data, SLR quantifies the strength and direction of the linear association between the two variables. This straight line is called the SLR **model**. It enables us to make predictions and infer the impact of the predictor on the target variable.

Specifically, in an SLR model, we assume a linear relationship between the target variable (y) and the input variable (x). The model can be represented mathematically as follows:

$$y = \beta_0 + \beta_1 x + \epsilon$$

Here, y is called the response variable, dependent variable, explained variable, predicted variable, target variable, or regressand. x is called the explanatory variable, independent variable, control variable, predictor variable, or regressor. β_0 is the intercept of the linear line that represents the expected value of y when x is 0. β_1 is the slope that represents the change in y for a one-unit increase in x. Finally, ϵ is the random error term that accounts for the variability in the target, y, that the predictor, x, cannot explain.

The main objective of SLR is to estimate the β_0 and β_1 parameters. An optimal set of β_0 and β_1 would minimize the total squared deviations between the observed target values, y, and the predicted values, $\hat{y}$, using the model. This is called the **least squares method**, where we seek the optimal β_0 and β_1 parameters that correspond to the minimum **sum of squared error** (**SSR**):

$$minSSR = min\sum_{i=1}^{n}u_i^2 = min(y_i - \hat{y}_i)^2$$

Here, each residual, u_i, is the difference between the observation, y_i, and its fitted value, $\hat{y}_i$. In simple terms, the objective is to locate the straight line that is closest to the data points given. *Figure 12.1* illustrates a collection of data points (in blue) and the linear model (in red):

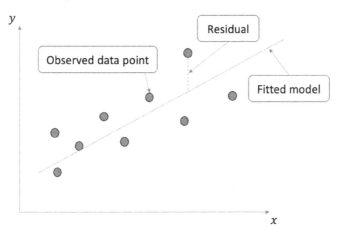

Figure 12.1 – The SLR model, where the linear model appears as a line and is trained by minimizing the SSR

Once we have estimated the model coefficients, β_0 and β_1, we can use the model to make predictions and inferences on the intensity of the linear relationship between the variables. Such a linear relationship indicates the goodness of fit, which is often measured using the coefficient of determination, or R^2. R^2 ranges from 0 to 1 and quantifies the proportion of the total variation in y that can be explained by x. It is defined as follows:

$$R^2 = 1 - \frac{\sum_i (y_i - \hat{y}_i)^2}{\sum_i (y_i - \bar{y})^2}$$

Here, $\bar{y}$ denotes the average value of the observed target variable, y.

Besides this, we can also use hypothesis testing to test the significance of the resulting coefficients, β_0 and β_1, thus helping us determine whether the observed relationship between the variables is statistically significant.

Let's go through an example of building a simple linear model using a simulated dataset.

Exercise 12.1 – building an SLR model

In this exercise, we will demonstrate the implementation of an SLR model in R. We'll be using a combination of built-in functions and packages to accomplish this task using a simulated dataset:

1. Simulate a dataset such that the response variable, Y, is linearly dependent on the explanatory variable, X, with some added noise:

```
# Set seed for reproducibility
set.seed(123)
# Generate independent variable X
X = runif(100, min = 1, max = 100) # 100 random uniform numbers
between 1 and 100
# Generate some noise
noise = rnorm(100, mean = 0, sd = 10) # 100 random normal
numbers with mean 0 and standard deviation 10
# Generate dependent variable Y
Y = 5 + 0.5 * X + noise
# Combine X and Y into a data frame
data = data.frame(X, Y)
```

Here, we use the `runif()` function to generate the independent variable, X, which is a vector of random uniform numbers. Then, we add some "noise" to the dependent variable, Y, making the observed data more realistic and less perfectly linear. This is achieved using the `rnorm()` function, which creates a vector of random normal numbers. The target variable, Y, is then created as a function of X, plus the noise.

Besides this, we use a seed (`set.seed(123)`) at the beginning to ensure reproducibility. This means that we will get the same set of random numbers every time we run this code. Each run will produce a different list of random numbers if we don't set a seed.

In this simulation, the true intercept (β_0) is 5, the true slope (β_1) is 0.5, and the noise is normally distributed with a mean of 0 and a standard deviation of 10.

2. Train a linear regression model based on the simulated dataset using the lm() function:

```
# Fit a simple linear regression model
model = lm(Y ~ X, data = data)
# Print the model summary
>>> summary(model)
Call:
lm(formula = Y ~ X, data = data)

Residuals:
     Min       1Q    Median       3Q      Max
-22.3797   -6.1323   -0.1973   5.9633  22.1723

Coefficients:
             Estimate Std. Error t value Pr(>|t|)
(Intercept)   4.91948    1.99064   2.471   0.0152 *
X             0.49093    0.03453  14.218   <2e-16 ***
---
Signif. codes:
0 '***' 0.001 '**' 0.01 '*' 0.05 '.' 0.1 ' ' 1

Residual standard error: 9.693 on 98 degrees of freedom
Multiple R-squared:  0.6735,  Adjusted R-squared:  0.6702
F-statistic: 202.2 on 1 and 98 DF,  p-value: < 2.2e-16
```

Here, we use the lm() function to fit the data, where lm stands for "linear model." This function creates our SLR model. The Y ~ X syntax is how we specify our model: it tells the function that Y is being modeled as a function of X.

The summary() function provides a comprehensive overview of the model, including the estimated coefficients, the standard errors, the t-values, and the p-values, among other statistics. Since the resulting p-value is extremely low, we can conclude that the input variable is predictive with strong statistical significance.

3. Use the plot() and abline() functions to visualize the data and the fitted regression line:

```
# Plot the data
plot(data$X, data$Y, main = "Simple Linear Regression", xlab =
"X", ylab = "Y")
# Add the fitted regression line
abline(model, col = "red")
```

Running this code generates *Figure 12.2*:

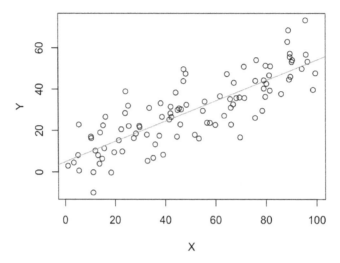

Figure 12.2 – Visualizing the data and the fitted regression line

Here, the `plot()` function creates a scatter plot of our data, and the `abline()` function adds the regression line to this plot. Such a visual representation is very useful for understanding the quality of the fitting.

We'll move on to the **multiple linear regression (MLR)** model in the next section.

Introducing multiple linear regression

MLR expands the single predictor in SLR to predict the target outcome based on multiple predictor variables. Here, the term "multiple" in MLR refers to the multiple predictors in the model, where each feature is given a coefficient. A specific coefficient, β, represents the change in the outcome variable for a single unit change in the associated predictor variable, assuming all other predictors are held constant.

One of the great advantages of MLR is its ability to include multiple predictors, allowing for a more complex and realistic (linear) representation of the real world. It can provide a holistic view of the connection between the target and all input variables. This is particularly useful in fields where the outcome variable is likely influenced by more than one predictor variable. It is modeled via the following formula:

$$y = \beta_0 + \beta_1 x_1 + \beta_2 x_2 + \ldots + \beta_p x_p + \epsilon$$

Here, we have a total of p features, and therefore, $(p + 1)$ coefficients due to the intercept term. ϵ is the usual noise term that represents the unexplained part. In other words, our prediction using MLR is as follows:

$$\hat{y} \ = \ \beta_0 + \beta_1 x_1 + \beta_2 x_2 + \ldots + \beta_p x_p$$

We can perform ceteris paribus analysis with this formulation, which is a Latin way of saying all other things are equal, and we only change one input variable to assess its impact on the outcome variable. In other words, MLR allows us to explicitly control (that is, keep unchanged) many other factors that simultaneously affect the target variable and observe the impact of only one factor.

For example, suppose we add a small increment, Δx_j, to the feature, x_j, and keep all other features unchanged. The new prediction, $\hat{y}_{new}$, is obtained by the following formula:

$$\hat{y}_{new} \ = \ \beta_0 + \beta_1 x_1 + \ldots + \beta_j \left(x_j + \Delta x_j \right) + \ldots + \beta_p x_p$$

We know that the original prediction is as follows:

$$\hat{y}_{old} \ = \ \beta_0 + \beta_1 x_1 + \ldots + \beta_j x_j + \ldots + \beta_p x_p$$

The difference between these two gives us the change in the output variable:

$$\Delta \hat{y} \ = \ \hat{y}_{new} - \hat{y}_{old} \ = \ \beta_j \Delta x_j$$

What we are doing here is essentially controlling all other input variables but only bumping x_j to see the impact on the prediction, $\hat{y}$. So, the coefficient, β_j, measures the sensitivity of the outcome to a specific feature. When we have a unit change, with $\Delta x_j \ = \ 1$, the change is exactly the coefficient itself, giving us $\Delta \hat{y} \ = \ \beta_j$.

The next section discusses the measure of the predictiveness of the MLR model.

Seeking a higher coefficient of determination

MLR tends to perform better than SLR due to the multiple predictors used in the model, such as a higher coefficient of determination (R^2). However, a regression model with more input variables and a higher R^2 does not necessarily mean that the model is a better fit and can predict better for the test set.

A higher R^2, as a result of more input features, could likely be due to overfitting. Overfitting occurs when a model is excessively complex, including too many predictors or even interaction terms between predictors. In such cases, the model may fit the observed data well (thus leading to a high R^2), but it may perform poorly when applied to new, unseen test data. This is because the model might have learned not only the underlying structure of the training data but also the random noise specific to the dataset.

Let's look at the metric of R^2 more closely. While R^2 measures how well the model explains the variance in the outcome variable, it has a major limitation: it tends to get bigger as more predictors enter the model, even if those predictors are irrelevant. As a remedy, we can use the adjusted R^2. Unlike R^2, the adjusted R^2 explicitly considers the number of predictors and adjusts the resulting statistic accordingly. If a predictor improves the model substantially, the adjusted R^2 will increase, but if a predictor does not improve the model by a significant amount, the adjusted R^2 may even decrease.

When building statistical models, simpler models are usually preferred when they perform similarly to more complex models. This principle of parsimony, also known as Occam's razor, suggests that among models with similar predictive power, the simplest one should be chosen. In other words, adding more predictors to the model makes it more complex, harder to interpret, and more likely to overfit.

More on adjusted R^2

The adjusted R^2 improves upon R^2 by adjusting for the number of features in the selected model. Specifically, the value of the adjusted R^2 only increases if adding this feature is worth more than what would have been expected from adding a random feature. Essentially, the additional predictors that are added to the model must be meaningful and predictive to lead to a higher adjusted R^2. These additional predictors, however, would always increase R^2 when added to the model.

The adjusted R^2 addresses this issue by incorporating the model's degree of freedom. Here, the degree of freedom refers to the number of values in a statistical calculation that is free to vary. In the context of regression models, this typically means the number of predictors. The adjusted R^2 can be expressed as follows:

$$\text{Adjusted } R^2 = 1 - \left(1 - R^2\right)\frac{(n-1)}{n-p-1}$$

Here, n denotes the number of observations and p represents the number of features in the model.

The formula works by adjusting the scale of R^2 based on the count of observations and predictors. The term $\frac{(n-1)}{n-p-1}$ is a ratio that reflects the degrees of freedom in the model, where $(n-1)$ represents the total degrees of freedom in the model. We subtract by 1 because we are estimating the mean of the dependent variable from the data. $(n-p-1)$ represents the degrees of freedom for the error, which, in turn, represents the number of observations left over after estimating the model parameters. The whole term, $\left(1 - R^2\right)\frac{(n-1)}{n-p-1}$, denotes the error variance that's been adjusted for the count of predictors.

Subtracting this from 1 results in the proportion of the total variance explained by the model, after adjusting for the number of predictors in the model. In other words, it's a version of R^2 that penalizes the addition of unnecessary predictors. This helps to prevent overfitting and makes adjusted R^2 a more balanced measure of a model's explanatory power when comparing models with different numbers of predictors.

Let's look at how to develop an MLR model in R.

Developing an MLR model

In this section, we will develop an MLR model using the same lm() function in R based on the mtcars dataset, which comes preloaded with R and was used in previous exercises. Again, the mtcars dataset contains measurements for 32 vehicles from a 1974 Motor Trend issue. These measurements include attributes such as miles per gallon (mpg), number of cylinders (cyl), horsepower (hp), and weight (wt).

Exercise 12.2 – building an MLR model

In this exercise, we will develop an MLR model to predict mpg using cyl, hp, and wt. We will then interpret the model results:

1. Load the mtcars dataset and build an MLR that predicts mpg based on cyl, hp, and wt:

    ```
    # Load the data
    data(mtcars)
    # Build the model
    model = lm(mpg ~ cyl + hp + wt, data = mtcars)
    ```

    Here, we first load the mtcars dataset using the data() function, then construct the MLR model using the lm() function. The mpg ~ cyl + hp + wt formula is used to specify the model. This formula tells R that we want to model mpg as a function of cyl, hp, and wt. The data = mtcars argument tells R to look for these variables in the mtcars dataset. The lm() function fits the model to the data and returns a model object, which we store in the variable model.

2. View the summary of the model:

    ```
    # Print the summary of the model
    >>> summary(model)
    Call:
    lm(formula = mpg ~ cyl + hp + wt, data = mtcars)

    Residuals:
        Min      1Q  Median      3Q     Max
    -3.9290 -1.5598 -0.5311  1.1850  5.8986

    Coefficients:
                Estimate Std. Error t value Pr(>|t|)
    (Intercept) 38.75179    1.78686  21.687  < 2e-16 ***
    cyl         -0.94162    0.55092  -1.709 0.098480 .
    hp          -0.01804    0.01188  -1.519 0.140015
    wt          -3.16697    0.74058  -4.276 0.000199 ***
    ---
    Signif. codes:  0 '***' 0.001 '**' 0.01 '*' 0.05 '.' 0.1 ' ' 1

    Residual standard error: 2.512 on 28 degrees of freedom
    Multiple R-squared:  0.8431,   Adjusted R-squared:  0.8263
    F-statistic: 50.17 on 3 and 28 DF,  p-value: 2.184e-11
    ```

    The summary includes the model's coefficients (the intercept and the slopes for each predictor), the residuals (differences between the actual observations and predicted values for the target), and several statistics that tell us how well the model fits the data, including R^2 and the adjusted R^2.

Let's interpret the output. Each coefficient represents the expected change in mpg for a single unit increase in the associated predictor, assuming all other predictors are constant. The R^2 value, which is 0.8431, denotes the proportion of variance (over 84%) in mpg that can be explained by the predictors together. Again, the adjusted R^2 value, which is 0.8263, is a modified R^2 that accounts for the number of features in the model.

In addition, the p-values for each predictor test the null hypothesis that the true value of the coefficient is zero. If a predictor's p-value is smaller than a preset significance level (such as 0.05), we would reject this null hypothesis and conclude that the predictor is statistically significant. In this case, wt is the only statistically significant factor compared with others using a significance level of 5%.

In the MLR model, all coefficients are negative, indicating a reverse direction of travel between the input variable and the target. However, we cannot conclude that all the predictors negatively correlate with the target variable. The correlation between the individual predictor and the target variable could be positive or negative in SLR.

The next section provides more context on this phenomenon.

Introducing Simpson's Paradox

Simpson's Paradox says that a trend appears in different data groups but disappears or changes when combined. In the context of regression analysis, Simpson's Paradox can appear when a variable that seems positively correlated with the outcome might be negatively correlated when we control other variables.

Essentially, this paradox illustrates the importance of considering confounding variables and not drawing conclusions from aggregated data without understanding the context. The confounding variables are those not among the explanatory variables under consideration but are related to both the target variable and the predictors.

Let's consider a simple example through the following exercise.

Exercise 12.3 – illustrating Simpson's Paradox

In this exercise, we will look at two scenarios with opposite signs of coefficient values for the same feature in both SLR and MLR:

1. Create a dummy dataset with two predictors and one output variable:

    ```
    set.seed(123)
    x1 = rnorm(100)
    x2 = -3 * x1 + rnorm(100)
    y = 2 + x1 + x2 + rnorm(100)
    df = data.frame(y = y, x1 = x1, x2 = x2)
    ```

Here, x1 is a set of 100 numbers randomly generated from a standard normal distribution. x2 is a linear function of x1 but with a negative correlation, and some random noise is added (via rnorm(100)). y is then generated as a linear function of x1 and x2, again with some random noise added. All three variables are compiled into a DataFrame, df.

2. Train an SLR model with y as the outcome and x1 as the input features. Check the summary of the model:

```
# Single linear regression
single_reg = lm(y ~ x1, data = df)
>>> summary(single_reg)
Call:
lm(formula = y ~ x1, data = df)

Residuals:
    Min      1Q  Median      3Q     Max
-2.7595 -0.8365 -0.0564  0.8597  4.3211

Coefficients:
            Estimate Std. Error t value Pr(>|t|)
(Intercept)   2.0298     0.1379   14.72   <2e-16 ***
x1           -2.1869     0.1511  -14.47   <2e-16 ***
---
Signif. codes:  0 '***' 0.001 '**' 0.01 '*' 0.05 '.' 0.1 ' ' 1

Residual standard error: 1.372 on 98 degrees of freedom
Multiple R-squared:  0.6813,   Adjusted R-squared:  0.678
F-statistic: 209.5 on 1 and 98 DF,  p-value: < 2.2e-16
```

The result shows that x1 is negatively correlated with y due to a negative coefficient of -2.1869.

3. Train an SLR model with y as the target and x1 and x2 as the input features. Check the summary of the model:

```
# Multiple linear regression
multi_reg = lm(y ~ x1 + x2, data = df)
>>> summary(multi_reg)
Call:
lm(formula = y ~ x1 + x2, data = df)

Residuals:
    Min      1Q  Median      3Q     Max
-1.8730 -0.6607 -0.1245  0.6214  2.0798

Coefficients:
```

```
             Estimate Std. Error t value Pr(>|t|)
(Intercept)   2.13507    0.09614  22.208  < 2e-16 ***
x1            0.93826    0.31982   2.934  0.00418 **
x2            1.02381    0.09899  10.342  < 2e-16 ***
---
Signif. codes:  0 '***' 0.001 '**' 0.01 '*' 0.05 '.' 0.1 ' ' 1

Residual standard error: 0.9513 on 97 degrees of freedom
Multiple R-squared:  0.8484,  Adjusted R-squared:  0.8453
F-statistic: 271.4 on 2 and 97 DF,  p-value: < 2.2e-16
```

The result shows that the estimated coefficient for x1 is now a positive quantity. Does this suggest that x1 is suddenly positively correlated with y? No, since there are likely other confounding variables that lead to a positive coefficient.

The key takeaway is that we can only make inferences on the (positive or negative) correlation between a predictor and a target outcome in an SLR setting. For example, if we build an SLR model to regress y against x, we can conclude that x and y are positively correlated if the resulting coefficient is positive ($\beta > 0$). Similarly, if $\beta > 0$, we can conclude that x and y are positively correlated. The same applies to the case of negative correlation.

However, such inference breaks in an MLR setting – that is, we cannot conclude a positive correlation if $\beta > 0$, and vice versa.

Let's take this opportunity to interpret the results. The Estimate column shows the estimated regression coefficients. These values indicate how much the y variable is expected to increase when the corresponding predictor variable increases by one unit while holding all other features constant. In this case, for each unit increase in x1, y is expected to increase by approximately 0.93826 units, and for each unit increase in x2, y is expected to increase by approximately 1.02381 units. The (Intercept) row shows the estimated value of y when all predictor variables in the model are zero. In this model, the estimated intercept is 2.13507.

Std. Error represents the standard errors for the estimates. Smaller values here indicate more precise estimates. The t value column shows the t-statistics for the hypothesis test that the corresponding population regression coefficient is zero, given that the other predictors are in the model. A larger absolute value of the t-statistic indicates stronger evidence against the null hypothesis. The Pr(>|t|) column gives the p-values for the hypothesis tests. In this case, both x1 and x2 have p-values below 0.05, indicating that both are statistically significant predictors of y at the 5% significance level.

Finally, the multiple R-squared and adjusted R-squared values provide measures of how well the model fits the data. The multiple R-squared value is 0.8484, indicating that this model explains approximately 84.84% of the variability in y. The adjusted R-squared value adjusts this measure for the number of features in the model. As discussed, it is a better measure when comparing models with different numbers of predictors. Here, the adjusted R-squared value is 0.8453. The F-statistic and its associated p-value are used to test the hypothesis that all population regression coefficients are zero. A small p-value (less than 0.05) indicates that we can reject this hypothesis, and conclude that at least one of the predictors is useful in predicting y.

The next section looks at the situation when we have a categorical predictor in the MLR model.

Working with categorical variables

In MLR, the process of including a binary predictor is similar to including a numeric predictor. However, the interpretation differs. Consider a dataset where y is the target variable, x_1 is a numeric predictor, and x_2 is a binary predictor:

$$y = \beta_0 + \beta_1 x_1 + \beta_2 x_2 + \epsilon$$

In this model, x_2 is coded as 0 and 1, and its corresponding coefficient, β_2, represents the difference in the mean values of y between the two groups identified by x_2.

For example, if x_2 is a binary variable representing sex (0 for males, 1 for females), and y is salary, then β_2 represents the average difference in salary between females and males, after accounting for the value of x_1.

Note that the interpretation of the coefficient of a binary predictor is dependent on the other variables in the model. So, in the preceding example, β_2 is the difference in salary between females and males for given values of x_1.

On the implementation side, R automatically creates dummy variables when a factor is used in a regression model. So, if x_2 were a factor with levels of "male" and "female," R would handle the conversion to 0 and 1 internally when fitting the model.

Let's look at a concrete example. In the following code, we're building an MLR model to predict mpg using qsec and am:

```
# Fit the model
model <- lm(mpg ~ qsec + am, data = mtcars)
# Display the summary of the model
>>> summary(model)
Call:
lm(formula = mpg ~ qsec + am, data = mtcars)

Residuals:
    Min      1Q  Median      3Q     Max
-6.3447 -2.7699  0.2938  2.0947  6.9194

Coefficients:
            Estimate Std. Error t value Pr(>|t|)
(Intercept) -18.8893     6.5970  -2.863  0.00771 **
qsec          1.9819     0.3601   5.503 6.27e-06 ***
```

```
am                8.8763      1.2897    6.883 1.46e-07 ***
---
Signif. codes:   0 '***' 0.001 '**' 0.01 '*' 0.05 '.' 0.1 ' ' 1

Residual standard error: 3.487 on 29 degrees of freedom
Multiple R-squared:  0.6868,  Adjusted R-squared:  0.6652
F-statistic:  31.8 on 2 and 29 DF,  p-value: 4.882e-08
```

Note that the am variable is treated as a numeric variable. Since it represents the type of transmission in the car (0 = automatic, 1 = manual), it should have been treated as a categorical variable. This can be achieved by converting it into a factor, as shown here:

```
# Convert am to categorical var
mtcars$am_cat = as.factor(mtcars$am)
# Fit the model
model <- lm(mpg ~ qsec + am_cat, data = mtcars)
# Display the summary of the model
>>> summary(model)
Call:
lm(formula = mpg ~ qsec + am_cat, data = mtcars)

Residuals:
    Min      1Q  Median      3Q     Max
-6.3447 -2.7699  0.2938  2.0947  6.9194

Coefficients:
            Estimate Std. Error t value Pr(>|t|)
(Intercept) -18.8893     6.5970  -2.863  0.00771 **
qsec          1.9819     0.3601   5.503 6.27e-06 ***
am_cat1       8.8763     1.2897   6.883 1.46e-07 ***
---
Signif. codes:   0 '***' 0.001 '**' 0.01 '*' 0.05 '.' 0.1 ' ' 1

Residual standard error: 3.487 on 29 degrees of freedom
Multiple R-squared:  0.6868,  Adjusted R-squared:  0.6652
F-statistic:  31.8 on 2 and 29 DF,  p-value: 4.882e-08
```

Note that only one variable, am_cat1, is created for the categorical variable, am_cat. This is because am_cat is binary, thus we only need one dummy column (keeping only am_cat1 and removing am_cat0 in this case) to represent the original categorical variable. In general, for a categorical variable with k categorical, R will automatically create $(k - 1)$ dummy variables in the model.

This process is called **one-hot encoding**, which involves converting categorical data into a format that can be used by the regression model. For categorical variables where there is no ordinal relationship, the integer encoding (as originally used) may not be sufficient. It could even be misleading to some extent. For these cases, one-hot encoding can be applied, where each unique original value is replaced with a binary variable. Specifically, each of these new variables (also known as "dummy" variables) would have a value of 1 for the observations where am was equal to the corresponding level, and 0 otherwise. This essentially creates a set of indicators that capture the presence or absence of each category. Finally, since we can infer the last dummy variable based on the values of the previous (k-1) dummy variables, we can remove the last dummy variable in the resulting one-hot encoded set of variables.

Using a categorical variable introduces a vertical shift to the model estimate, as discussed in the following section. To see this, let's look more closely at the impact of the categorical variable, am_cat1. Our MLR model now assumes the following form:

$$\hat{y} = \beta_0 + \beta_1 x_{qsec} + \beta_2 x_{am\_cat}$$

We know that x_{am_cat} is a binary variable. When $x_{am\_cat} = 0$, the prediction becomes as follows:

$$\hat{y} = \beta_0 + \beta_1 x_{qsec}$$

When $x_{am\_cat} = 1$, the prediction is as follows:

$$\hat{y} = \beta_0 + \beta_1 x_{qsec} + \beta_2 = (\beta_0 + \beta_2) + \beta_1 x_{qsec}$$

By comparing these two quantities, we can see that they are two linear models parallel to each other since the slope is the same and the only difference is β_2 in the intercept term.

A visual illustration helps here. In the following code snippet, we first create a new DataFrame, newdata, that covers the range of qsec values in the original data, for each of the am_cat values (0 and 1). Then, we use the predict() function to get the predicted mpg values from the model for this new data. Next, we plot the original data points with geom_point() and add two regression lines with geom_line(), where the lines are based on the predicted values in newdata. The color = am_cat aesthetic setting adds different colors to the points and lines for the different am_cat values, and the labels are adjusted in scale_color_discrete() so that 0 corresponds to "Automatic" and 1 corresponds to "Manual":

```
# Load required library
library(ggplot2)
# Create new data frame for the predictions
newdata = data.frame(qsec = seq(min(mtcars$qsec), max(mtcars$qsec),
length.out = 100),
                     am_cat = c(rep(0, 100), rep(1, 100)))
newdata$am_cat = as.factor(newdata$am_cat)
# Get predictions
newdata$mpg_pred = predict(model, newdata)
# Plot the data and the regression lines
ggplot(data = mtcars, aes(x = qsec, y = mpg, color = am_cat)) +
  geom_point() +
```

```
    geom_line(data = newdata, aes(y = mpg_pred)) +
    labs(title = "mpg vs qsec by Transmission Type",
         x = "Quarter Mile Time (qsec)",
         y = "Miles per Gallon (mpg)",
         color = "Transmission Type") +
    scale_color_discrete(labels = c("Automatic", "Manual")) +
    theme(text = element_text(size = 16),   # Default text size
          title = element_text(size = 15),   # Title size
          axis.title = element_text(size = 18),   # Axis title size
          legend.title = element_text(size = 16),   # Legend title size
          legend.text = element_text(size = 16),  # Legend text size
          legend.position = "bottom")   # Legend position
```

Running this code generates *Figure 12.3*:

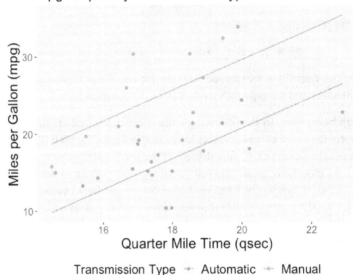

Figure 12.3 – Visualizing the two linear regression models based on different transmission types. These two lines are parallel to each other due to a shift in the intercept term

What this figure suggests is that manual transmission cars have the same miles per gallon (mpg) more than automatic transmission cars given the same quarter-mile time (qsec). However, this is unlikely in practice since different car types (manual versus automatic) will likely have different quarter-mile times. In other words, there is an interaction effect between these two variables.

The following section introduces the interaction term as a remedy to this situation.

Introducing the interaction term

In regression analysis, an interaction occurs when the effect of one predictor on the target variable differs depending on the level of another predictor variable. In our running example, we are essentially looking at whether the relationship between mpg and qsec is different for different values of am. In other words, we are testing whether the slope of the line relating mpg and qsec differs for manual (am=1) and automatic (am=0) transmissions.

For example, if there is no interaction effect, then the impact of qsec on mpg is the same, regardless of whether the car has a manual or automatic transmission. This would mean that the lines depicting the relationship between mpg and qsec for manual and automatic cars would be parallel.

If there is an interaction effect, then the effect of qsec on mpg differs for manual and automatic cars. This would mean that the lines depicting the relationship between mpg and qsec for manual and automatic cars would not be parallel. They could either cross or, more commonly, just have different slopes.

To depict these differences in relationships, we can add an interaction term to the model, which is done using the * operator. For example, the formula for a regression model with an interaction between qsec and am_cat would be mpg ~ qsec * am_cat. This is equivalent to mpg ~ qsec + am_cat + qsec:am_cat, where qsec:am_cat represents the interaction term. The following code shows the details:

```
# Adding interaction term
model_interaction <- lm(mpg ~ qsec * am_cat, data = mtcars)
# Print model summary
>>> summary(model_interaction)
Call:
lm(formula = mpg ~ qsec * am_cat, data = mtcars)

Residuals:
    Min      1Q  Median      3Q     Max
-6.4551 -1.4331  0.1918  2.2493  7.2773

Coefficient s:
             Estimate Std. Error t value Pr(>|t|)
(Intercept)   -9.0099     8.2179  -1.096  0.28226
qsec           1.4385     0.4500   3.197  0.00343 **
am_cat1      -14.5107    12.4812  -1.163  0.25481
qsec:am_cat1   1.3214     0.7017   1.883  0.07012 .
Signif. code s:  0 '***' 0.001 '**' 0.01 '*' 0.05 '.' 0.1 ' ' 1

Residual standard error: 3.343 on 28 degrees of freedom
Multiple R-squared:  0.722,  Adjusted R-squared:  0.6923
F-statistic: 24.24 on 3 and 28 DF,  p-value: 6.129e-08
```

Let's also plot the updated model, which consists of two intersecting lines due to the interaction effect. In the following code snippet, `geom_smooth(method =""l"", se = FALSE)` is used to fit different linear lines to each group of points (automatic and manual cars). `as.factor(am_cat)` is used to treat `am_cat` as a factor (categorical) variable so that a separate line is fit for each category:

```
# Create scatter plot with two intersecting lines
ggplot(mtcars, aes(x = qsec, y = mpg, color = as.factor(am_cat))) +
  geom_point() +
  geom_smooth(method =""l"", se = FALSE) + # fit separate regression
lines per group
  scale_color_discrete(name =""Transmission Typ"",
                       labels = c""Automati"",""Manua"")) +
  labs(x =""Quarter mile time (seconds"",
       y =""Miles per gallo"",
       title =""Separate regression lines fit for automatic and manual
car"") +
  theme(text = element_text(size = 16),
        title = element_text(size = 15),
        axis.title = element_text(size = 20),
        legend.title = element_text(size = 16),
        legend.text = element_text(size = 16))
```

Running this code generates *Figure 12.4*:

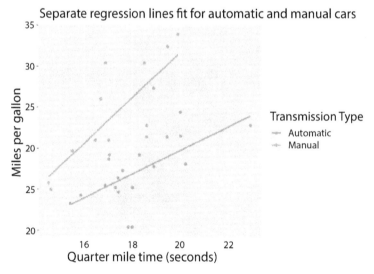

Figure 12.4 – Two intersecting lines due to the intersection term
between quarter-mile time and transmission type

The next section focuses on another related topic: nonlinear terms.

Handling nonlinear terms

Linear regression is a widely used model for understanding the linear relationships between a response and explanatory variables. However, not all underlying relationships in the data are linear. In many situations, a feature and a response variable may not have a straight-line relationship, thus necessitating the handling of nonlinear terms in the linear regression model to increase its flexibility.

The simplest way to incorporate nonlinearity, and therefore build a curve instead of a straight line, is by including polynomial terms of predictors in the regression model. In polynomial regression, some or all predictors are raised to a specific polynomial term – for example, transforming a feature, x, into x^2 or x^3.

Let's go through an exercise to understand the impact of adding polynomial features to a linear regression model.

Exercise 12.4 – adding polynomial features to a linear regression model

In this exercise, we will create a simple dataset where the relationship between the input feature, x, and the target variable, y, is not linear, but quadratic. First, we will fit an SLR model, then add a quadratic term and compare the results:

1. Generate a sequence of x values ranging from -10 to 10. For each x, compute the corresponding y as the square of x, plus some random noise to show a (noisy) quadratic relationship. Put x and y in a DataFrame:

    ```
    # Create a quadratic dataset
    set.seed(1)
    x = seq(-10, 10, by = 0.5)
    y = x^2 + rnorm(length(x), sd = 5)
    # Put it in a dataframe
    df = data.frame(x = x, y = y)
    Fit a simple linear regression to the data and print the
    summary.
    lm1 <- lm(y ~ x, data = df)
    >>> summary(lm1)
    Call:
    lm(formula = y ~ x, data = df)

    Residuals:
        Min      1Q  Median      3Q     Max
    -43.060 -29.350  -5.451  19.075  64.187

    Coefficient s:
    ```

```
                 Estimate Std. Error t value Pr(>|t|)
(Intercept) 35.42884     5.04627     7.021 2.01e-08 ***
x               -0.04389     0.85298    -0.051    0.959
Signif. code s:    0 '***' 0.001 '**' 0.01 '*' 0.05 '.' 0.1 ' ' 1


Residual standard error: 32.31 on 39 degrees of freedom
Multiple R-squared:  6.787e-05,  Adjusted R-squared:  -0.02557
F-statistic: 0.002647 on 1 and 39 DF,  p-value: 0.9592
```

The result suggests a not-so-good model fitting to the data, which possesses a nonlinear relationship.

2. Fit a quadratic model to the data by including x^2 as a predictor using the I() function. Print the summary of the model:

```
lm2 <- lm(y ~ x + I(x^2), data = df)
>>> summary(lm2)
Call:
lm(formula = y ~ x + I(x^2), data = df)

Residuals:
     Min       1Q  Median       3Q      Max
-11.700   -2.134   -0.078    2.992    7.247

Coefficients:
                 Estimate Std. Error t value Pr(>|t|)
(Intercept)   0.49663     1.05176     0.472    0.639
x             -0.04389     0.11846    -0.370    0.713
I(x^2)         0.99806     0.02241    44.543   <2e-16 ***
Signif. code s:    0 '***' 0.001 '**' 0.01 '*' 0.05 '.' 0.1 ' ' 1


Residual standard error: 4.487 on 38 degrees of freedom
Multiple R-squared:   0.9812,  Adjusted R-squared:  0.9802
F-statistic: 992.1 on 2 and 38 DF,  p-value: < 2.2e-16
```

The result shows that the polynomial model fits the data better than the simple linear model. Thus, adding nonlinear terms can improve model fit when the relationship between predictors and the response is not strictly linear.

3. Plot the linear and quadratic models together with the data:

```
ggplot(df, aes(x = x, y = y)) +
  geom_point() +
  geom_line(aes(y = linear_pred), color =""blu"", linetype
=""dashe"") +
  geom_line(aes(y = quadratic_pred), color =""re"") +
  labs(title =""Scatter plot with linear and quadratic fit"",
```

```
    x ="""",
    y ="""") +
theme(text = element_text(size = 15)) +
scale_color_discrete(name =""Mode"",
                     labels = c"«Linear Mode"»,"«Quadratic
Mode"»)) +
  annotate""tex"", x = 0, y = 40, label =""Linear Mode"", color
=""blu"") +
  annotate""tex"", x = 6, y = 80, label =""Quadratic Mode"",
color =""re"")
```

Here, we first calculate the predicted values for both models and add them to the DataFrame. Then, we create a scatter plot of the data and add two lines representing the predicted values from the linear model (in blue) and the quadratic model (in red).

Running this code generates *Figure 12.5*. The result suggests that adding a polynomial feature could extend the flexibility of a linear model:

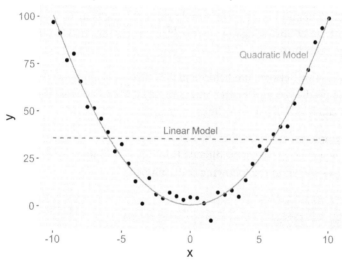

Figure 12.5 – Visualizing the linear and quadratic fits to the nonlinear data

Other common ways to introduce nonlinearity include the logarithmic transformation ($\log x$) or a square root transformation ($\sqrt{x}$). These transformations can also be applied to the target variable, y, and we can have multiple transformed features in the same linear model.

Note that the model with transformed features remains a linear model. If there is a nonlinear transformation to the coefficients, the model would be a nonlinear one.

The next section sheds more light on a widely used type of transformation: the logarithmic transformation.

More on the logarithmic transformation

The logarithmic transformation, or log transformation, maps an input to the corresponding output based on the logarithmic function, giving $y = \log x$. A popular reason behind using such a transformation is to introduce nonlinearity in the linear regression model. When the relationship between the input features and the target output is nonlinear, applying a transformation can sometimes linearize the relationship, making it possible to model the relationship with a linear regression model. For the logarithmic transformation, it can help when the rate of change in the outcome variable increases or decreases as the value of the predictor increases.

To be specific, the rate of change decreases as the input becomes more extreme. The natural consequence of such a transformation is that potential outliers in the input data are squeezed so that they appear less extreme in the transformed column. In other words, the resulting linear regression model will be less sensitive to the original outliers due to the log transformation.

Another side benefit of using log transformation is its ability to deal with heteroscedasticity. Heteroscedasticity is when the variability of the error term in a regression model is not constant across all levels of the predictors. This violates one of the assumptions of linear regression models and can lead to inefficient and biased estimates. In this case, log transformations can stabilize the variance of the error term by shrinking the potential big error terms, making it more constant across different levels of the predictors.

Lastly, when the relationship between predictors and the outcome is multiplicative rather than additive, taking the log of the predictors and/or the outcome variable can convert the relationship into an additive one, which can be modeled using linear regression.

Let's consider an example where we predict the miles per gallon (mpg) from horsepower (hp). We'll compare the model where we predict mpg directly from hp and another model where we predict the log of mpg from hp, as shown in the following code snippet:

```
# Fit the original model
model_original = lm(mpg ~ hp, data = mtcars)

# Fit the log-transformed model
mtcars$log_mpg = log(mtcars$mpg)
model_log = lm(log_mpg ~ hp, data = mtcars)

# Predictions from the original model
mtcars$pred_original = predict(model_original, newdata = mtcars)

# Predictions from the log-transformed model (back-transformed to the
original scale using exp)
mtcars$pred_log = exp(predict(model_log, newdata = mtcars))
```

```r
library(tidyr)
library(dplyr)

# Reshape data to long format
df_long <- mtcars %>%
  gather(key ="""Mode""", value ="""Predictio""", pred_original, pred_log)

# Create plot
ggplot(df_long, aes(x = hp, y = mpg)) +
  geom_point(data = mtcars, aes(x = hp, y = mpg)) +
  geom_line(aes(y = Prediction, color = Model)) +
  labs(
    x ="""Horsepower (hp""",
    y ="""Miles per gallon (mpg""",
    color ="""Mode"""
  ) +
  scale_color_manual(values = c""pred_origina"" ="""blu""",""pred_lo""
="""re""")) +
  theme(
    legend.position ="""botto""",
    text = element_text(size = 16),
    legend.title = element_text(size = 16),
    axis.text = element_text(size = 16),  # control the font size of
axis labels
    legend.text = element_text(size = 16)  # control the font size of
legend text
  )
```

Running this code generates *Figure 12.6*, where we can see a slight curvature in the blue line:

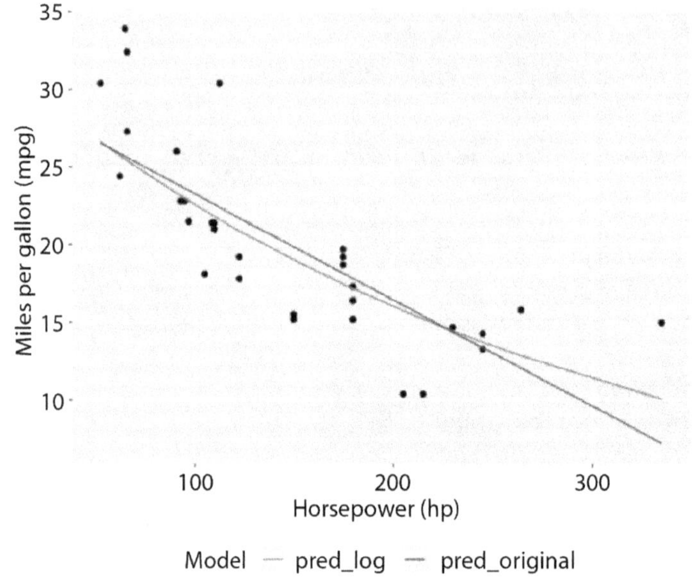

Figure 12.6 – Visualizing the original and log-transformed model

Note that the log transformation can only be applied to positive data. In the case of `mtcars$mpg`, all values are positive, so we can safely apply the log transformation. If the variable included zero or negative values, we would need to consider a different transformation or approach.

The next section focuses on deriving and using the closed-form solution to the linear regression model.

Working with the closed-form solution

When developing a linear regression model, the available training set (X, y) is given, and the only unknown parameters are the coefficients, β. Here, a bold lowercase letter means a vector (such as β and y), and a bold uppercase letter denotes a matrix (such as X). It turns out that the closed-form solution to a linear regression model can be derived using the concept of the **ordinary least squares** (**OLS**) estimator, which aims to minimize the sum of the squared residuals in the model. Having the closed-form solution means we can simply plug in the required elements (in this case, X and y) and perform the calculation to obtain the solution, without resorting to any optimization procedure.

Specifically, given a data matrix, X (which includes a column of ones for the intercept term and is in bold to indicate more than one feature), of dimensions $n \times p$ (where n is the number of observations and p is the number of predictors) and a response vector, y, of length n, the OLS estimator for the coefficient vector, β, is given by the following formula:

$$\beta = (X^T X)^{-1} X^T y$$

This solution assumes that the term (X^TX) is invertible, meaning it should be a full-rank matrix. If this is not the case, the solution either does not exist or is not unique.

Now, let's look at how to derive this solution. We start with the minimization problem for the least squares: minimizing $(y - X\beta)^T(y - X\beta)$ over β. This quadratic form can be expanded to $y^Ty - \beta^T X^Ty - y^TX\beta + \beta^TX^TX\beta$. Note that $\beta^TX^Ty = y^TX\beta$ since both terms are scalars and therefore are equal to each other after the transpose operation. We can write the **residual sum of squares** (**RSS**) expression as $y^Ty - 2\beta^TX^Ty + \beta^TX^TX\beta$.

Here, we apply the first-order condition to solve for the value of β that minimizes this expression (recall that the point that either minimizes or maximizes a graph has a derivative of 0). This means that we would set its first derivative to zero, leading to the following formula:

$$\frac{\partial(y^Ty - 2\beta^TX^Ty + \beta^TX^TX\beta)}{\partial\beta} = -2X^Ty + 2X^TX\beta = 0$$

$$X^TX\beta = X^Ty$$

$$\beta = (X^TX)^{-1}X^Ty$$

Thus, we have derived the closed-form solution of β that minimizes the sum of the squared residuals. Let's go through an example to see how it can be implemented.

Implementing the closed-form solution

Let's look at implementing the OLS estimation in R for an SLR model. An example that uses synthetic data is shown in the following code snippet:

```
# Set seed for reproducibility
set.seed(123)
# Generate synthetic data
n = 100 # number of observations
x = runif(n, -10, 10) # predictors
beta0 = 2 # intercept
beta1 = 3 # slope
epsilon = rnorm(n, 0, 2) # random error term
y = beta0 + beta1*x + epsilon # response variable
# Design matrix X
X = cbind(1, x)
```

Here, we generate 100 observations with a single input feature, where the observation is noise-perturbed and follows a process given by $y = \beta_0 + \beta_1 x + \epsilon$. The error term assumes a normal distribution that's parameterized by a mean of 0 and a standard deviation of 2.

Before proceeding to the estimation, note that we also appended a column of 1s on the left of the input feature, x, to form a matrix, X. This column of 1s is used to indicate the intercept term and is often referred to as the bias trick. That is, the coefficient, β_0, for the intercept term will be part of the coefficient vector, and there is no need to create a separate coefficient just for the intercept.

Let's calculate the result using the closed-form solution:

```
beta_hat = solve(t(X) %*% X) %*% t(X) %*% y
>>> print(beta_hat)
        [,1]
   1.985344
x  3.053152
```

Here, `%*%` is used for matrix multiplication, `t(X)` is the transpose of X, and `solve()` is used to calculate the inverse of a matrix.

We can also run the linear regression procedure using the `lm()` function for comparison:

```
# Fit linear regression model for comparison
model = lm(y ~ x)
>>> print(coef(model))
(Intercept)          x
   1.985344    3.053152
```

The results are the same as the ones that we obtained via the manual approach.

The next two sections cover two common issues in linear regression settings: multicollinearity and heteroskedasticity.

Dealing with multicollinearity

Multicollinearity refers to the case when two (or more) predictors are highly correlated in a multiple regression model. This means that one independent variable can be linearly predicted from the others with a high degree of accuracy. This is a situation that we do not want to fall into. In other words, we would like to see a high degree of correlation between the predictors and the target variable, while a low degree of correlation among these predictors themselves.

In the face of multicollinearity in a linear regression model, the resultant model tends to generate unreliable and unstable estimates of the regression coefficients. It can inflate the coefficients of the parameters, making them statistically insignificant, even though they might be substantively important. In addition, multicollinearity makes it difficult to assess the effect of each independent variable on the dependent variable as the effects are intertwined. However, it does not affect the predictive power or interpretability of the model; instead, it only changes the calculations for individual features.

Detecting any potential multicollinearity among the predictors can be performed by examining the pair-wise correlation. Alternatively, we can resort to a particular test statistic called the **variance inflation factor** (**VIF**), which quantifies how much the variance is increased due to multicollinearity. A VIF of 1 indicates that two variables are not correlated, while a VIF greater than 5 (in many fields) would suggest a problematic amount of multicollinearity.

When multicollinearity exists in the linear regression model, we could choose to keep one predictor only and remove all other highly correlated predictors. We can also combine these correlated variables into a few uncorrelated ones via **principle component analysis** (**PCA**), a widely used technique for dimension reduction. Besides this, we can resort to ridge regression to control the magnitude of the coefficients; this will be introduced later in this chapter.

To check multicollinearity using VIF, we can use the `vif()` function from the `car` package, as shown in the following code snippet:

```
# install the package if not already installed
if(!require(car)) install.packages('car')
# load the package
library(car)
# fit a linear model
model = lm(mpg ~ hp + wt + disp, data = mtcars)
# calculate VIF
vif_values = vif(model)
>>> print(vif_values)
      hp        wt      disp
2.736633 4.844618 7.324517
```

Looking at the result, `disp` seems to have high multicollinearity (VIF = 7.32 > 5), suggesting that it has a strong correlation with `hp` and `wt`. This implies that `disp` is not providing much information that is not already contained in the other two predictors.

To handle the multicollinearity here, we can consider removing `disp` from the model since it has the highest VIF, applying PCA to combine the three predictors, or using ridge or lasso regression (more on this in the last two sections of this chapter).

The next section focuses on the issue of heteroskedasticity.

Dealing with heteroskedasticity

Heteroskedasticity (or heteroscedasticity) refers to the situation in which the variability of the error term, or residuals, is not the same across all levels of the independent variables. This violates one of the key assumptions of OLS linear regression, which assumes that the residuals have a constant variance – in other words, the residuals are homoskedastic. Violating this assumption could lead to incorrect inferences on the statistical significance of the coefficients since the resulting standard errors of the regression coefficients could be larger or smaller than they should be.

There are a few ways to handle heteroskedasticity. First, we can transform the outcome variable using the logarithmic function, as introduced earlier. Other functions, such as taking the square root or inverse of the original outcome variable, could also help reduce heteroskedasticity. Advanced regression models such as **weighted least squares** (**WLS**) or **generalized least squares** (**GLS**) may also be explored to reduce the impact of heteroskedasticity.

To formally test for heteroskedasticity, we can conduct a Breusch-Pagan test using the `bptest()` function from the `lmtest` package. In the following code snippet, we fit an MLR model to predict mpg using `wt` and `hp`, followed by performing the Breusch-Pagan test:

```
# Load library
library(lmtest)
# Fit a simple linear regression model on mtcars dataset
model = lm(mpg ~ wt + hp, data = mtcars)
# Perform a Breusch-Pagan test to formally check for
heteroskedasticity
>>> bptest(model)
    studentized Breusch-Pagan test

data:  model
BP = 0.88072, df = 2, p-value = 0.6438
```

Since the p-value (0.6438) is greater than 0.05, we do not reject the null hypothesis of the Breusch-Pagan test. This suggests that there is not enough evidence to say that heteroskedasticity is present in the regression model. So, we would conclude that the variances of the residuals are not significantly different from being constant, or homoskedastic.

The next section shifts to looking at regularized linear regression models and ridge and lasso penalties.

Introducing penalized linear regression

Penalized regression models, such as ridge and lasso, are techniques that are used to handle problems such as multicollinearity, reduce overfitting, and even perform variable selection, especially when dealing with high-dimensional data with multiple input features.

Ridge regression (also called L2 regularization) is a method that adds a penalty equivalent to the square of the magnitude of coefficients. We would add this term to the loss function after weighting it by an additional hyperparameter, often denoted as λ, to control the strength of the penalty term.

Lasso regression (L1 regularization), on the other hand, is a method that, similar to ridge regression, adds a penalty for non-zero coefficients, but unlike ridge regression, it can force some coefficients to be exactly equal to zero when the penalty tuning parameter is large enough. The larger the value of the hyperparameter, λ, the greater the amount of shrinkage. The penalty on the size of coefficients helps reduce model complexity and multicollinearity, leading to a model that can generalize better on unseen data. However, ridge regression includes all the features in the final model, so it doesn't induce any sparsity. Therefore, it's not particularly useful for variable selection.

In summary, ridge and lasso regression are both penalized linear regression methods that add a constraint regarding the magnitude of the estimated coefficients to the model optimization process, which helps prevent overfitting, manage multicollinearity, and reduce model complexity. However, ridge tends to include all predictors in the model and helps reduce their effect, while lasso can exclude predictors from the model altogether, leading to a simpler and more interpretable model.

Let's start with ridge regression and look at its loss function more closely.

Working with ridge regression

Ridge regression, also referred to as L2 regularization, is a commonly used technique to alleviate overfitting in linear regression models by penalizing the magnitude of the estimated coefficients in the resulting model.

Recall that in an SLR model, we seek to minimize the sum of the squared differences between our predicted and actual values, which we refer to as the least squares method. The loss function we wish to minimize is the RSS:

$$RSS = \sum_{i=1}^{n}\left(y_i - \left(\beta_0 + \sum_{j=1}^{p}\beta_j x_{ij}\right)\right)^2$$

Here, y_i is the actual target value, β_0 is the intercept term, $\{\beta_j\}$ are the coefficient estimates for each predictor, x_{ij}, and the summations are overall observations and predictors.

Purely minimizing the RSS would give us an overfitting model, as represented by the high magnitude of the resulting coefficients. As a remedy, we could apply ridge regression by adding a penalty term to this loss function. This penalty term is the sum of the squares of each coefficient multiplied by a tuning parameter, λ. So, the ridge regression loss function (also known as the **cost function**) is as follows:

$$L_{ridge} = RSS + \lambda \sum_{j=1}^{p}\beta_j^2$$

Here, the λ parameter is a user-defined tuning parameter. A larger λ means a higher penalty and a smaller λ means less regularization effect. $\lambda = 0$ gives the ordinary least squares regression result, while as λ approaches infinity, the impact of the penalty term dominates, and the coefficient estimates approach zero.

By adding this penalty term, ridge regression tends to decrease the size of the coefficients, which can help mitigate the problem of multicollinearity (where predictors are highly correlated). It does this by spreading the coefficient estimates of correlated predictors across each other, which can lead to a more stable and interpretable model.

However, it's important to note that ridge regression does not typically produce sparse solutions and does not perform variable selection. In other words, it will not result in a model where some coefficients are exactly zero (unless λ is infinite), thus all predictors are included in the model. If feature selection is important, methods such as lasso (L1 regularization) or elastic net (a combination of L1 and L2 regularization) might be more appropriate.

Note that we would often penalize the intercept, β_0.

Let's go through an exercise to learn how to develop a ridge regression model.

Exercise 12.5 – implementing ridge regression

In this exercise, we will implement a ridge regression model and compare the estimated coefficients with the OLS model. Our implementation will be based on the glmnet package:

1. Install and load the glmnet package:

    ```
    # install the package if not already installed
    if(!require(glmnet)) install.packages('glmnet')
    library(glmnet)
    ```

    Here, we use an if-else statement to detect if the glmnet package is installed.

2. Store all columns other than mpg as predictors in X and mpg as the target variable in y:

    ```
    # Prepare data
    data(mtcars)
    X = as.matrix(mtcars[, -1]) # predictors
    y = mtcars[, 1] # response
    ```

3. Fit a ridge regression model using the glmnet() function:

    ```
    # Fit ridge regression model
    set.seed(123) # for reproducibility
    ridge_model = glmnet(X, y, alpha = 0)
    ```

    Here, the alpha parameter controls the type of model we fit. alpha = 0 fits a ridge regression model, alpha = 1 fits a lasso model, and any value in between fits an elastic net model.

4. Use cross-validation to choose the best lambda value:

    ```
    # Use cross-validation to find the optimal lambda
    cv_ridge = cv.glmnet(X, y, alpha = 0)
    best_lambda = cv_ridge$lambda.min
    >>> best_lambda
    2.746789
    ```

    Here, we use the cross-validation approach to identify the optimal lambda that gives the lowest error on the cross-validation set on average. All repeated cross-validation steps are completed via the cv.glmnet() function.

5. Fit a new ridge regression model using the optimal `lambda` and extract its coefficients without the intercept:

```
# Fit a new ridge regression model using the optimal lambda
opt_ridge_model = glmnet(X, y, alpha = 0, lambda = best_lambda)
# Get coefficients
ridge_coefs = coef(opt_ridge_model)[-1]  # remove intercept
>>> ridge_coefs
[1] -0.371840170 -0.005260088 -0.011611491  1.054511975
-1.233657799  0.162231830
 [7]   0.771141047  1.623031037  0.544153807 -0.547436697
```

6. Fit a linear regression model and extract its coefficients:

```
# Ordinary least squares regression
ols_model = lm(mpg ~ ., data = mtcars)
# Get coefficients
ols_coefs = coef(ols_model)[-1] # remove intercept
>>> ols_coefs
        cyl         disp          hp         drat
wt           qsec           vs
-0.11144048  0.01333524 -0.02148212  0.78711097 -3.71530393
0.82104075  0.31776281
          am         gear         carb
 2.52022689  0.65541302 -0.19941925
```

7. Plot the coefficients of both models on the same graph:

```
plot(1:length(ols_coefs), ols_coefs, type="b", col="blue",
pch=19, xlab="Coefficient", ylab="Value", ylim=c(min(ols_coefs,
ridge_coefs), max(ols_coefs, ridge_coefs)))
lines(1:length(ridge_coefs), ridge_coefs, type="b", col="red",
pch=19)
legend("bottomright", legend=c("OLS", "Ridge"), col=c("blue",
"red"), pch=19)
```

Running this code generates *Figure 12.7*:

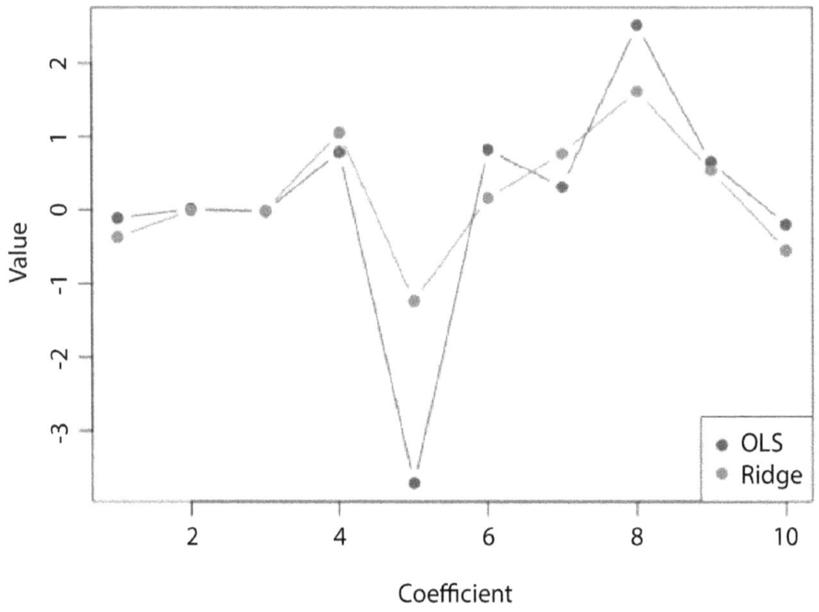

Figure 12.7 – Visualizing the estimated coefficients from the ridge and OLS models

This plot shows that the estimated coefficients from the ridge regression model are, in general, smaller than those from the OLS model.

The next section focuses on lasso regression.

Working with lasso regression

Lasso regression is another type of regularized linear regression. It is similar to ridge regression but differs in terms of the specific process of calculating the magnitude of the coefficients. Specifically, it uses the L1 norm of the coefficients, which consists of the total sum of absolute values of the coefficients, as the penalty that's added to the OLS loss function.

The lasso regression cost function can be written as follows:

$$L_{lasso} = RSS + \lambda \sum_{j=1}^{p} |\beta_j|$$

The key characteristic of lasso regression is that it can reduce some coefficients exactly to 0, effectively performing variable selection. This is a consequence of the L1 penalty term and is not the case for ridge regression, which can only shrink coefficients close to 0. Therefore, lasso regression is particularly useful when we believe that only a subset of the predictors matters when it comes to predicting the outcome.

In addition, in contrast to ridge regression, which can't perform variable selection and therefore may be less interpretable, lasso regression automatically selects the most important features and discards the rest, which can make the final model easier to interpret.

Like ridge regression, the lasso regression penalty term is also subject to a tuning parameter, λ. The optimal λ parameter is typically chosen via cross-validation or a more intelligent search policy such as Bayesian optimization.

Let's go through an exercise to understand how to develop a lasso regression model.

Exercise 12.6 – implementing lasso regression

To implement a lasso regression model, we can follow a similar process as we did for the ridge regression model:

1. To fit a lasso regression model, set `alpha = 1` in the `glmnet()` function:

    ```
    lasso_model = glmnet(X, y, alpha = 1)
    ```

2. Use the same cross-validation procedure to identify the optimal value of `lambda`:

    ```
    # Use cross-validation to find the optimal lambda
    cv_lasso = cv.glmnet(X, y, alpha = 1)
    best_lambda = cv_lasso$lambda.min
    >>> best_lambda
    0.8007036
    ```

3. Fit a new lasso regression model using the optimal `lambda`:

    ```
    # Fit a new lasso regression model using the optimal lambda
    opt_lasso_model = glmnet(X, y, alpha = 1, lambda = best_lambda)
    ```

    The resulting coefficients can also be extracted using the `coef()` function, followed by `[-1]` to remove the intercept term:

    ```
    # Get coefficients
    lasso_coefs = coef(opt_lasso_model)[-1]  # remove intercept
    >>> lasso_coefs
    [1] -0.88547684   0.00000000 -0.01169485   0.00000000
    -2.70853300   0.00000000   0.00000000
     [8]   0.00000000   0.00000000   0.00000000
    ```

4. Plot the estimated coefficients together with the previous two models:

    ```
    plot(1:length(ols_coefs), ols_coefs, type="b", col="blue",
    pch=19, xlab="Coefficient", ylab="Value", ylim=c(min(ols_coefs,
    ridge_coefs), max(ols_coefs, ridge_coefs)))
    lines(1:length(ridge_coefs), ridge_coefs, type="b", col="red",
    pch=19)
    ```

```
lines(1:length(lasso_coefs), lasso_coefs, type="b", col="green",
pch=19)
legend("bottomright", legend=c("OLS", "Ridge", "Lasso"),
col=c("blue", "red", "green"), pch=19)
```

Running this code generates *Figure 12.8*, which suggests that only two variables are kept in the resultant model. So, the lasso regression model can produce a sparse solution by setting the coefficients of some features equal to zero:

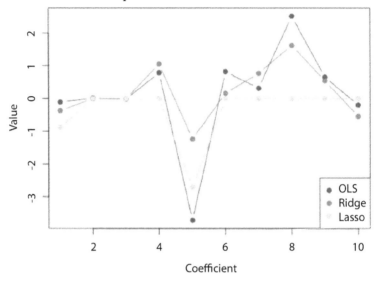

Figure 12.8 – Visualizing the estimated coefficients from the ridge, lasso, and OLS regression models

In summary, the lasso regression model gives us a sparse model by setting the coefficients of non-significant variables to zero, thus achieving feature selection and model estimation at the same time.

Summary

In this chapter, we covered the nuts and bolts of the linear regression model. We started by introducing the SLR model, which consists of only one input variable and one target variable, and then extended to the MLR model with two or more predictors. Both models can be assessed using R^2, or more preferably, the adjusted R^2 metric. Next, we discussed specific scenarios, such as working with categorical variables and interaction terms, handling nonlinear terms via transformations, working with the closed-form solution, and dealing with multicollinearity and heteroskedasticity. Lastly, we introduced widely used regularization techniques such as ridge and lasso penalties, which can be incorporated into the loss function as a penalty term and generate a regularized model, and, additionally, a sparse solution in the case of lasso regression.

In the next chapter, we will cover another type of widely used linear model: the logistic regression model.

13

Logistic Regression in R

In this chapter, we will introduce logistic regression, covering its theoretical construct, connection with linear regression, and practical implementation. As it is an important classification model that is widely used in areas where interpretability matters, such as credit risk modeling, we will focus on its modeling process in different contexts, along with extensions such as adding regularization to the loss function and predicting more than two classes.

By the end of this chapter, you will understand the fundamentals of the logistic regression model and its comparison with linear regression, including extended concepts such as the `sigmoid` function, odds ratio, and **cross-entropy loss** (**CEL**). You will also have grasped the commonly used evaluation metrics in the classification setting, as well as enhancements that deal with imbalanced datasets and multiple classes in the target variable.

In this chapter, we will cover the following:

- Introducing logistic regression
- Comparing logistic regression with linear regression
- More on log odds and odds ratio
- Introducing the cross-entropy loss
- Evaluating a logistic regression model
- Dealing with an imbalanced dataset
- Penalized logistic regression
- Extending to multi-class classification

Technical requirements

To run the code in this chapter, you will need to have the latest versions of the following packages:

- `caret` – 6.0.94

- `tibble` – 3.2.1

- `dplyr` – 1.0.10

- `pROC` – 1.18.2

- `nnet` – 7.3.18

- `glmnet` – 4.1.7

The versions mentioned along with the packages in the preceding list are the latest ones while I am writing this book.

All the code and data for this chapter is available at `https://github.com/PacktPublishing/` `The-Statistics-and-Machine-Learning-with-R-Workshop/blob/main/` `Chapter_13/working.R`.

Introducing logistic regression

Logistic regression is a binary classification model. It is still a linear model, but now the output is constrained to be a binary variable, taking the value of 0 or 1, instead of modeling a continuous outcome as in the case of linear regression. In other words, we will observe and model the outcome $y = 1$ or $y = 0$. For example, in the case of credit risk modeling, $y = 0$ refers to a non-default loan application, while $y = 1$ indicates a default loan.

However, instead of directly predicting the binary outcome, the logistic regression model predicts the probability of y taking a specific value, such as $P(y = 1)$. The probability of assuming the other category is $P(y = 0) = 1 - P(y = 1)$, since the total probability should always sum to 1. The final prediction would be the winner of the two, taking the value of 1 if $P(y = 1) > P(y = 0)$, and 0 otherwise. In the credit risk example, $P(y = 1)$ would be interpreted as the probability of a loan defaulting.

In logistic regression, the term *logistic* is related to *logit*, which refers to the log odds. The odds are another way to describe the probability; instead of specifying the individual $P(y = 1)$ and $P(y = 0)$, it refers to the ratio of $P(y = 1)$ to $P(y = 0)$. Thus the log odds are calculated via $\log \frac{P(y = 1)}{P(y = 0)}$. Therefore, we can simply use the term *odds* to describe the probability of an event happening ($y = 1$) over the probability of it not occurring ($y = 0$).

First, let us look at how the logistic regression model transforms a continuous output (as in linear regression) into a probability score, which is a number bounded between 0 and 1.

Understanding the sigmoid function

The `sigmoid` function is the key ingredient that maps any continuous number (from negative infinity to positive infinity) to a probability. Also known as the logistic function, the `sigmoid` function is characterized by an S-shaped curve, taking any real number as input and mapping it into a score between 0 and 1, which happens to be the range of a valid probability score.

The standard `sigmoid` function takes the following form:

$$f(x) = \frac{1}{1 + e^{-x}}$$

Note that this is a nonlinear function. That is, the input values will get disproportionate scaling when going through the transformation using this function. It is also a continuous function (thus differentiable) and monotone ($f(x)$ will increase as x increases), thus enjoying high popularity as the go-to activation function to be used at the last layer of a typical neural network model for binary classification tasks.

Let us try to visualize this function. In the following code snippet, we use `seq(-10, 10, by = 0.1)` to create a sequence of equally spaced numbers from -10 to 10, with a step size of 0.1. For each number, we calculate the corresponding output using the `sigmoid` function. Here, we directly pass all the numbers of the function, which then calculates all the output in a parallel mode called vectorization. Here, vectorization refers to the process of applying an operation to an entire vector simultaneously instead of looping over each element one by one as in a `for` loop. Finally, we plot the function to show the characteristic S-shaped curve of the `sigmoid` function and add the gridlines using the `grid()` function:

```
# Create a range of equally spaced numbers between -10 and 10
x = seq(-10, 10, by = 0.1)

# Calculate the output value for each number using sigmoid function
sigmoid = 1 / (1 + exp(-x))

# Plot the sigmoid function
plot(x, sigmoid, type = "l", lwd = 2,
     main = "Sigmoid Function",
     xlab = "x",
     ylab = "f(x)",
     col = "red")

# Add grid lines
grid()
```

Running the preceding code generates the output shown in *Figure 13.1*. The plot shows a different level of steepness across the whole domain, where the function is more sensitive in the middle region and becomes more saturated at the two extremes.

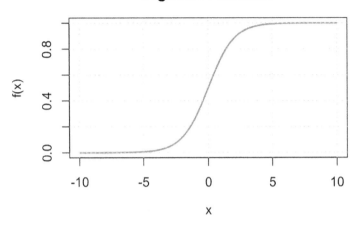

Figure 13.1 – Visualizing the sigmoid function

Now that we understand the `sigmoid` function, let us look at the mathematical construct of the logistic regression model.

Grokking the logistic regression model

The logistic regression model is essentially a linear regression model generalized to the setting where the dependent outcome variable is binary. In other words, it is a linear regression model that models the log odds of the probability of an event.

To see this, let us first recall the following linear regression model, where we use a total of p features to model the target output variable, z:

$$z = \beta_0 + \beta_1 x_1 + \beta_2 x_2 + \ldots + \beta_p x_p$$

Here, z is interpreted as the log odds, or logit, of the event of $y = 1$. We are interested in estimating the parameters from β_0 to β_p.

Now, we know that the z variable is unbounded, meaning it can vary from negative infinity to positive infinity. We need a way to bound this output and convert it into a probability score valued between 0 and 1. This is achieved via an additional transformation using the `sigmoid` function, which happens to satisfy all our needs. Mathematically, we have the following:

$$P(y = 1) = \frac{1}{1 + e^{-z}}$$

Plugging in the definition of z gives the full logistic regression model:

$$P(y = 1) = \frac{1}{1 + e^{-(\beta_0 + \beta_1 x_1 + \beta_2 x_2 + \ldots + \beta_p x_p)}}$$

Here, $P(y = 1)$ refers to the probability of having a success of $y = 1$ (this is a general statement), and correspondingly, $P(y = 0)$ indicates the probability of having a failure.

Note that we can equivalently express the model as the following:

$$\log\frac{P(y = 1)}{P(y = 0)} = \beta_0 + \beta_1 x_1 + \beta_2 x_2 + \ldots + \beta_p x_p$$

Here, the term $\log\frac{P(y = 1)}{P(y = 0)}$ stands for the log odds.

A key change here is the introduction of the `sigmoid` function. This makes the relationship between the predictors and the resulting probability no longer linear, but sigmoidal instead. To observe the subtlety here, we can look at the different regions of the `sigmoid` function across the domain. For example, when looking at the region around 0, a small change in the input would result in a relatively large change in the resulting probability output. However, the same change in the input will cause a very small change in the output when located on the two extreme sides of the function. Also, as the input becomes more extreme, either toward the negative or positive side, the resulting probability will gradually approach 0 or 1. *Figure 13.2* recaps the characteristics of the logistic regression model.

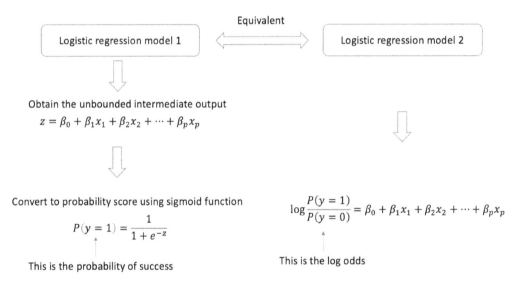

Figure 13.2 – Summarizing the logistic regression model

Note that a logistic regression model shares similar assumptions with the linear regression model. Specifically, it assumes that the observations are independent of each other, and the target outcome follows a Bernoulli distribution parameterized by p – that is, $y \sim Bernoulli(p)$. Given this, we do not assume a linear regression between the output variable and the input predicts; instead, we use the logistic link function to transform and introduce nonlinearity to the input variables.

The following section further compares logistic regression with linear regression.

Comparing logistic regression with linear regression

In this section, we will focus on a binary credit classification task using the *German Credit* dataset, which contains 1,000 observations and 20 columns. Each observation denotes a customer who had a loan application from the bank and is labeled as either good or bad in terms of credit risk. The dataset is available in the caret package in R.

For our study, we will attempt to predict the target binary variable, Class, based on Duration, and compare the difference in the prediction outcome between linear regression and logistic regression. We specifically choose one predictor only so that we can visualize and compare the decision boundaries of the resultant model in a two-dimensional plot.

Exercise 13.1 – comparing linear regression with logistic regression

In this exercise, we will demonstrate the advantage of using a logistic regression model in producing a probabilistic output compared to the unbounded output using a linear regression model:

1. Load the *German Credit* dataset from the caret package. Convert the target variable (Class) to numeric:

    ```
    # install the caret package if you haven't done so
    install.packages("caret")
    # load the caret package
    library(caret)
    # load the German Credit dataset
    data(GermanCredit)
    GermanCredit$Class_num = ifelse(GermanCredit$Class == "Bad", 1,
    0)
    ```

    Here, we create a new target variable called Class_num to map the original Class variable to 1 if it takes on the value of "Bad", and 0 otherwise. This is necessary as both linear regression and logistic regression models cannot accept a string-based variable as the target (or predictor).

2. Build a linear regression model to regress Class_num against Duration:

    ```
    lm_model = lm(Class_num ~ Duration, data=GermanCredit)
    coefs = coefficients(lm_model)
    intercept = coefs[1]
    slope = coefs[2]
    ```

    Here, we use the lm() function to build the linear regression model and coefficients() to extract the model coefficients, including the intercept and slope.

3. Visualize the prediction and the target:

```
ggplot(GermanCredit,
        aes(Duration, Class_num)) +
    geom_point() +
    geom_abline(intercept=intercept, slope=slope) +
    theme(axis.title.x = element_text(size = 18),
          axis.title.y = element_text(size = 18))
```

Here, we plot the observed target variable as a scatter plot and use the geom_abline() function to plot the model as a straight line based on the estimated slope and intercept.

Running the preceding code generates *Figure 13.3*.

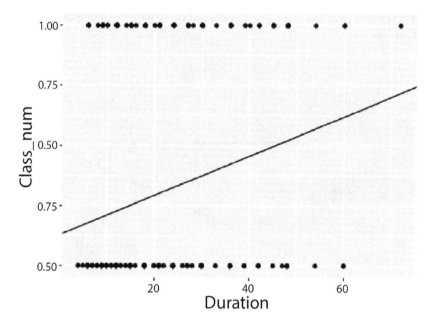

Figure 13.3 – Visualizing the linear regression model

Since all target values are 0 or 1, we can think of the predictions as probabilities valued between 0 and 1. However, as we zoom out, the problem of having an unbounded probability would surface, as shown in the following steps.

4. Re-plot the graph by zooming out to a wide domain of (-30, 120) for the *x* axis and (-0.5, 1.5) for the *y* axis:

```
ggplot(GermanCredit,
        aes(Duration, Class_num)) +
    geom_point() +
    geom_abline(intercept=intercept, slope=slope) +
```

```
xlim(-30, 120) +
ylim(-0.5, 1.5) +
theme(axis.title.x = element_text(size = 18),
    axis.title.y = element_text(size = 18))
```

Here, we enlarged the range of possible values for the *x* axis and *y* axis using the `xlim()` and `ylim()` functions.

Running the preceding code generates the output shown in *Figure 13.4*, which shows that the predicted values are outside the range of [0,1] when the value of `Duration` becomes more extreme, a situation called extrapolation beyond the observed range of values. This means that the predicted probabilities would be smaller than 0 or bigger than 1, which is obviously an invalid output. This calls for a generalized linear regression model called logistic regression, where the response will follow a logistic, S-shaped curve based on the transformation of the `sigmoid` function.

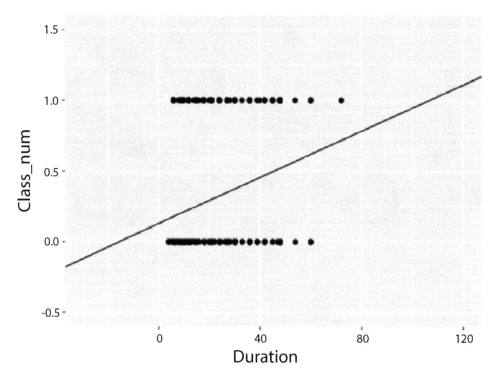

Figure 13.4 – Visualizing the linear regression model with extended range

5. Build a logistic regression model using the `glm()` function:

```
glm_model = glm(Class_num ~ Duration, data=GermanCredit,
family=binomial)
>>> glm_model
Call:  glm(formula = Class_num ~ Duration, family = binomial,
data = GermanCredit)

Coefficients:
(Intercept)      Duration
   -1.66635       0.03754

Degrees of Freedom: 999 Total (i.e. Null);  998 Residual
Null Deviance:      1222
Residual Deviance: 1177    AIC: 1181
```

The result shows the estimated intercept and slope, along with the residual deviance, a measure of goodness of fit for the logistic regression model.

6. Plot the estimated logistic curve on the previous plot:

```
ggplot(GermanCredit,
       aes(Duration, Class_num)) +
  geom_point() +
  geom_abline(intercept=intercept, slope=slope) +
  geom_smooth(
    method = "glm",
    se = FALSE,
    method.args = list(family=binomial)
  ) +
  theme(axis.title.x = element_text(size = 18),
        axis.title.y = element_text(size = 18))
```

Running the preceding code will generate the output shown in *Figure 13.5*, which suggests a slight deviation between the linear regression line and the logistic regression curve.

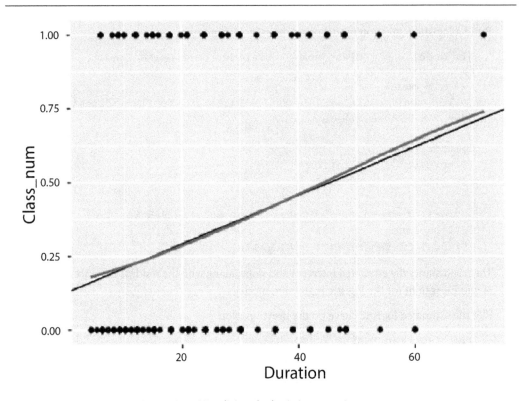

Figure 13.5 – Visualizing the logistic regression curve

Again, we can zoom out the figure and focus on the difference when going beyond the observed range of possible values in the dataset.

7.   Plot the logistic curve within a wider range of values beyond the observed range:

```
# Get coefficients from logistic model
intercept_glm = coef(glm_model)[1]
slope_glm = coef(glm_model)[2]

# Generate sequence of x-values
x_values = seq(from = min(GermanCredit$Duration) - 150,
               to = max(GermanCredit$Duration) + 150,
               by = 0.1)

# Compute probabilities using logistic function
y_values = 1 / (1 + exp(-(intercept_glm + slope_glm * x_
values)))

# Data frame for plot
plot_df = data.frame(x = x_values, y = y_values)
```

```
# Plot
ggplot() +
  geom_point(data = GermanCredit, aes(Duration, Class_num)) +
  geom_abline(intercept=intercept, slope=slope) +
  geom_line(data = plot_df, aes(x, y), color = "blue") +
  theme_minimal() +
  xlim(-30, 120) +
  ylim(-0.5, 1.5) +
  theme(axis.title.x = element_text(size = 18),
        axis.title.y = element_text(size = 18))
```

Here, we first extract the coefficients for the logistic regression model, followed by generating a sequence of values for the input and the corresponding output using the sigmoid function transformation. Lastly, we plot the logistic curve together with the linear fit in the same plot with the observed data.

Running the preceding code generates the output shown in *Figure 13.6*, which shows that the logistic regression curve gets gradually saturated as the input value becomes more extreme. In addition, all values are now bounded in the range of [0, 1], making it a valid candidate for interpretation as probabilities instead of an unbounded value.

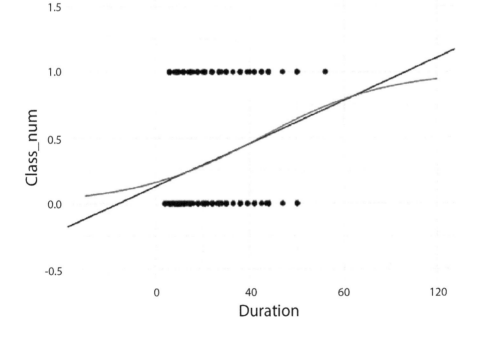

Figure 13.6 – Visualizing the logistic regression model with extended range

The next section looks at how to make predictions using a logistic regression model.

Making predictions using the logistic regression model

As discussed in the previous section, the direct predictions from a logistic regression model take the form of probabilities valued between 0 and 1. To convert them into binary predictions, we could take the most probable prediction by rounding the probability using a threshold of 0.5. For example, if the predicted probability is $P(y = 1) = 0.8$, the rounding operation will lead to a final binary prediction of $y = 1$. On the other hand, if $P(y = 1) = 0.3$, rounding will result in $y = 0$.

Let us go through the following exercise to understand how to perform predictions using the logistic regression model.

Exercise 13.2 – performing predictions using the logistic regression model

We have seen how to perform predictions using the explicit sigmoid transformation after extracting the slope and intercept of the logistic regression model. In this exercise, we will explore a more convenient approach using the `predict()` function:

1. Generate a sequence of `Duration` values ranging from 5 to 80 with a step size of 2 and predict the corresponding probabilities for the sequence using the `predict()` function based on the previous logistic regression model:

    ```
    library(tibble)
    library(dplyr)

    # making predictions
    pred_df = tibble(
      Duration = seq(5, 80, 2)
    )

    pred_df = pred_df %>%
      mutate(
        pred_prob = predict(glm_model, pred_df, type="response")
      )
    ```

    Here, we use the `seq()` function to create the equally spaced vector and store it in a `tibble` object called `pred_df`. We then use `predict()` to predict the corresponding probabilities by specifying `type="response"`.

2. Visualize the predicted probabilities together with the raw data:

    ```
    ggplot() +
      geom_point(data = GermanCredit, aes(Duration, Class_num)) +
    ```

```
    geom_point(data = pred_df, aes(Duration, pred_prob) ,
  color="blue") +
    theme(axis.title.x = element_text(size = 18),
          axis.title.y = element_text(size = 18))
```

The preceding code will generate the following output:

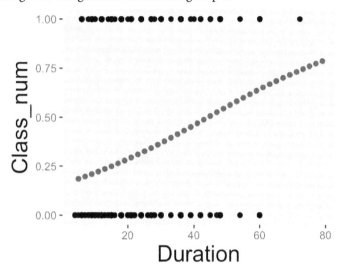

Figure 13.7 – Visualizing the predicted probabilities

3. Convert the probabilities to binary outcomes using the round() function:

```
# getting the most likely outcome
pred_df = pred_df %>%
  mutate(
    most_likely_outcome = round(pred_prob)
  )
```

Here, we round the predicted probabilities using a default threshold of 0.5.

4. Add the binary outcomes as green points to the previous graph:

```
ggplot() +
  geom_point(data = GermanCredit, aes(Duration, Class_num)) +
  geom_point(data = pred_df, aes(Duration, pred_prob) ,
color="blue") +
  geom_point(data = pred_df, aes(Duration, most_likely_outcome)
, color="green") +
  theme(axis.title.x = element_text(size = 18),
        axis.title.y = element_text(size = 18))
```

Running the preceding code generates the following output, which suggests that all predicted probabilities above 0.5 are converted to 1, and those below 0.5 are converted to 0.

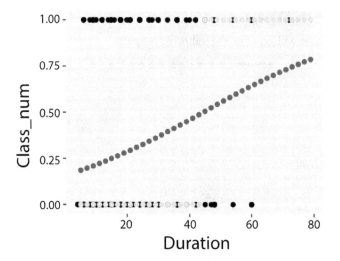

Figure 13.8 – Visualizing the predicted binary outcomes

The next section discusses the log odds further.

More on log odds and odds ratio

Recall that the odds refer to the ratio of the probability of an event happening over its complement:

$$odds = \frac{probability\ of\ event\ happening}{probability\ of\ event\ not\ happening} = \frac{p}{1-p} = \frac{P(y = 1)}{P(y = 0)}$$

Here, the probability is calculated as follows:

$$p = P(y = 1) = \frac{1}{1 + e^{-z}}$$

$$1 - p = 1 - \frac{1}{1 + e^{-z}} = \frac{e^{-z}}{1 + e^{-z}}$$

Plugging in the definition of p and $1 - p$ gives us the following:

$$odds = \frac{p}{1-p} = e^z$$

Instead of directly working with the odds, we often use the log odds or logit. This term is typically modeled as a linear combination of predictors in a logistic regression model via the following:

$$\log\frac{P(y = 1)}{P(y = 0)} = z = \beta_0 + \beta_1 x_1 + \dots + \beta_p x_p$$

Here, we can interpret each coefficient β_j as the expected change in the log odds for a one-unit increase in the jth predictor x_j, while keeping all other predictors constant. This equation essentially says that the log odds of the target value y being 1 is linearly related to the input variables.

Now suppose x_i is a binary input variable, making $x_i = 1$ or 0. We can calculate the odds of $x_i = 1$ as follows:

$$\frac{p_1}{1 - p_1} = e^{z_1}$$

This measures the chance of an event for $x_i = 1$, over the chance of a non-event. Similarly, we can calculate the odds of $x_i = 0$ as follows:

$$\frac{p_0}{1 - p_0} = e^{z_0}$$

This measures the chance of an event for $x_i = 0$ over the chance of a non-event.

We can then calculate the odds ratio of x_i, which is the ratio of the odds $x_i = 1$ to the odds of $x_i = 0$:

$$\frac{\frac{p_1}{1-p_1}}{\frac{p_0}{1-p_0}} = \frac{e^{\beta_0 + \beta_1 x_1 + \dots \beta_i^* 1 + \dots + \beta_p x_p}}{e^{\beta_0 + \beta_1 x_1 + \dots \beta_i^* 0 + \dots + \beta_p x_p}} = e^{\beta_i}$$

Here, e^{β_i} measures the quantified impact of the binary input variable, x_i, on the odds of the outcome y being 1, while all other input variables remain unchanged. This gives us a way to measure the impact of any predictor in a logistic regression model, covering both categorical and numerical input variables.

For categorical input variables, we may use gender (0 for male and 1 for female) to predict whether insurance is purchased (1 for yes and 0 for no). We set the base categorical as 0 for male. If the estimated coefficient $\beta_{gender} = 0.2$ for gender, its odds ratio is calculated as $e^{0.2} \approx 1.22$. Therefore, the odds of female customers purchasing the insurance is 1.22 times the odds of their male counterparts purchasing the insurance, assuming all other variables remain unchanged.

For numerical input variables, we may use age to predict whether insurance is purchased. There is no need to set the base category in this case. If the estimated coefficient $\beta_{age} = 0.3$, the corresponding odds ratio is calculated as $e^{0.3} \approx 1.35$. This means that the odds of a client purchasing is 1.35 times the odds of similar people who are one year younger, assuming *ceteris paribus*.

Note that we can calculate the log odds using the predicted probabilities, as shown in the following code snippet:

```
# calculate log odds using predicted probabilities
pred_df = pred_df %>%
  mutate(
    log_odds = log(pred_prob / (1 - pred_prob))
  )
```

The next section introduces more on the loss function of the logistic regression model.

Introducing the cross-entropy loss

The binary CEL, also called the **log loss**, is often used as the cost function in logistic regression. This is the loss that the logistic regression model will attempt to minimize by moving the parameters. This function takes the predicted probabilities and the corresponding targets as the input and outputs a scalar score, indicating the goodness of fit. For a single observation with a target of y_i and predicted probability of p_i, the loss is calculated as follows:

$$Q_i(y_i, p_i) = -\left[y_i \log p_i + (1 - y_i)\log(1 - p_i)\right]$$

Summing up all individual losses gives the total binary CEL:

$$Q(y, p) = \frac{1}{N}\sum_i^N Q_i = \frac{1}{N}\sum_{i=1}^N -\left[y_i \log p_i + (1 - y_i)\log(1 - p_i)\right]$$

The binary CEL function is a suitable choice for binary classification problems because it heavily penalizes confident but incorrect predictions. For example, as p_i approaches 0 or 1, the resulting CEL will go to infinity if the prediction is incorrect. This property thus encourages the learning process to output probabilities that are close to the true probabilities of the targets.

More generally, we use the CEL to model a classification problem with two or more classes in the target variable. For the ith observation, x_i, the classification function would produce a probability output, denoted as $p_{i,k} = f(x_i; w)$, to indicate the likelihood of belonging to the kth class. When we have a classification task with a total of C classes, the CEL for x_i is defined as $Q_i(w) = -\sum_{k=1}^C y_{i,k} \log(p_{i,k})$, which essentially sums across all classes. Again, $y_{i,k} = 1$ if the target label for the ith observation belongs to the kth class, and $y_{i,k} = 0$ otherwise.

The summation means that the class-wise evaluation (i.e., the term $y_{i,k} \log(p_{i,k})$) is performed for all classes and summed together to produce the total cost for the ith observation. For each observation, there are a total of C predictions corresponding to the respective probability of belonging to each class. The CEL thus aggregates the matrix of predicted probabilities by summing them into a single number. In addition, the result is negated to produce a positive number, since $\log(x)$ is negative if x is a probability between 0 and 1.

Note that the target label in the CEL calculation needs to be one-hot encoded. This means that a single categorical label is converted into an array of binary numbers that contains 1 for the class label and 0 otherwise. For example, for a digit image on number 8, instead of passing the original class as the target output, the resulting one-hot-encoded target $[0, 0, 0, 0, 0, 0, 0, 1, 0, 0]$ would be used, where the 8th position is activated (i.e., hot) and the rest disabled. The target array would also have the same length as the probability output array, whose elements correspond to each of the classes.

Intuitively, we would hope for the predicted probability for the correct class to be close to 1 and for the incorrect class to be 0. That is, the loss should increase as the predicted probabilities diverge from the actual class label. The CEL is designed to follow this intuition. Specifically, we can look at the following four scenarios for the ith observation:

- When the target label belongs to the kth class (i.e., $y_{i,k} = 1$) and the predicted probability for the kth class is very strong (i.e., $p_{i,k} \approx 1$), the cost should be low.

- When the target label belongs to the kth class (i.e., $y_{i,k} = 1$) and the predicted probability for the kth class is very weak (i.e., $p_{i,k} \approx 0$), the cost should be high.

- When the target label does not belong to the kth class (i.e., $y_{i,k} = 0$) and the predicted probability for the kth class is very strong (i.e., $p_{i,k} \approx 1$), the cost should be high.

- When the target label does not belong to the kth class (i.e., $y_{i,k} = 0$) and the predicted probability for the kth class is very weak (i.e., $p_{i,k} \approx 0$), the cost should be low.

To calculate the CEL, we first calculate the weighted sum between the vector of binary labels and the vector of predicted probabilities across all classes for each observation. The results of all observations are added together, followed by a minus sign to reverse the cost to a positive number. The CEL is designed to match the intuition for the cost: the cost would be low when the prediction and the target closely match, and high otherwise. In other words, to calculate the total cost, we would simply sum individual costs, leading to $Q(\mathbf{w}) = -\sum_{i=1}^{N}\sum_{k=1}^{C} y_{i,k} \log(p_{i,k})$. *Figure 13.9* summarizes the preceding discussion on the CEL.

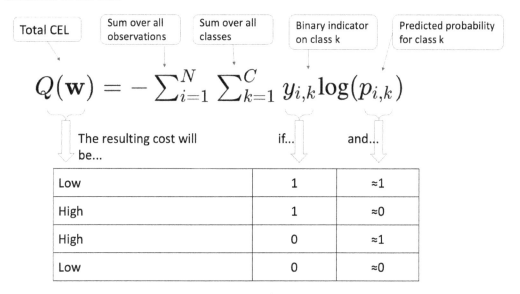

Figure 13.9 – Illustrating the CEL

Note that the last two scenarios do not contribute at all to the loss calculation since the target value is equal to 0; any value multiplied by 0 becomes 0. Also, observe that the predicted probabilities of a specific observation for these two classes need to add up to 1, making it sufficient to only focus on the predicted probability of one class (mostly class 1).

The next section introduces how to evaluate a logistic regression model.

Evaluating a logistic regression model

There are multiple metrics we can use to evaluate a logistic regression model. These are the metrics we use to determine the goodness of fit (over the test set), which needs to be differentiated from the CEL we use to train the model (over the training set).

The following list provides the commonly used metrics:

- **Accuracy rate**: This is the proportion of the number of correctly predicted observations made by the model out of the count of all observations. Since a correct prediction can be either a true positive or a true negative, the accuracy is calculated by summing up the true positives and true negatives and dividing the total number of observations.

- **Error rate**: This is the proportion of incorrectly predicted observations made by the model over the total observations. An incorrect prediction can be a false positive or a false negative. It is calculated as *1 - accuracy rate*; that is, the error rate is the complement of the accuracy rate. In other words, it is calculated as *(false positives + false negatives) / total observations*.

- **Precision**: Precision is the proportion of correct positive predictions among all positive predictions. This measure essentially tells us out of all predicted positive instances, how many of them are correct. Thus it indicates the model's accuracy in predicting positive observations and is calculated as *true positives / (true positives + false positives)*. In the denominator, we note that among all positive instances, some are true positives and the rest are false positives.

- **Recall**: Recall refers to the proportion of actual positive instances that the model correctly predicts. Also called sensitivity or **true positive rate** (**TPR**), the recall measures the model's ability to detect positive observations. It is calculated as *true positives / (true positives + false negatives)*, where the formula differs in the denominator compared with precision.

- **Specificity**: Also called **true negative rate** (**TNR**), specificity measures the proportion of actual negative instances correctly predicted by the model. This is the opposite of sensitivity, which focuses on the model's ability to capture the true positives. In other words, specificity measures the model's ability to identify negative instances or non-events correctly. It is calculated as *true negatives / (true negatives + false positives)*.

- **Area under the curve** (**AUC**): The AUC is directly determined by the **receiver operating characteristic** (**ROC**) curve, which plots the TPR (sensitivity) against the false positive rate (*1 - specificity*) at different thresholds for binary classification. As the name suggests, AUC measures the area under the ROC curve in the form of a proportion valued between 0 and 1, indicating the degree of separability between two classes. A perfect model with 100% correct predictions has an AUC of 1, while a model with completely wrong predictions has an AUC of 0. A model that performs random guesses (choosing 0 or 1 with 50% probability) corresponds to an AUC of 0.5, suggesting no class separation capacity.

Note that we would exercise the principle of parsimony upon assessing two models with equally good evaluation metrics. The principle of parsimony says that if two competing models provide a similar level of fit to the data, the one with fewer input variables should be picked, thus preferring simplicity over complexity. The underlying assumption is that the most accurate model is not necessarily the best model.

Figure 13.10 describes the process of laying out the confusion matrix that captures the prediction results in different scenarios, along with the details on calculating the aforementioned evaluation metrics.

Confusion matrix

	Predicted $\hat{y} = 0$	Predicted $\hat{y} = 1$	Total
Non-event $y = 0$	a	c	$a + c$
Event $y = 0$	b	d	$b + d$
Total	$a + b$	$c + d$	n

$$\text{Accuracy} = \frac{a+d}{n} \qquad \text{Error rate} = \frac{b+c}{n} \qquad \text{Recall} = \frac{d}{b+d}$$

$$\text{Precision} = \frac{d}{c+d} \qquad \text{Specificity} = \frac{a}{a+c}$$

Figure 13.10 – Illustrating the confusion matrix and common
evaluation metrics for binary classification tasks

Note that we will only be able to calculate these metrics after selecting a threshold to cut off predicted probabilities. Specifically, an observation with a predicted probability greater than the cutoff threshold will be classified as positive, and negative otherwise.

Precision and recall usually have an inverse relationship with respect to the adjustment of the classification threshold. Reviewing both precision and recall is useful for cases where there is a huge imbalance in the target variable's values. As precision and recall are two related but different metrics, which should we optimize for?

To answer this question, we need to assess the relative impact of making a false negative prediction. This is measured by the false negative rate, which is the opposite (or complement) of recall. As *Figure 13.11* suggests, failing to spot a spam email is less risky than missing a fraudulent transaction or a positive cancer patient. We should aim to optimize for precision (so that the model's predictions are more precise and targeted) for the first case and recall (so that we minimize the chance of missing a potentially positive case) for the second case.

	Positive class	Impact of having a False Negative (FN)	How bad is FN?	Optimize for?
Spam filter	• Spam	• Spam goes to the inbox	• Acceptable	• Precision
Fraudulent transaction	• Fraud	• Fraudulent transactions pass through	• Very bad	• Recall
Cancer diagnose	• Cancer	• Patient walks away without knowing they have cancer	• Very bad	• Recall

Figure 13.11 – Three cases with different impacts of having a false negative prediction

As for the AUC, or area under the ROC curve, there is no need to select a specific threshold as it is calculated by evaluating a sequence of thresholds from 0 to 1. The ROC curve plots the sensitivity on the *y* axis and *1 - specificity* on the *x* axis. This also corresponds to plotting TPR versus 1-TNR, or TPR versus FPR. As the classification threshold goes up, FPR goes down, leading to a leftward movement of the curve.

A perfect binary classifier has an AUC score of 1. This means that FPR, or *1 – specificity*, is 0. That is, there is no false positive, and all negative cases are not predicted as positive. In addition, the sensitivity, or TPR, is 1, meaning all positive cases are predicted as positive correctly.

Figure 13.12 illustrates three different AUC curves. The topmost curve (in green) corresponds to a better model due to the highest AUC. Both models, represented by the green and red curves, perform better than random guessing, as indicated by the straight off-diagonal line in blue.

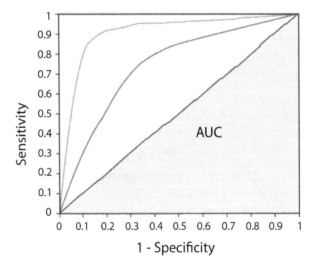

Figure 13.12 – Illustrating three different AUC curves

Continuing the previous exercise, we can now calculate the corresponding evaluation metrics. First, we use the trained logistic regression model to score all observations in the training set and obtain the corresponding probabilities using the `predict()` function and setting `type="response"`, as shown in the following code snippet. Note that we need to pass in a DataFrame with the corresponding feature names as input to the model:

```
# Create new data frame with all durations
new_data = data.frame(Duration = GermanCredit$Duration)
# Calculate predicted classes based on predicted probabilities
predicted_probs = predict(glm_model, new_data, type="response")
```

Next, we set a single cutoff threshold (`0.5`, in this case) to convert the predicted probabilities into the corresponding binary outcomes using the `ifelse()` function:

```
# Convert to binary outcomes
predicted_classes = ifelse(predicted_probs > 0.5, 1, 0)
```

With the binary outcomes and true target labels, we can obtain the confusion matrix as follows:

```
# Create confusion matrix
conf_matrix = table(predicted = predicted_classes, actual =
GermanCredit$Class_num)
>>> conf_matrix
         actual
predicted   0    1
        0 670  260
        1  30   40
```

Here, the confusion matrix provides a breakdown of the correct and incorrect classifications from the model. Within the confusion matrix, the top-left cell means true negatives, the top-right cell means false positives, the bottom-left cell means false negatives, and the bottom-right cell means true positives.

Based on the confusion matrix, we can calculate the evaluation metrics as follows:

```
# Accuracy
accuracy = sum(diag(conf_matrix)) / sum(conf_matrix)
>>> print(paste("Accuracy: ", accuracy))
"Accuracy:  0.71"
# Error rate
error_rate = 1 - accuracy
>>> print(paste("Error rate: ", error_rate))
"Error rate:  0.29"
# Precision
precision = conf_matrix[2,2] / sum(conf_matrix[2,])
print(paste("Precision: ", precision))
```

```
"Precision:   0.571428571428571"
# Recall / Sensitivity
recall = conf_matrix[2,2] / sum(conf_matrix[,2])
print(paste("Recall: ", recall))
>>> "Recall:   0.133333333333333"
# Specificity
specificity = conf_matrix[1,1] / sum(conf_matrix[,1])
print(paste("Specificity: ", specificity))
>>> "Specificity:   0.957142857142857"
```

Here, we extract the corresponding items of the confusion matrix and plug in the definitions of different evaluation metrics to complete the calculations.

We can also calculate the AUC, starting with calculating the ROC curve using the pROC package:

```
library(pROC)
# Calculate ROC curve
roc_obj = roc(GermanCredit$Class_num, predicted_probs)
# Plot ROC curve
>>> plot(roc_obj)
```

Running the code generates *Figure 13.13*, which suggests that the model is doing marginally better than random guessing.

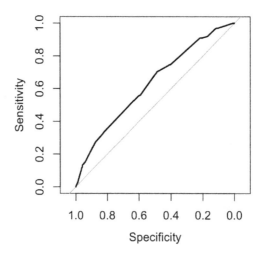

Figure 13.13 – Visualizing the ROC curve

To calculate the AUC, we can call the auc() function:

```
# Calculate AUC
auc = auc(roc_obj)
```

```
>>> print(paste(«AUC: «, auc))
«AUC:    0.628592857142857»
```

Note that when we have no preference for precision or recall, we can use the F1 score, defined as follows:

$$\text{F1 score } = \frac{2(\text{Precision} \times \text{Recall})}{\text{Precision} + \text{Recall}}$$

The next section discusses a challenging modeling situation when we have an imbalanced dataset to begin with.

Dealing with an imbalanced dataset

When building a logistic regression model using a dataset whose target is a binary outcome, it could be the case that the target values are not equally distributed. This means that we would observe more non-events ($y = 0$) than events ($y = 1$), as is often the case in applications such as fraudulent transactions in banks, spam/phishing emails for corporate employees, identification of diseases such as cancer, and natural disasters such as earthquakes. In these situations, the classification performance may be dominated by the majority class.

Such domination can result in misleadingly high accuracy scores, which correspond to poor predictive performance. To see this, suppose we are developing a default prediction model using a dataset that consists of 1,000 observations, where only 10 (or 1%) of them are default cases. A naive model would simply predict every observation as non-default, resulting in a 99% accuracy.

When we encounter an imbalanced dataset, we are often more interested in the minority class, which represents the outcome to be detected in a classification problem. Since the signal on the minority class is relatively weak, we would need to rely on good modeling techniques to recognize a good pattern so that the signal can be correctly detected.

There are multiple techniques we can use to address the challenge of an imbalanced dataset. We will introduce a popular approach called data resampling, which requires oversampling and/or undersampling of the original dataset to make the overall distribution less imbalanced. Resampling includes oversampling the minority class, undersampling the majority class, or using a combination of them, as represented by **Synthetic Minority Oversampling TEchnique (SMOTE)**. However, such a remedy does not come without risk. Here, oversampling may lead to overfitting due to more samples being added to the original dataset, while undersampling may result in loss of information due to some majority observations being removed from the original dataset.

Figure 13.14 illustrates the process of oversampling the minority class (the left panel) and undersampling the majority class (the right panel). Note that once the model is built based on the balanced dataset, we would still need to calibrate it on a new test set so that it performs well on a new real dataset as well.

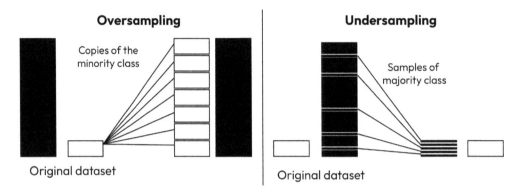

Figure 13.14 – Illustrating the process of oversampling the minority and undersampling the majority

Let us go through an exercise to understand how to perform undersampling or oversampling in a logistic regression setting.

Exercise 13.3 – performing undersampling and oversampling

In this exercise, we will create two artificial datasets based on undersampling and oversampling. We will then assess the performance of the resulting logistic regression model using the confusion matrix:

1. Divide the raw dataset into training (70%) and test (30%) sets:

    ```
    set.seed(2)
    index = sample(1:nrow(GermanCredit), nrow(GermanCredit)*0.7)
    train = GermanCredit[index, ]
    test = GermanCredit[-index, ]
    ```

    Here, we randomly sample a set of indexes used to select observations for the training set and allocate the rest to the test set.

2. Check the class ratio in the training set:

    ```
    >>> table(train$Class_num)
      0   1
    504 196
    ```

    The result shows that the majority (class 0) is more than twice the size of the minority (class 1).

3. Separate the training set into a majority set and a minority set based on the class label:

    ```
    # separate the minority and majority classes
    table(train$Class_num)
    minority_data = train[train$Class_num == 1,]
    majority_data = train[train$Class_num == 0,]
    ```

We will then use these two datasets to perform undersampling and oversampling.

4. Undersample the majority class and combine the undersampled majority class with the minority class. Check the resulting class ratio:

```
# undersample the majority class
undersampled_majority = majority_data[sample(1:nrow(majority_
data), nrow(minority_data)),]
# combine undersampled majority class and minority class
undersampled_data = rbind(minority_data, undersampled_majority)
>>> table(undersampled_data$Class_num)
  0   1
196 196
```

The class ratio is balanced now. Let us perform oversampling for the minority class.

5. Oversample the minority class and combine the oversampled minority class with the majority class. Check the resulting class ratio:

```
# oversample the minority class
oversampled_minority = minority_data[sample(1:nrow(minority_
data), nrow(majority_data), replace = TRUE),]
# combine majority class and oversampled minority class
oversampled_data = rbind(majority_data, oversampled_minority)
>>> table(oversampled_data$Class_num)
  0   1
504 504
```

6. Fit logistic regression models on both the undersampled and oversampled datasets:

```
# fit logistic regression models on undersampled and oversampled
data
undersampled_model = glm(Class_num ~ Duration, family =
binomial(link = 'logit'), data = undersampled_data)
oversampled_model = glm(Class_num ~ Duration, family =
binomial(link = 'logit'), data = oversampled_data)
```

7. Obtain the predicted probabilities on the test set:

```
# get the predicted probabilities on the test set
undersampled_pred = predict(undersampled_model, newdata = test,
type = "response")
oversampled_pred = predict(oversampled_model, newdata = test,
type = "response")
```

8. Apply a threshold of 0.5 to convert the probabilities into binary classes:

```
# apply threshold to convert the probabilities into binary
classes
```

```
undersampled_pred_class = ifelse(undersampled_pred > 0.5, 1, 0)
oversampled_pred_class = ifelse(oversampled_pred > 0.5, 1, 0)
```

9. Calculate the confusion matrix:

```
# calculate the confusion matrix
undersampled_cm = table(predicted = undersampled_pred_class,
actual = test$Class_num)
oversampled_cm = table(predicted = oversampled_pred_class,
actual = test$Class_num)
>>> undersampled_cm
          actual
predicted   0   1
        0 117  59
        1  79  45
>>> oversampled_cm
          actual
predicted   0   1
        0 115  59
        1  81  45
```

The result shows that both models deliver a similar performance using the undersampled or oversampled training dataset. Again, we would use another validation dataset, which can be taken from the original training set, to calibrate the model parameters further so that it performs better on the test set.

It turns out that we can also add a lasso or ridge penalty in a logistic regression model, as discussed in the next section.

Penalized logistic regression

As the name suggests, a penalized logistic regression model includes an additional penalty term in the loss function of the usual logistic regression model. Recall that a standard logistic regression model seeks to minimize the negative log-likelihood function (or equivalently, maximize the log-likelihood function), defined as follows:

$$Q(\beta) = \frac{1}{N}\sum_{i=1}^{N} -\left[y_i \log p_i + (1 - y_i)\log(1 - p_i)\right]$$

Here, $p_i = \frac{1}{1 + e^{-(\beta_0 + \beta_1 x_i^{(1)} + \beta_2 x_i^{(2)} + \dots + \beta_p x_i^{(p)})}}$ is the predicted probability for input $x^{(i)}$, y_i is the corresponding target label, and $\beta = \{\beta_0, \beta_1, \dots, \beta_p\}$ are model parameters to be estimated. Note that we now express the loss as a function of the coefficient vector as it is directly determined by the set of parameters used in the model.

Since the penalty term aims to shrink the magnitude of the estimated coefficients, we would add it to the loss function so that the penalty terms will be relatively small (subject to a tuning hyperparameter, λ). For the case of ridge penalty, we would add up the squared coefficients, resulting in the following penalized negative log-likelihood function:

$$Q_{ridge}(\beta) \;=\; Q(\beta) + \lambda\,||\beta||_2^2 \;=\; Q(\beta) + \lambda\sum_{j=1}^{p}\beta_j^2$$

Correspondingly, the penalized negative log-likelihood function using the lasso regularization term takes the following form:

$$Q_{lasso}(\beta) \;=\; Q(\beta) + \lambda\,\big|\beta\big| \;=\; Q(\beta) + \lambda\sum_{j=1}^{p}\big|\beta_j\big|$$

In both cases, the penalty term has the potential effect of shrinking the magnitude of the coefficients toward 0 relative to the original maximum likelihood estimates. This can help to prevent overfitting by controlling the complexity of the model estimation process. The tuning hyperparameter, λ, controls the amount of shrinkage. In particular, a larger λ adds more weight to the penalty term and thus leads to more shrinkage effect, while a smaller λ puts less weight on the overall magnitude of the estimated coefficients.

Let us illustrate the process of development of penalized logistic regression models. The task can be achieved using the `glmnet` package, which supports both lasso and ridge penalties. In the following code snippet, we use the first nine columns as predictors to model the binary outcome:

```
# Create a matrix of predictors and a response vector
# For glmnet, we need to provide our data as matrices/vectors
X = GermanCredit[1:nrow(GermanCredit), 1:9]
y = GermanCredit$Class_num
# Define an alpha value: 0 for ridge, 1 for lasso, between 0 and 1 for
elastic net
alpha_value = 1 # for lasso
# Run the glmnet model
fit = glmnet(X, y, family = "binomial", alpha = alpha_value)
```

Here, we set `alpha=1` to enable the lasso penalty. Setting `alpha=0` enables the ridge penalty, and setting alpha between 0 and 1 corresponds to the elastic net penalty.

Note that this procedure would evaluate a sequence of values for the λ hyperparameter, giving us an idea of the level of impact on the resulting coefficients based on the penalty. In particular, we can plot out the coefficient paths, as shown in the following, indicating the resulting coefficients for different values of λ:

```
# plot coefficient paths
>>> plot(fit, xvar = "lambda", label = TRUE)
```

Running the code generates *Figure 13.15*, which suggests that more parameters shrink to 0 when λ gets big.

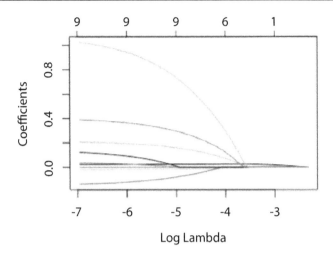

Figure 13.15 – Visualizing the coefficient paths using lasso penalized logistic regression

The next section discusses a more general setting: the multinomial logistic regression, a model class for multi-class classification.

Extending to multi-class classification

Many problems feature more than two classes. For example, the **Standard and Poor's (S&P)** bond rating includes multiple classes, such as AAA, AA, A, and more like these. Corporate client accounts in a bank are categorized into good credit, past due, overdue, doubtful, or loss. Such settings require the multinomial logistic regression model, which is a generalization of the binomial logistic regression model in the multi-class classification context. Essentially, the target variable, y, can take more than two possible discrete outcomes and allows for more than two categorical values.

Assume that the target variable can take on three values, giving $y \in \{0,1,2\}$. Let us choose class 0 as the pivot value or the baseline. We will model the odds of the probabilities of the other categories (classes 1 and 2) relative to this baseline. In other words, we have the following:

$$\frac{p(y = 1)}{p(y = 0)} = e^{z_1}$$
$$\frac{p(y = 2)}{p(y = 0)} = e^{z_2}$$

Therefore, the relative ratio of the predicted probabilities for each class is as follows:

$$p(y = 2){:}p(y = 1){:}p(y = 0) = e^{z_2}{:}e^{z_1}{:}1$$

We know the following:

$$p(y = 2) + p(y = 1) + p(y = 0) = 1$$

So we have the following:

$$p(y = 2) = \frac{e^{z_2}}{e^{z_2} + e^{z_1} + 1}$$

$$p(y = 1) = \frac{e^{z_1}}{e^{z_2} + e^{z_1} + 1}$$

$$p(y = 0) = \frac{1}{e^{z_2} + e^{z_1} + 1}$$

Again, one of the main assumptions in a multinomial logistic regression model is that the log odds consist of a linear combination of the predictor variables. This is the same assumption as in binary logistic regression. The corresponding interpretation of the coefficients in a multinomial logistic regression will also change slightly. In particular, each coefficient now represents the change in the log odds of the corresponding category relative to the baseline category for a one-unit change in the corresponding predictor variable, while holding all other predictors constant.

We can rely on the multinom() function from the nnet package to create a multinomial logistic regression model. In the following code snippet, we use the previous mtcars dataset and convert the gear variable into a factor, which will serve as the target variable:

```
library(nnet)
# convert gear to factor
mtcars$gear = as.factor(mtcars$gear)
>>> table(mtcars$gear)

 3   4   5
15  12   5
```

The frequency count shows a total of three categories in the variable. Next, we use mpg, hp, and disp to predict gear in a multinomial logistic regression model:

```
# fit the model
multinom_model = multinom(gear ~ mpg + hp + disp, data = mtcars)
# weights:  15 (8 variable)
initial  value 35.155593
iter  10 value 10.945783
iter  20 value 9.011992
iter  30 value 8.827997
iter  40 value 8.805003
iter  50 value 8.759821
iter  60 value 8.742738
iter  70 value 8.737492
iter  80 value 8.736569
final  value 8.735812
converged
```

The output message suggests that we have a total of eight variables in the model. This makes sense since we have four variables (`intercept`, `mpg`, `hp`, and `disp`) to model the difference between four-gear and three-gear cars in one submodel, and another four variables to model the difference between five-gear and three-gear cars in another submodel.

Let us view the summary of the model:

```
# view summary of the model
>>> summary(multinom_model)
Call:
multinom(formula = gear ~ mpg + hp + disp, data = mtcars)

Coefficients:
   (Intercept)       mpg        hp         disp
4    0.3892548 0.2707320 0.02227133 -0.04428756
5  -17.6837050 0.6115097 0.15511207 -0.08815984

Std. Errors:
   (Intercept)       mpg        hp       disp
4    17.30456 0.5917790 0.05813736 0.02735148
5    15.46373 0.5754793 0.08651377 0.06060359


Residual Deviance: 17.47162
AIC: 33.47162
```

As expected, the summary includes two sets of coefficients for the two submodels (indexed by 4 and 5, respectively).

Lastly, let us make predictions using the multinomial logistic regression model and calculate the confusion matrix:

```
# make prediction
predicted_gears = predict(multinom_model, newdata = mtcars)
# view the confusion matrix
>>> table(Predicted = predicted_gears, Actual = mtcars$gear)
          Actual
Predicted  3  4  5
        3 14  0  0
        4  1 12  1
        5  0  0  4
```

The result suggests a decent classification performance with only two misclassifications.

Summary

In this chapter, we delved into the world of logistic regression, its theoretical underpinnings, and its practical applications. We started by exploring the fundamental construct of logistic regression and its comparison with linear regression. We then introduced the concept of the sigmoid transformation, a crucial element in logistic regression, which ensures the output of our model is bounded between 0 and 1. This section helped us better understand the advantages of logistic regression for binary classification tasks.

Next, we delved into the concept of log odds and odds ratio, two critical components of the logistic regression model. Understanding these allowed us to comprehend the real-world implications of the model's predictions and to interpret its parameters effectively. The chapter then introduced the CEL, the cost function used in logistic regression. Specifically, we discussed how this loss function ensures our model learns to predict accurate probabilities for the binary classes.

When it comes to evaluating a logistic regression, we learned about various metrics, including accuracy, error rate, precision, recall, sensitivity, specificity, and AUC. This understanding will allow us to assess the performance of our logistic regression model accurately.

An important discussion revolved around the handling of imbalanced datasets, a common scenario in real-world data. We understood the effects of data imbalance on our model and learned about strategies, such as resampling techniques, to handle such situations effectively. Further, we discussed penalized logistic regression where we incorporate L1 (lasso) or L2 (ridge) regularization into our logistic regression model. This penalization technique helps us prevent overfitting by keeping the magnitude of the model weights small and creating simpler models when dealing with high-dimensional data.

Finally, we touched upon multinomial logistic regression, an extension of logistic regression used for multi-class classification problems. This part provided insight into handling situations where the target variable consists of more than two classes.

At the end of this chapter, we gained an extensive understanding of logistic regression, its implementation, and its nuances. This knowledge lays the groundwork for diving deeper into more complex classification methods and strategies.

In the next chapter, we will cover Bayesian statistics, another major branch of statistical modeling.

14

Bayesian Statistics

In this chapter, we will introduce the Bayesian inference framework, covering its core components and implementation details. Bayesian inference introduces a useful framework that provides an educated guess on the predictions of the target outcome as well as quantified uncertainty estimates. Starting from a prior distribution that embeds domain expertise, the Bayesian inference approach allows us to continuously learn updated information from the data and update the posterior distribution to form a more realistic view of the underlying parameters.

By the end of this chapter, you will have grasped essential skills when working with the Bayesian inference framework. You will learn the core theory behind Bayes' theorem and its use in the Bayesian linear regression model.

We will cover the following main topics in this chapter:

- Introducing Bayesian statistics
- Diving deeper into Bayesian inference
- The full Bayesian inference procedure
- Bayesian linear regression with a categorical variable

Technical requirements

To run the code in this chapter, you will need to have the latest versions of the following packages:

- `ggplot2`, 3.4.0
- `ggridges`, 0.5.4
- `rjags`, 4.13
- `coda`, 0.19.4

Please note that the versions mentioned along with the packages in the preceding list are the latest ones while I am writing this chapter.

All the code and data for this chapter is available at `https://github.com/PacktPublishing/` `The-Statistics-and-Machine-Learning-with-R-Workshop/blob/main/` `Chapter_14/working.R`.

Introducing Bayesian statistics

The Bayesian approach to statistics and **machine learning** (**ML**) provides a logical, transparent, and interpretable framework. This is a uniform framework that can build problem-specific models for both statistical inference and prediction. In particular, Bayesian inference offers a method to figure out unknown or unobservable quantities given known facts (observed data), employing probability to describe the uncertainty over the possible values of unknown quantities—namely, random variables of interest.

Using Bayesian statistics, we are able to express our prior assumption about unknown quantities and adjust this based on the observed data. It provides the Bayesian versions of common statistical procedures such as hypothesis testing and linear regression, covered in *Chapters 11*, *Statistics estimation*, and *12*, *Linear Regression in R*. Compared to the frequentist approach, which we have adopted in all the models covered so far, the Bayesian approach additionally allows us to construct problem-specific models that can make the best use of the data in a continuous learning fashion (via the Bayesian posterior update, to be covered in the following section).

For example, the unknown quantities correspond to the parameters we are trying to estimate in a linear or logistic regression model. Instead of treating them as fixed quantities and using the principle of maximum likelihood to estimate their values, the Bayesian approach treats them as moving variables with their respective probability distribution of possible values.

Let us get a first glimpse of the famous Bayesian theorem.

A first look into the Bayesian theorem

The Bayesian theorem describes the relationship between conditional probabilities of statistical quantities. In the context of linear regression, we would treat the parameter β as a random variable instead of fixed quantities as we did with linear regression in *Chapter 12*. For example, in a simple linear regression model $y = \beta x$, instead of obtaining the single best parameter β^* by minimizing the **ordinary least squares** (**OLS**) given the available data (x, y), we would instead treat β as a random variable that follows a specific distribution.

Doing this involves two distributions about β, the parameter of interest, as follows:

- The first distribution is the **prior distribution** $P(\beta)$, which corresponds to a subjective distribution we assign to β before we observe any data. This distribution encapsulates our prior belief about the probabilities of possible values of β before we observe any actual data.

- The second distribution is the **posterior distribution** $P(\beta|x,y)$, which corresponds to the updated belief about this distribution after we observe the data. This is the distribution we want to estimate through the update. Such an update is necessary in order to conform our prior belief to what we actually observe in reality.

Naturally, we would hope that the posterior distribution $P(\beta|x,y)$ stays closer to what the data reflects as the training size gets large, and correspondingly stays further away from the prior belief. Here, the data refers to (x,y). To proceed with the update, the data would enter as what we call the likelihood function $P(y|x,\beta)$, also referred to as the **generative model**. That is, $P(y|x,\beta)$ represents the likelihood (similar to probability, although in an unnormalized way) of observing the target y given the input feature x and parameter value β, where we have treated β as a specific value instead of a random variable. In other words, we would first sample from the distribution $P(\beta)$ to obtain a concrete value of β, and then follow the specific observation model to obtain the actual data point y given the input feature x. For example, we would assume a normal distribution for the observation model if the errors are assumed to follow a normal distribution.

Now, we are ready to compile these three quantities together via the following Bayesian theorem:

$$P(\beta|x,y) = \frac{P(y|x,\beta)P(\beta)}{P(y|x)}$$

Here, $P(y|x)$ is referred to as the evidence, which acts as a normalizing constant to ensure that the posterior distribution is a valid probability distribution, meaning each probability is non-negative and sums (or integrates) to 1.

Figure 14.1 illustrates the Bayesian theorem:

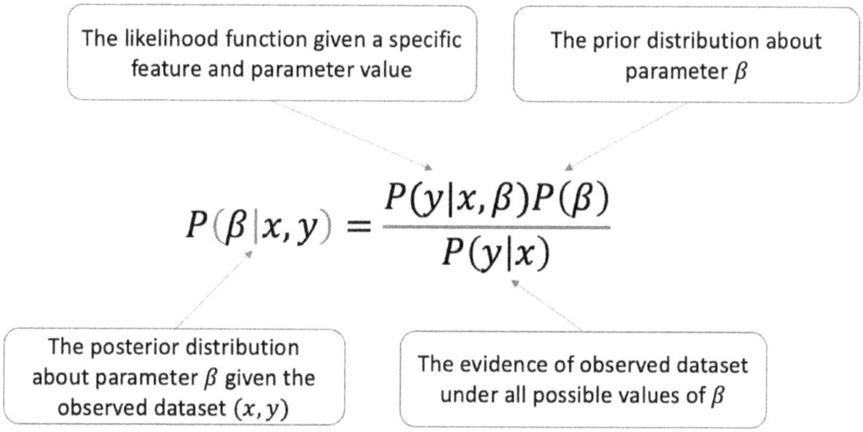

| The likelihood function given a specific feature and parameter value | The prior distribution about parameter β |

$$P(\beta|x,y) = \frac{P(y|x,\beta)P(\beta)}{P(y|x)}$$

| The posterior distribution about parameter β given the observed dataset (x,y) | The evidence of observed dataset under all possible values of β |

Figure 14.1 – Illustrating the Bayesian theorem that calculates the posterior distribution $P(\beta|x,y)$

Note that the prior distribution $P(\beta)$ in Bayesian linear regression could be chosen to model our prior belief about the parameter β, which is something not available when using the frequentist framework in OLS-based linear regression. In practice, we often go with a normal prior distribution, but it could be any distribution that captures the prior belief about the probabilities of possible values of β before we observe any data. The Bayesian framework thus allows us to incorporate the prior knowledge into the modeling in a principled manner. For example, if we believe that all features should have similar effects, we can then configure the prior distributions of the coefficients to be centered around the same value.

Since the parameter β follows a posterior distribution $P(\beta|x, y)$, the resulting prediction y^* given a new input data x^* will not be a single number, as in the case of the frequentist approach. Instead, we will obtain a series of possible values of y^*, which follows a posterior predictive distribution $P(y^*|x^*, \beta)$. That is, the prediction y^* is treated as a random variable due to the randomness in the parameter β. We can then use this distribution to understand the uncertainty in the resulting predictions. For example, if the posterior predictive distribution $P(y^*|x^*, \beta)$ is wide, the resulting predictions, which are sampled from $P(y^*|x^*, \beta)$, contain a higher degree of uncertainty. On the other hand, if the distribution is narrow, the resulting predictions are more concentrated and thus more confident. The posterior distribution $P(\beta|x, y)$ will also continue to evolve as new data becomes available.

The next section introduces more on the generative model.

Understanding the generative model

In Bayesian inference, the generative model specifies the probability distribution that governs how the data is generated. For example, when the available target data is binary, we could assume it is generated following a Bernoulli distribution with a parameter p that represents the probability of success. To get a list of binary outcomes, we would first assign a probability value to p and then use this Bernoulli distribution to generate binary labels by repeated sampling from this distribution.

Let us go through an exercise to understand the generative process.

Exercise 14.1 – Generating binary outcomes

In this exercise, we will generate a list of binary outcomes based on a Bernoulli distribution. This involves comparing a random sample from a uniform distribution valued between 0 and 1 to the preset probability of success. Follow the next steps:

1. We begin with generating a random number in the range of 0 and 1 using a uniform distribution:

    ```
    set.seed(1)
    random_prob = runif(1, min = 0, max = 1)
    >>> random_prob
    0.2655087
    ```

    Again, remember to set the random seed for reproducibility purposes.

2. Next, we compare the number to a preset probability of 0.2:

```
prop_success = 0.2
>>> random_prob < prop_success
FALSE
```

This completes the generation of a single binary number. Now, let us expand it to 10 numbers.

3. With the help of the following code, we'll generate 10 binary numbers:

```
n_samples = 10
data = c()
for(sample_idx in 1:n_samples) {
  data[sample_idx] <- runif(1, min = 0, max = 1) < prop_success
}
>>> data
FALSE FALSE FALSE FALSE FALSE FALSE FALSE FALSE  TRUE FALSE
```

Here, we used a `for` loop to repeatedly generate a uniform random number, compare it with the preset probability of success, and then store the result in a vector.

4. Convert the vector to numbers 0 and 1, as follows:

```
data = as.numeric(data)
>>> data
0 0 0 0 0 0 0 0 1 0
```

Note that we have observed 1 instance of success out of 10 draws, despite a 20% probability of success. As the sample size increases, we would expect the empirical probability of success (10% in this case) to be close to the theoretical value (20%).

It turns out that this generative model corresponds to a binomial process or a binomial distribution, which allows us to generate binary outcomes in one shot. Specifically, we can use the `rbinom()` function to simulate data from a binomial distribution, as shown in the following code snippet:

```
set.seed(1)
>>> rbinom(n = n_samples, size = 1, prob = prop_success)
```

Here, n is the number of samples to be generated from the generative model, `size` is the number of trials to run, and `prob` is the underlying probability of success valued between 0.0 and 1.0.

Note that we are essentially working with a known parameter p, which is the probability of success. In practice, this would be an unknown parameter, something we are interested in estimating from the data. Bayesian inference would allow us to do that, with the assistance of a prior distribution and the likelihood function, as introduced in the following section.

Understanding prior distributions

A prior distribution, an essential component of Bayesian inference, represents the prior knowledge or belief about the underlying parameter before we observe the actual data. It essentially specifies the probability distribution of the parameters based on domain-specific preference or expertise. If we have a valid reason to believe that certain values of the parameters are more likely, we can choose a prior distribution that reflects this preference.

The prior distribution, denoted as $P(\beta)$, treats β as a random variable and specifies its probability distribution. That is, it tells us which values of β are more likely than others. In our running example on Bayesian linear regression, the prior distribution for β is often chosen to be a multivariate Gaussian distribution. This is mainly for mathematical convenience, as the Gaussian distribution has nice properties that make it easier to work with. However, if we have no prior preference, a uniform distribution (which gives the same probability to all possible choices and is thus uninformed) could be a good candidate.

Note that we can also use the prior to impose a form of regularization on the model. By choosing a prior distribution that favors smaller values of the parameters (such as a Gaussian distribution centered at 0), we can discourage the model from finding solutions with large coefficients, thus preventing overfitting.

In the following code snippet, we randomly generate 10 samples of $P(\beta)$ following a uniform distribution between 0 and 0.2:

```
set.seed(1)
prop_successes = runif(n_samples, min = 0.0, max = 0.2)
>>> prop_successes
0.05310173 0.07442478 0.11457067 0.18164156 0.04033639 0.17967794
0.18893505 0.13215956 0.12582281 0.01235725
```

When we model the probability of success $p \in [0,1]$ as a random variable, we often assign a beta distribution as the prior. For example, in the following code snippet, we generate 1,000 samples of p from the beta distribution $p \sim Beta(35,55)$ using the `rbeta()` function, followed by showing its density plot after converting it to a DataFrame format:

```
library(ggplot2)
# Sample 1000 draws from Beta(35,55) prior
prior_A = rbeta(n = 1000, shape1 = 35, shape2 = 55)
# Store the results in a data frame
prior_sim = data.frame(prior_A)
# Construct a density plot of the prior sample
ggplot(prior_sim, aes(x = prior_A)) +
  geom_density()
```

Running the preceding code will generate the output shown in *Figure 14.2*. It shows a prior concentration of the probability of success close to 0.4, within the range of 0 and 1:

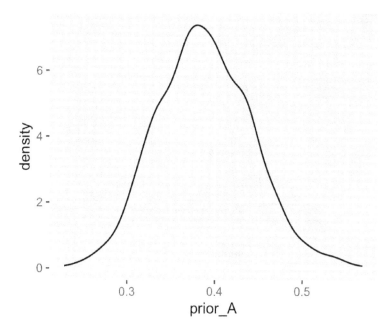

Figure 14.2 – Visualizing the density plot of the prior distribution

The *Beta(a, b)* distribution is defined on the interval from 0 to 1, thus providing a natural and flexible prior to the probability random variable. We can tune the `Beta` shape parameters *a* and *b* to produce alternative prior models. In the following code snippet, we compare the original `Beta(35, 55)` prior distributions with two alternatives: `Beta(1, 1)` and `Beta(100, 100)`. We then plot all three prior distributions together:

```
# Sample draws from the Beta(1,1) prior
prior_B = rbeta(n = 1000, shape1 = 1, shape2 = 1)
# Sample draws from the Beta(100,100) prior
prior_C = rbeta(n = 1000, shape1 = 100, shape2 = 100)
# Combine the results in a single data frame
prior_all = data.frame(samples = c(prior_A, prior_B, prior_C),
                       priors = rep(c("A","B","C"), each = 1000))
# Plot the 3 priors
ggplot(prior_all, aes(x = samples, fill = priors)) +
  geom_density(alpha = 0.5)
```

The preceding code returns the output shown in *Figure 14.3*. Each prior distribution has a different preference region. For example, distribution B is close to a uniform distribution and has no specific preference, while distribution C places a strong preference for 0.5, as indicated by the peak value around 0.5 and a narrow spread:

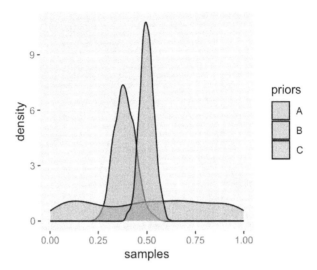

Figure 14.3 – Visualizing three different prior distributions

In the next section, we'll introduce you to the likelihood function.

Introducing the likelihood function

The likelihood function describes how likely the observed data is, given a set of fixed model parameters. In a parametric model (a model that assumes a certain set of parameters), the likelihood is the probability of the observed data as a function of the parameters. The specific form of the likelihood function depends on the distribution (more specifically, the observation model) assumed for the data. For example, if we assume the data follows a normal distribution, the likelihood function would take the form of a normal probability density function.

Let us look at a concrete example. Suppose we are developing a simple linear regression model with standard normal errors, expressed as $y = \beta x + \epsilon$, where $\epsilon \sim N(0,1)$. Here, we have ignored the intercept term and only considered the slope. For a specific data point $\left(x_i, y_i\right)$, we can express the likelihood l_i as the probability evaluated at the probability density function of the error term:

$$l_i = \frac{1}{\sqrt{2\pi}} e^{\frac{-(y_i - \beta x_i)^2}{2}}$$

Since the dataset consists of a total of n input-output pairs, the joint likelihood of all data points can be expressed as follows:

$$L = \Pi_{i=1}^{n} l_i = \left(\frac{1}{\sqrt{2\pi}}\right)^{n} e^{\frac{-\Sigma_{i=1}(y_i - \beta x_i)^2}{2}}$$

In practice, we would often work with the log-likelihood after introducing the log transformation, as shown here:

$$\log L = \log\left(\frac{1}{\sqrt{2\pi}}\right)^{n} e^{\frac{-\Sigma_{i=1}(y_i - \beta x_i)^2}{2}} = -0.5 n \log 2\pi - 0.5 \sum_{i=1}^{n} (y_i - \beta x_i)^2$$

Compared to the objective function used in OLS-based linear regression, we find that the exponent term $\sum_{i=1}^{n} (y_i - \beta x_i)^2$ is exactly the sum of squared errors. When using the maximum likelihood estimation procedure, these two different objective functions become equivalent to each other. In other words, we have the following:

$$\log L \approx -\sum_{i=1}^{n} (y_i - \beta x_i)^2$$

Here, we ignored the constant term $\left(\frac{1}{\sqrt{2\pi}}\right)^n$ at the last step.

Let us go through an example of how to calculate the joint likelihood of a set of observed data. In the following code listing, we create a list of data points in x and y and a simple linear regression model with coefficient b to generate the predicted values in y_pred, along with the residual terms in residuals:

```
# observed data
x = c(1, 2, 3, 4, 5)
y = c(2, 3, 5, 6, 7)
# parameter value
b = 0.8
# calculate the predicted values
y_pred = b * x
# calculate the residuals
residuals = y - y_pred
```

We can then plug in the closed-form expression for the joint likelihood and calculate the total log-likelihood, as follows:

```
log_likelihood = -0.5 * length(y) * log(2 * pi) - 0.5 *
sum(residuals^2)
log_likelihood
-18.09469
```

Let us look at another example of the binomial model. As discussed earlier, when the underlying parameter p represents the probability of success, we can use the rbinom() function to obtain the probability of observing a certain outcome (number of successes) in a total number of draws and with a specific probability of success. In the following code snippet, we first create a vector of probabilities to indicate different probabilities of success and calculate the corresponding likelihood in a total of 1,000 trials of sampling from a binomial model $Bin(p, 10)$, where 10 is the total number of draws. Lastly, we visualize all likelihood functions via a stacked density plot using the geom_density_ridges() function from the ggridges package:

```
library(ggridges)
# Define a vector of 1000 p values
p_grid = seq(from = 0, to = 1, length.out = 1000)
# Simulate 10 trials for each p in p_grid, each trial has 1000 samples
sim_result = rbinom(n = 1000, size = 10, prob = p_grid)
# Collect results in a data frame
```

```
likelihood_sim = data.frame(p_grid, sim_result)
# Density plots of p_grid grouped by sim_result
ggplot(likelihood_sim, aes(x = p_grid, y = sim_result, group = sim_
result)) +
   geom_density_ridges()
```

The preceding code results in the following output:

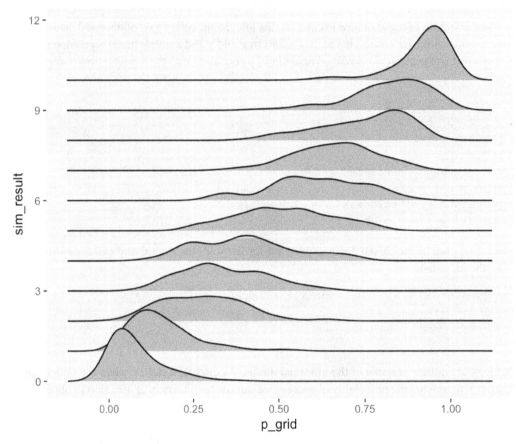

Figure 14.4 – Visualizing the stacked density plot as the likelihood functions
of different sampling from the binomial distribution

The next section introduces the posterior model.

Introducing the posterior model

A posterior distribution represents what we know about unknown parameters after observing the available data. It combines the prior beliefs, expressed via the prior distribution, with the evidence presented by the data, expressed in the likelihood function, to form a new distribution over the possible parameter values, which can be either discrete or continuous.

Based on Bayes' theorem, the posterior distribution is proportional to the product of the prior distribution and the likelihood function:

$$P(\beta|x,y) \ \propto \ P(y|x,\beta)P(\beta)$$

Note that we do not need to know the evidence term in the denominator when solving the optimal value of the parameter β, since $P(y|x)$ is totally independent of β. As we gather more data and update our beliefs, the posterior distribution would often become more sharply peaked around the true parameter value, indicating an increased level of confidence.

However, when we need to know $P(y|x)$ in order to calculate $P(\beta|x,y)$, the task is not so straightforward since a closed-form solution may not be available, or the parameters are multi-dimensional and prohibit a direct calculation of nested integration. In such cases, we would often resort to numerical methods such as **Markov chain Monte Carlo** (**MCMC**) or approximate methods such as variational inference.

Let us go through an exercise to understand the overall inference process. We will use the `rjags` package to do the calculations based on our running example.

Exercise 14.2 – Obtaining the posterior distribution

In this exercise, we will use the `rjags` package to perform Bayesian inference, including specifying the model architecture and obtaining the posterior distribution for the underlying parameter. Follow the next steps:

1. We begin with defining the likelihood function as a binomial distribution (using `dbin`) with parameters p for the probability of success and n for the total number of samples. We will also, define the prior distribution for p as a beta distribution (using `dbeta`) with parameters a and b:

```
library(rjags)
# define the model
bayes_model = "model{
    # Likelihood model for X
    X ~ dbin(p, n)

    # Prior model for p
    p ~ dbeta(a, b)
}"
Compile the model using the bayes.model() function.
# compile the model
bayes_jags = jags.model(textConnection(bayes_model),
                    data = list(a = 1, b = 1, X = 3, n =
10))
```

Here, we specify textConnection (bayes_model) to pass the model specification string to jags. The data argument is a list specifying the observed data and the parameters for the prior distribution.

2. Next, we draw samples from the posterior distribution:

```
# simulate the posterior
bayes_sim = coda.samples(model = bayes_jags, variable.names =
c("p"), n.iter = 10000)
```

Here, the coda.samples() function is used to run MCMC simulations and draw samples of the parameter. The n.iter argument specifies the number of iterations for the MCMC simulation.

3. Finally, we plot the posterior:

```
# plot the posterior
plot(bayes_sim, trace = FALSE, xlim = c(0,1), ylim = c(0,3))
```

Running the preceding code will result in the output shown in *Figure 14.5*:

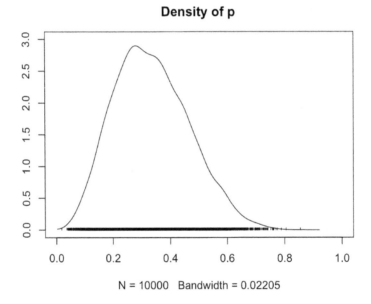

Figure 14.5 – Visualizing the posterior distribution for the underlying parameter (probability of success)

In the next section, we'll dive deeper into Bayesian inference, starting by introducing the normal-normal model, a commonly used type of model in Bayesian inference.

Diving deeper into Bayesian inference

Bayesian inference is a statistical method that makes use of conditional probability to update the prior beliefs about the parameters of a statistical model given the observed data. The output of Bayesian inference is a posterior distribution, which is a probability distribution that represents our updated beliefs about the parameter after observing the data.

When calculating the exact posterior distribution is difficult, we would often resort to MCMC, which is a technique for estimating the distribution of a random variable. It's a method commonly used to generate samples from the posterior distribution in Bayesian inference, especially when the dimensionality of the model parameters is high, making an analytical solution intractable.

The following section introduces the normal-normal model and uses MCMC to estimate its posterior distribution.

Introducing the normal-normal model

The normal-normal model is another foundational model in Bayesian inference. It refers to the case when the likelihood is normally distributed and the prior is also normally distributed, both following a bell-shaped curve. This type of model is often used in Bayesian statistics as it has a closed-form solution for the posterior distribution, which also happens to be normally distributed.

Let us look at a concrete exercise of the normal-normal model using MCMC-based Bayesian inference.

Exercise 14.3 – Working with the normal-normal model

In this exercise, we will define a normal likelihood function whose mean follows a normal prior and whose standard deviation follows a uniform prior. We will then use `rjags` to obtain the posterior estimates for the mean and the standard deviation. Follow the next steps:

1. Let's start with simulating 100 data points that follow a normal distribution with a true mean of 2 and a true standard deviation of 1:

```
library(coda)
set.seed(1)
mu_true = 2
sd_true = 1
n = 100
data = rnorm(n, mean = mu_true, sd = sd_true)
```

2. Now, we specify the model architecture, including a normal likelihood function for the data, a normal prior distribution for the mean variable, and a uniform prior for the standard deviation variable. The normal prior is parameterized by `0` and `0.1` for the mean and standard deviation, respectively, and the uniform prior ranges from `0` to `10`:

```
model_string = "model {
    for (i in 1:n) {
        y[i] ~ dnorm(mu, prec)
    }

    mu ~ dnorm(0, 0.1)
    sigma ~ dunif(0, 10)

    prec <- pow(sigma, -2)
}"
```

The preceding model states that each observation in our data (`y[i]`) follows a normal distribution with mean `mu` and precision, `prec`, (defined as the reciprocal of the variance, hence `prec <- pow(sigma, -2)`). The mean `mu` follows a normal distribution with mean `0` and precision close to `0`, which corresponds to a relatively large variance and, therefore, a weak prior belief about `mu`. The standard deviation, `sigma`, follows a uniform distribution from `0` to `10`, which expresses complete uncertainty about its value between these two bounds.

3. Next, compile the model in `jags` and burn in the Markov chain, as follows:

```
data_jags = list(y = data, n = n)
model = jags.model(textConnection(model_string), data = data_
jags)
update(model, 1000)   # burn-in
```

Here, the `burn-in` period is a number of initial iterations that we discard when performing Bayesian inference, under the assumption that the resulting chain may not have converged during this period. The idea is to let the Markov chain burn in until it reaches a distribution that is stable and reflective of the posterior distribution we are interested in. In this case, we would discard the first 1,000 iterations of the Markov chain, and these are not used in the subsequent analysis.

4. Here, we generate samples from the posterior distribution:

```
params = c("mu", "sigma")
samples = coda.samples(model, params, n.iter = 10000)
# print summary statistics for the posterior samples
>>> summary(samples)
Iterations = 2001:12000
Thinning interval = 1
Number of chains = 1
Sample size per chain = 10000
```

1. Empirical mean and standard deviation for each variable,
   plus standard error of the mean:

```
        Mean      SD  Naive SE Time-series SE
mu    2.1066 0.09108 0.0009108      0.0009108
sigma 0.9097 0.06496 0.0006496      0.0008996
```

2. Quantiles for each variable:

```
        2.5%    25%     50%     75% 97.5%
mu     1.929 2.0453 2.1064 2.1666 2.287
sigma 0.792 0.8652 0.9055 0.9513 1.046
```

Here, the `summary()` function provides useful summary statistics for the posterior samples of `mu` and `sigma`, including their means, medians, and credible intervals.

5. Finally, let's visualize the results by running the following command:

```
>>> plot(samples)
```

Running the preceding line of code will generate the output shown in *Figure 14.6*, which shows trace plots and density plots for the posterior samples of `mu` and `sigma`:

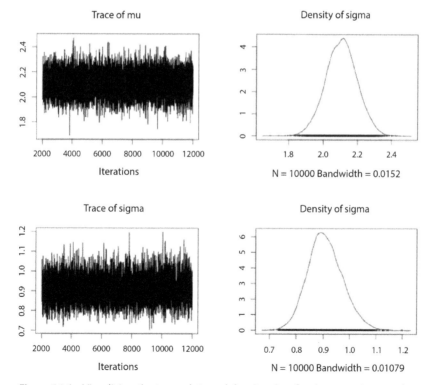

Figure 14.6 – Visualizing the trace plots and density plots for the posterior samples

In the next section, we will discuss more on the MCMC.

Introducing MCMC

A Markov chain is a mathematical model that transits from one state to another within a finite or countable number of possible states. It is a sequence of random variables where the future state depends only on the present state and not on the sequence of events before it. This property is known as the **Markov property**, or **memorylessness**.

In the context of statistical inference, we can sample from complex probability distributions and create models of sequence data via Monte Carlo simulations, and thus the term MCMC. MCMC algorithms construct a Markov chain of parameter values where the stationary distribution of the chain is the posterior distribution of the underlying parameters. The chain is generated by iteratively proposing new parameter values and accepting or rejecting these candidate values based on a preset rule, ensuring that the samples converge to the posterior distribution.

Let us analyze the details of the previous MCMC chain. In the following code snippet, we convert the chain to a DataFrame and print its first few rows:

```
# Store the chains in a data frame
mcmc_chains <- data.frame(samples[[1]], iter = 1:10000)
# Check out the head
>>> head(mcmc_chains)
         mu       sigma iter
1 2.159540 0.8678513    1
2 2.141280 0.8719263    2
3 1.975057 0.8568497    3
4 2.054670 0.9313297    4
5 2.144810 1.0349093    5
6 2.001104 1.0597861    6
```

These are MCMC samples generated to approximate the posterior distribution for mu and sigma, respectively. Each sample depends on the previous sample only and is unrelated to other prior samples. We can plot these samples in a line plot called a trace plot, as shown in the following code snippet:

```
# Use plot() to construct trace plots
>>> plot(samples, density = FALSE)
```

Running the preceding code generates the output shown in *Figure 14.7*. A trace plot is a commonly used diagnostic tool in MCMC. It is a graphical representation of the values of the samples at each iteration or step of the MCMC algorithm. In this case, the trace plot shows no obvious trend, suggesting that both chains have converged stably:

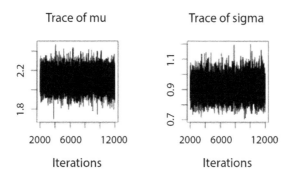

Figure 14.7 – Visualizing the trace plots of MCMC chains

Let us observe the first 100 samples using `ggplot()` for mu:

```
# Trace plot the first 100 iterations of the mu chain
>>> ggplot(mcmc_chains[1:100, ], aes(x = iter, y = mu)) +
  geom_line() +
  theme(axis.title.x = element_text(size = 20),  # Increase x-axis
label size
        axis.title.y = element_text(size = 20))  # Increase y-axis
label size
```

Running the preceding code generates the output shown in *Figure 14.8*. We can see that the sampler moves to another region after sufficiently exploring the previous region:

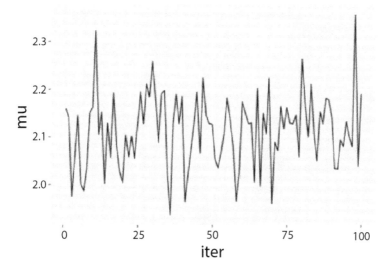

Figure 14.8 – Visualizing the first 100 iterations of the mu chain

Note that we can also display the density plot only by setting `trace = FALSE`, as follows:

```
# Use plot() to construct density plots
>>> plot(samples, trace = FALSE)
```

The preceding code returns the output shown in *Figure 14.9*:

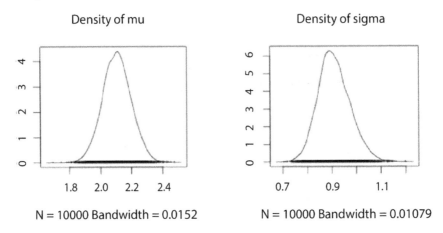

Figure 14.9 – Visualizing the density plot of both chains

In practice, we would often run multiple chains in MCMC. This allows us to check whether all chains converge to the same distribution, which is a critical step to ensure that MCMC sampling is done correctly. Since each chain in MCMC starts at a different initial point, running multiple chains could help check whether the posterior distribution is dependent on the starting values. When all chains converge to the same distribution regardless of the initial points, we have a higher level of confidence in the stability and robustness of the MCMC process.

The following code snippet runs MCMC over four chains:

```
model2 = jags.model(textConnection(model_string), data = data_jags,
n.chains = 4)
# simulate the posterior
samples2 <- coda.samples(model = model2, variable.names = params,
n.iter = 1000)
```

We can check the trace plot as follows:

```
# Construct trace plots
>>> plot(samples2, density = FALSE)
```

As shown in *Figure 14.10*, all four chains have more or less converged to a stable state:

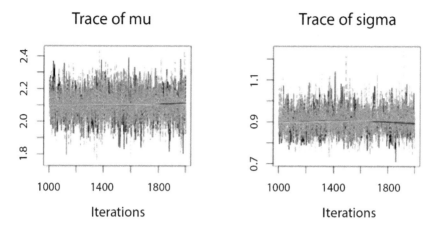

Figure 14.10 – Visualizing the density plot of both chains

Finally, we can check the summary of the MCMC samples, as follows:

```
>>> summary(samples2)
Iterations = 1001:2000
Thinning interval = 1
Number of chains = 4
Sample size per chain = 1000


1. Empirical mean and standard deviation for each variable,
   plus standard error of the mean:

        Mean       SD Naive SE Time-series SE
mu    2.1052 0.09046  0.00143        0.00144
sigma 0.9089 0.06454  0.00102        0.00134


2. Quantiles for each variable:

        2.5%    25%    50%    75% 97.5%
mu    1.9282 2.0456 2.104 2.1655 2.282
sigma 0.7952 0.8626 0.906 0.9522 1.041
```

In the next section, we will cover the full Bayesian inference procedure, including quantifying the posterior uncertainty and making predictions based on the posterior distribution of the parameters.

The full Bayesian inference procedure

The full Bayesian inference starts by specifying the model architecture, including the prior distribution for unknown (unobserved) parameters and the likelihood function that determines how the data is generated. We can then perform MCMC to infer the posterior distribution of these parameters given the observed dataset. Finally, we can use the posterior distribution to either quantify the uncertainty about these parameters or make predictions for new input data with quantified uncertainty about the predictions.

The following exercise illustrates this process using the mtcars dataset.

Exercise 14.4 – Performing full Bayesian inference

In this exercise, we will perform Bayesian linear regression with a single feature and two unknown parameters: intercept and slope. The model looks at the relationship between car weight (wt) and horsepower (hp) in the mtcars dataset. Follow the next steps:

1. Specify a Bayesian inference model where each target wt is modeled as a realization of a random variable following a normal distribution. The mean parameter is a linear combination of two parameters (a for intercept and b for slope) with the corresponding input feature. Both a and b follow a normal distribution, and the variance parameter follows a uniform distribution. The code is illustrated in the following snippet:

    ```
    # load the necessary libraries
    library(rjags)
    library(coda)

    # define the model
    model = "model{
        # Define model for data Y[i]
        for(i in 1:length(Y)) {
          Y[i] ~ dnorm(m[i], s^(-2))
          m[i] <- a + b * X[i]
        }

        # Define the a, b, s priors
        a ~ dnorm(0, 0.5^(-2))
        b ~ dnorm(1, 0.5^(-2))
        s ~ dunif(0, 20)
    }"
    ```

2. Compile the model with three chains and control the random seed for model reproducibility:

    ```
    # compile the model
    model = jags.model(textConnection(model),
    ```

```
                    data = list(Y = mtcars$wt, X = mtcars$hp),
                    n.chains = 3,
                    inits = list(.RNG.name = "base::Wichmann-
Hill", .RNG.seed = 100))
```

3. Run a burn-in period of `three` iterations:

```
# burn-in
update(model, 1000)
```

4. MCMC samples from the posterior distribution of the parameters, as seen here:

```
# generate MCMC samples
samples = coda.samples(model, variable.names = c("a", "b", "s"),
n.iter = 5000)
```

5. Create a trace plot of the MCMC samples to assess convergence, as follows:

```
# check convergence using trace plot
>>> plot(samples)
```

Running the preceding code generates the output shown in *Figure 14.11*, suggesting a decent convergence for all three parameters:

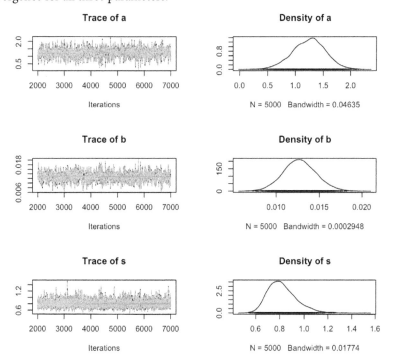

Figure 14.11 – Visualizing the trace plots of all three parameters

6. Calculate the mean of the posterior distributions for parameters a and b, and use these mean values to make point predictions and plot the prediction line on a scatterplot of the data, as follows:

```
# Get the posterior estimates
posterior_estimates = summary(samples)
# Calculate the mean for each parameter
a_mean = posterior_estimates$statistics["a", "Mean"]
b_mean = posterior_estimates$statistics["b", "Mean"]
# Plot the prediction line
ggplot(mtcars, aes(x = hp, y = wt)) +
  geom_point() +
  geom_abline(intercept = a_mean, slope = b_mean) +
  labs(title = "Bayesian Linear Regression",
       x = "Horsepower",
       y = "Weight") +
  theme(plot.title = element_text(hjust = 0.5))
```

Here, we use the mean of the posterior distribution as the parameter value to make a point prediction for each input feature. The commands generate the output shown in *Figure 14.12*:

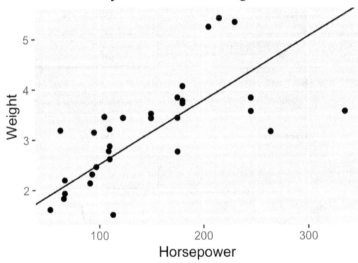

Figure 14.12 – Making points predictions using the mean value
of the posterior distributions for each parameter

It is worth noting that Bayesian linear regression offers quantified uncertainty compared to frequentist linear regression covered in previous chapters. Such uncertainty comes in the form of credible intervals, which differ from the confidence interval. Specifically, we obtain the credible interval by treating the parameter as a random variable and the data as a fixed quantity, which makes more sense as we only get to observe the data once in most cases.

7. Compute the **highest posterior density interval (HPDI)** for parameters a and b using the HPDinterval() function and plot the confidence intervals in the histogram, as follows:

```
# Extract samples
a_samples = as.matrix(samples[, "a"])
b_samples = as.matrix(samples[, "b"])
# Calculate credible intervals
a_hpd = coda::HPDinterval(coda::as.mcmc(a_samples))
b_hpd = coda::HPDinterval(coda::as.mcmc(b_samples))
# Plot histograms and credible intervals
par(mfrow=c(2,1))   # Create 2 subplots
# Parameter a
hist(a_samples, freq=FALSE, xlab="a", main="Posterior
distribution of a", col="lightgray")
abline(v=a_hpd[1,1], col="red", lwd=2)   # Lower limit of the
credible interval
abline(v=a_hpd[1,2], col="red", lwd=2)   # Upper limit of the
credible interval
# Parameter b
hist(b_samples, freq=FALSE, xlab="b", main="Posterior
distribution of b", col="lightgray")
abline(v=b_hpd[1,1], col="red", lwd=2)   # Lower limit of the
credible interval
abline(v=b_hpd[1,2], col="red", lwd=2)   # Upper limit of the
credible interval
```

Running the commands generates the output shown in *Figure 14.13*. Here, we calculate the 95% credible interval by default:

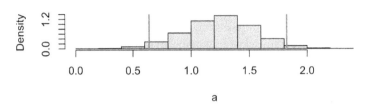

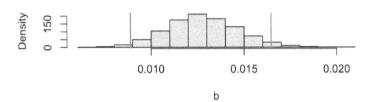

Figure 14.13 – Visualizing the posterior distribution and credible interval of the model parameters

8. Finally, make posterior predictions for a new input value, 120, and plot the posterior predictive distribution based on the list of posterior samples for the parameters, as follows:

```
# make posterior predictions
# Obtain the mean values of the MCMC samples for each parameter
a_mean = mean(samples[[1]][,"a"])
b_mean = mean(samples[[1]][,"b"])
# New input (e.g., horsepower = 120)
new_input = 120
# Prediction
predicted_weight = a_mean + b_mean * new_input
print(predicted_weight)
# Predictive distribution
predicted_weights = samples[[1]][,"a"] + samples[[1]][,"b"] *
new_input
# Plot the predictive distribution
>>> hist(predicted_weights, breaks = 30, main = "Posterior
predictive distribution", xlab = "Predicted weight")
```

Running the commands generates the output shown in *Figure 14.14*. This posterior predictive distribution captures the model's uncertainty about the prediction:

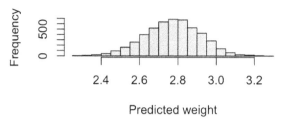

Posterior predictive distribution

Figure 14.14 – Visualizing the posterior predictive distribution for a new input feature

The next section covers a Bayesian linear regression model using a categorical input variable.

Bayesian linear regression with a categorical variable

When the predictor is categorical, such as a binary feature, we would set one parameter for each corresponding category. The following exercise demonstrates such an example.

Exercise 14.5 – Performing Bayesian inference with a categorical variable

In this exercise, we will examine the relationship between am (automatic or manual transmission, a categorical variable) and mpg (miles per gallon, a continuous variable). We will define the mean of the normal likelihood for mpg as a function of am, with a different mean mu[i] for each level of am. We'll also give mu a normal prior and standard deviation s a uniform prior. Follow the next steps:

1. Specify the aforementioned model architecture, as follows:

```
# define the model
model = "model{
    # Define model for data Y[i]
    for(i in 1:length(Y)) {
      Y[i] ~ dnorm(mu[am[i]+1], s^(-2))
    }
    # Define the mu, s priors
    for(j in 1:2){
      mu[j] ~ dnorm(20, 10^(-2))
    }
    s ~ dunif(0, 20)
}"
```

2. Compile the model, generate the posterior samples, and show the convergence plots:

```
# compile the model
model = jags.model(textConnection(model),
                   data = list(Y = mtcars$mpg, am = mtcars$am),
                   n.chains = 3,
                   inits = list(.RNG.name = "base::Wichmann-
Hill", .RNG.seed = 100))
# burn-in
update(model, 1000)
# generate MCMC samples
samples = coda.samples(model, variable.names = c("mu", "s"),
n.iter = 5000)
# check convergence using trace plot
>>> plot(samples)
```

Running the preceding code generates the output shown in *Figure 14.15*, suggesting a decent convergence of all model parameters:

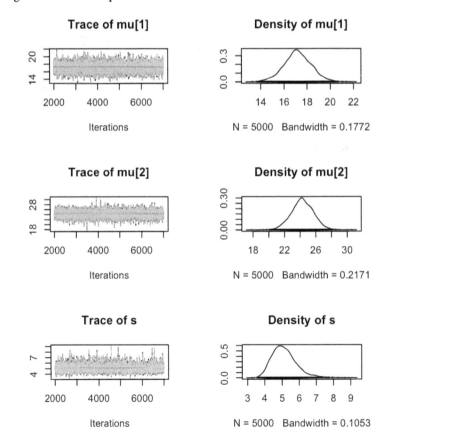

Figure 14.15 – Visualizing the convergence plots

3. Plot the distribution of the dataset along with the mean estimate of parameters for both levels of the categorical variable, as follows:

```
# Get the posterior estimates
posterior_estimates = summary(samples)
# Calculate the mean for each parameter
mu1_mean = posterior_estimates$statistics["mu[1]", "Mean"]
mu2_mean = posterior_estimates$statistics["mu[2]", "Mean"]
# Plot the prediction line
ggplot(mtcars, aes(x = as.factor(am), y = mpg)) +
  geom_jitter(width = 0.2) +
  geom_hline(aes(yintercept = mu1_mean, color = "Automatic"),
linetype = "dashed") +
  geom_hline(aes(yintercept = mu2_mean, color = "Manual"),
linetype = "dashed") +
  scale_color_manual(name = "Transmission", values =
c("Automatic" = "red", "Manual" = "blue")) +
  labs(title = "Bayesian Linear Regression",
       x = "Transmission (0 = automatic, 1 = manual)",
       y = "Miles Per Gallon (mpg)") +
  theme(plot.title = element_text(hjust = 0.5),
        legend.position = "bottom")
```

The preceding code will generate the output shown in *Figure 14.16*:

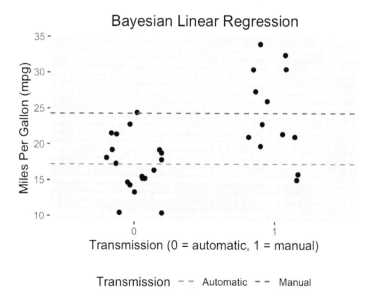

Figure 14.16 – Visualizing the distribution and the mean estimates of the dataset

Summary

This chapter provides a comprehensive introduction to Bayesian statistics, beginning with an exploration of the fundamental Bayes' theorem. We delved into its components, starting with understanding the generative model, which helps us simulate data and examine how changes in parameters affect the data generation process.

We then focused on understanding the prior distribution, an essential part of Bayesian statistics that represents our prior knowledge about an uncertain parameter. This was followed by an introduction to the likelihood function, a statistical function that determines how likely it is for a set of observations to occur given specific parameter values.

Next, we introduced the concept of the posterior model. This combines our prior distribution and likelihood to give a new probability distribution that represents updated beliefs after having seen the data. We also explored more complex models, such as the normal-normal model, wherein both the likelihood and the prior are normally distributed. We further investigated the mechanics of Bayesian inference through the MCMC method, a powerful tool for estimating the distribution of parameters and making predictions. A detailed walk-through of the full Bayesian inference procedure accompanied this.

Lastly, we discussed Bayesian linear regression with a categorical variable, which extends the methodology to models that include categorical predictors.

Congratulations! You've successfully navigated to the end of this book, a testament to your dedication and effort. This journey, hopefully enriching for you and certainly so for me, marks a significant milestone in your ongoing exploration of the dynamic field of statistics and ML. I'm honored that you chose this book as a companion in this voyage, and I trust it has laid a solid foundation for your future pursuits.

Index

C

Other Books You May Enjoy

If you enjoyed this book, you may be interested in these other books by Packt:

Building Statistical Models in Python

Huy Hoang Nguyen, Paul N Adams, Stuart J Miller

ISBN: 978-1-80461-428-0

- Explore the use of statistics to make decisions under uncertainty
- Answer questions about data using hypothesis tests
- Understand the difference between regression and classification models
- Build models with stats models in Python
- Analyze time series data and provide forecasts
- Discover Survival Analysis and the problems it can solve

Enhancing Deep Learning with Bayesian Inference

Matt Benatan, Jochem Gietema, Marian Schneider

ISBN: 978-1-80324-688-8

- Understand advantages and disadvantages of Bayesian inference and deep learning
- Understand the fundamentals of Bayesian Neural Networks
- Understand the differences between key BNN implementations/approximations
- Understand the advantages of probabilistic DNNs in production contexts
- How to implement a variety of BDL methods in Python code
- How to apply BDL methods to real-world problems
- Understand how to evaluate BDL methods and choose the best method for a given task
- Learn how to deal with unexpected data in real-world deep learning applications

Packt is searching for authors like you

If you're interested in becoming an author for Packt, please visit `authors.packtpub.com` and apply today. We have worked with thousands of developers and tech professionals, just like you, to help them share their insight with the global tech community. You can make a general application, apply for a specific hot topic that we are recruiting an author for, or submit your own idea.

Share your thoughts

Now you've finished *The Statistics and Machine Learning with R Workshop*, we'd love to hear your thoughts! Scan the QR code below to go straight to the Amazon review page for this book and share your feedback or leave a review on the site that you purchased it from.

https://packt.link/r/1-803-24030-X

Your review is important to us and the tech community and will help us make sure we're delivering excellent quality content.

Download a free PDF copy of this book

Thanks for purchasing this book!

Do you like to read on the go but are unable to carry your print books everywhere?

Is your eBook purchase not compatible with the device of your choice?

Don't worry, now with every Packt book you get a DRM-free PDF version of that book at no cost.

Read anywhere, any place, on any device. Search, copy, and paste code from your favorite technical books directly into your application.

The perks don't stop there, you can get exclusive access to discounts, newsletters, and great free content in your inbox daily

Follow these simple steps to get the benefits:

1. Scan the QR code or visit the link below

https://packt.link/free-ebook/9781803240305

2. Submit your proof of purchase
3. That's it! We'll send your free PDF and other benefits to your email directly

www.ingramcontent.com/pod-product-compliance
Lightning Source LLC
Chambersburg PA
CBHW060640060326
40690CB00020B/4459